WAR IS HELL

WAR IS HELL

THE RISE OF TOTAL WAR FROM NAPOLEON TO THE PRESENT

DANIEL E. LONG

STACKPOLE
BOOKS
Essex, Connecticut

STACKPOLE BOOKS
The Globe Pequot Publishing Group, Inc.
64 South Main St.
Essex, CT 06426
www.GlobePequot.com

British Library Cataloguing in Publication Information available

Library of Congress Cataloging-in-Publication Data available

ISBN 9780811777582 (cloth) | ISBN 9780811777599 (epub)

To my late father, Sergeant Forest E. Long,
1928th Ordnance Ammunition Company,
9th Air Force, March 1943–May 1946,
who unknowingly instilled in me a love of World War II history,
and my wife, Cara J. Spittle Long, who believed in me when I didn't.

CONTENTS

ACKNOWLEDGMENTS

A s is customary to note, I take sole responsibility for the contents of this book and any errors it may contain. Having said that, no writer works by themselves. I may write in the solitude of my office (except for the joyful sounds of grandchildren running around), but once that is done, I would be left flopping like a fish on land without the help of these marvelous people. First, I want to thank my family for putting up with me while I wrote this. Almost as important, I want to acknowledge Dave Reisch, who gave me the chance to write and publish this book in the first place. Both Alden Perkins and Jason Rossi deserve recognition for figuratively holding my hand as a first-time author through all the steps associated with publishing a book. Of all the other good people associated with Stackpole Books–Globe Pequot Publishing who worked on this book every step of the way, a special thank-you goes to my editor Niki Guinan—who knows how many red pencils she would have used editing my manuscript if it weren't for modern word processors. Sorry, Niki. I also want to thank Dr. Randall L. Bytwerk for the gracious use of his illustrations from the German Propaganda Archive (https://research.calvin.edu/german-propaganda-archive/). David Wardell deserves credit for verifying that this was a project worth pursuing. A tremendous shout-out goes to Marilyn Spittle for graciously offering pre-editing suggestions and services. And a very big thank-you to Sandy Knipmeyer of the Anchorage Public Library, Alaska, and Deanna Thomas of the Anchor Point Public Library, Alaska, for putting up with me and getting me all the material I requested. Living in the middle of nowhere is great, until you need something you can't get.

INTRODUCTION

"Those who cannot remember the past are condemned to repeat it."

—George Santayana

What is total war? How do we define it? While the actual phrase *total war* was first used by Carl von Clausewitz in his book *Vom Kriege* [*On War*] and later used as the title of General Erich Ludendorff's 1935 book *Der Totale Krieg*, the idea and concepts encompassed by the term—also known as *war of annihilation*—are nearly as old as man, encompassing many of the wars fought by mankind. However, this concept, as implemented by many of the combatants, reached new levels during World Wars I and II. Since the end of World War II, much time and effort has been spent trying to determine how such heinous acts could have occurred. From the end of the Thirty Years' War (1618–1648) to the beginning of the First World War (1914–1918), numerous steps were taken to reduce the involvement of civilian populations in war. However, in less than thirty years (1918–1945), most of these advances were negated, and civilians were once again targeted extensively during the Second World War (1939–1945) as well as during both the periods leading up to this conflict (1933–1939) and following it (1945–present).

In the introduction to his book *A War of Peoples, 1914–1919*, Adrian Gregory describes the problem of writing a book like this one. Referring to the sheer volume of documents available on any given historical topic, he writes, "The result is countless billions of pages of text in dozens of languages, far beyond the capability of any single historian to know anything but the tiniest fraction first hand."[1] He continues, "No historian can hope to develop first-hand knowledge of any but a small portion of them [the overwhelming number of records]. As a result any general history is reliant on the interpretations of experts, but at times an author must make choices on the basis of what is felt to be the balance of probabilities."[2] Such is the conundrum of determining what is presented in this book. As such, I made the best choices I believed possible in writing this work. Any omissions or errors contained herein are my sole responsibility.

Before further discussing the philosophy of total war, this definition, along with others, needs to be established. This concept has existed since early writings on military philosophy, though a precise definition has been difficult to obtain. Jeremy Black devotes the entire first chapter of his book *The Age of Total War, 1860–1945* to a philosophical discussion of the meaning of the term *total war* without reaching a definite conclusion. Instead, he is only able to reach a working definition of the concept suitable for his book.[3] Raymond Aron defines *total war* as the "merciless mobilization of the national resources and a race for inventions," while Bertrand de Jouvenel describes it as the "total identification of the nation with the army."[4] For this book, the definition of *total war* is the "deliberate use of military force against the enemy's civilian and non-combatant population."[5] Stig Förster writes, "The whole involvement of civilians in war, as active participants and its victims, is one of the most significant hallmarks of total war."[6] Quincy Wright further expounds,

> The object of war is the complete submission of the enemy. It is assumed that military methods aimed to destroy or control his armed forces and occupy his territories, economic *measures designed to starve his population and reduce his resources*, and diplomatic measures designed to destroy his hope of relief or support will, when sufficient, induce the enemy government and population to change their minds and submit to whatever terms are demanded.[7]

Similarly, Jan Philipp Reemtsma writes that a war of annihilation is a "war which is waged . . . in order to exterminate or merely to decimate a population . . . to kill the enemy in the greatest possible numbers."[8] Black, paraphrasing Arthur Marwick, further expounds on this view of by describing total war "in terms of greater destructiveness and disruption, a fundamental challenge to the socio-political order, greater popular participation due to the mobilization of national resources, and a major impact on value systems."[9] However, as Black points out, this description does not account for either political purpose or military means. As Lawrence H. Keeley writes, "War has always been a struggle between peoples, their societies, and their economies, not just warriors, war parties, armies, and navies."[10]

As this book deals with the effect of war on civilians, how does one define *civilians*? Often, they are defined by stating what they are not: They are not combatants, nor are they members of a viable political party.[11] Neither are they civilians who aid institutions that sanction guerrilla warfare.[12] In earlier times, civilians were easy to distinguish because wars were fought by professional armies clad in distinctive uniforms. However, in today's world, most conflict is waged not by professional armies but by ordinary people who are unable to be identified as combatants. This problem leads to the issue of how to target the combatant without involving innocent victims. Richard Rubenstein points out that the US public may prefer targeting specific individuals over a general war as a way of minimizing potential harm to not just the soldiers involved but also the civilians who would perish in the conflict. However, this attitude can have the paradoxically opposite effect; targeting the specific individual leads to the killing of civilians associated with said target, including family, friends, neighbors, and visitors to the area selected for attack.[13] Clearly, while the philosophical debate on the most appropriate way to protect the innocent may rage on, there are no easy answers.

Because I am dealing with definitions, one more is in order. How does one define *genocide*, along with other acts of mass violence? The United States Holocaust Memorial Museum defines *genocide* as

> any of the following acts committed with intent to destroy, in whole or in part, a national, ethnic, racial or religious group.
>
> - Killing members of the group.
> - Causing serious bodily or mental harm to members of the group.
> - Deliberately inflicting on the group conditions of life calculated to bring about its physical destruction in whole or in part.
> - Imposing measures intended to prevent births within the group.
> - Forcibly transferring children of the group to another group.[14]

Further definitions of other acts directed at large groups include the following:

Crimes against humanity are "any of the following acts when committed as part of a widespread or systematic attack directed against any civilian population." The acts include murder, extermination, enslavement, deportation, imprisonment, torture, rape (and other gender-based or sex crimes), group-based persecution, enforced disappearance, apartheid, and "other inhumane acts of a similar character intentionally causing great suffering or serious injury to body or to mental or physical health."

Ethnic cleansing refers to the forced removal of an ethnic group from a territory.

Mass atrocities are instances of "large-scale, systematic violence against civilian populations." Although the term *mass atrocities* has no formal legal definition, it usually refers to genocide (defined earlier), crimes against humanity, war crimes, and ethnic cleansing.

Mass killing is the deliberate actions of armed groups, including but not limited to state security forces, rebel armies, and other militias, that result in the deaths of at least one thousand noncombatant civilians targeted as part of a specific group over a period of one year or less.

War crimes are serious violations of international humanitarian law and occur in the state of armed conflict.[15]

Finally, World Without Genocide lists the eight stages of genocide, which it uses to determine the status of human rights violations:

1. Classification: The labeling of "us" versus "them" based on ethnicity, race, religion, or nationality
2. Symbolization: Names or symbols given to the classified categories
3. Dehumanization: The comparison of humans to animals, insects, or diseases, thereby facilitating the murder of the classified group
4. Organization: The unification of groups to train militias that are then used in the genocide
5. Polarization: The spread of propaganda, limiting the contact of the group with others, and/or the creation of laws, thereby excluding the group
6. Preparation: The identification of the intended victims
7. Extermination: The mass killing of the identified victims
8. Denial: The cover-up of the mass killings, along with the intimidation of any witnesses and placing the blame for the killings on the victims[16]

I explore these eight stages repeatedly throughout this book.

The casualties from genocide are enormous. Estimates of the number of deaths from genocide range from a minimum of 60 million to close to 170 million.[17] While war has always resulted in casualties, many of whom were innocents caught in the wrong place at the wrong time, the frequency of the occurrence of genocide and specifically targeting civilians has increased dramatically starting in the early twentieth century, and it has continued unabated. It is now time to study these occurrences.

So in reading and contemplating this introduction, one may be inclined to ask, "So what?" Well, all we have to do is not be oblivious to the world around us to know that the world is a mess. War, for all man has done to try to eliminate conflict in this world, remains present in our lives, either directly or indirectly. This leads us to wonder if man can ever eliminate war or conflict in general. Is man inherently good or evil? If he's good, then we'll keep on trying to eliminate the malevolence in the world. If he's evil, then, Houston, we have a problem.

This idea for this book has been incubating in my mind for some time. It is not an in-depth analysis of every conflict man has ever had. Such an undertaking would be nearly impossible and just as nearly suicidal. Instead, it looks at war throughout history and how it has affected those who were not fighting when they become involved. And not fighting is meant to include all noncombatants, whether they started out as one or not. Perusing the table of contents, you can get an idea of how this book is divided. Chapter 1 describes the development of warfare from ancient, prehistoric times until 1860. In there, I track the codification of war and its implications on the civilian. The remaining chapters follow suit as I progress until today. Finally, in chapter 7, I come back full circle to the questions I pose in the previous paragraph: Is man inherently good or evil? Does his ability to wage war affect that question or its outcome?

In philosophical debates that I have had on this topic in years past, the question would arise, Why do we have laws? Because this is not a philosophical text, I skip the discussion and come to my personal conclusions. I believe in a higher power. Otherwise, for me, laws are arbitrary and capricious, and I will obey them as long as it is convenient for me to do so. So what does this have to do with this book? I get there in that last chapter.

You are free to question why certain information is included or excluded. I had to make a decision on content, so I did. At the end of the day, my goal is to inform and to teach people. If I have done so with this book, then I am satisfied with it.

CHAPTER 1

ANTIQUITY–1860

"The supreme art of war is to subdue the enemy without fighting."

—Sun Tzu

Historically, war was waged to involve the entire population. The early history of mankind saw very little distinction between differing classes of people. As such, the entire population hunted and gathered food when necessary and protected itself as needed. Therefore, any conflict involving early man would be considered total war, as the entire population was involved.[1] The number of individuals in these tribes was so small that it was not possible to leave women and children behind while the men fought. Doing so would require some of the men to be left behind to guard those not fighting, thereby taking away from the number of combatants. With the relatively small number of tribesmen, all their members had to fight to have a chance of winning. Losing typically meant the elimination of the defeated tribe, either by the death of the combatants of the losing side or by taking the survivors and either incorporating them into the victorious tribe or using them as slaves.[2] The same is true today for small, isolated, Indigenous tribes. For example, in one band of Copper Inuit in Arctic Canada, every adult male had been involved in a "homicide."[3]

This field of thought leads to the conclusion that the larger the population involved in a conflict, the lower the percentage of the population that can be involved in the conflict and the further from total war the conflict becomes. As Jeremy Black further explains, the Zulus, as they fought the British in 1879, expected all men from age sixteen into their sixties to fight.[4] In this example, therefore, the Zulu were more closely engaged in total war than were the British due to the higher percentage of citizens involved.

War is almost universally viewed as an evil activity that should be avoided at nearly all costs. However, Georg W. F. Hegel in his book *Philosophy of Right* states, "War should not to be regarded as an absolute evil."[5] Vanhoutte further explains, "Far from being the absolute evil, . . . war can even be, and has been,

6

understood as the basis of civil society. Furthermore, war can truly function as a means to investigate and even understand civil society."[6] Black corroborates these sentiments.[7]

PREHISTORY–1500

As pointed out, the early history of mankind saw very little distinction between classes of people. Therefore, the entire population hunted and gathered food when necessary and protected itself as needed. In this aspect, as Vanhoutte states, war is not an evil; it is a fact of life of societies of early tribal man. Through early history, war was often waged in such a manner. Progressing through history, we saw the rise of discrete groups of people organized into distinct fighting parties, such as we see today: the Greeks against the Trojans in *The Iliad*, Hannibal and the Carthaginians against the Romans, the Mongols under Genghis Khan, Christians against Jews and Muslims during the Middle Ages, the Spaniards against the Aztecs, and the United States against Native Americans.[8]

Warfare in man's prehistory did not involve the large numbers of individuals commonly associated with modern (beginning in the seventeenth century) warfare. However, armed conflict did exist in these early eras. George R. Milner defines *warfare* during this time as "situations where separately constituted and spatially discrete groups of people engage in armed, often planned, potentially lethal, and culturally sanctioned confrontations that advance the shared interests of the members of the separate communities that take part in the fighting."[9]

Milner points out that due to archeological limitations, we do not know many of the details of these conflicts, such as motives, total numbers of combatants, or planning involved, but we do know that they occurred.[10] There is a train of thought among many people that these pre-Columbian civilizations all lived together in perfect harmony, with conflict between neighboring clans rarely, if ever, occurring.[11] This thought follows the writings of Jean-Jacques Rousseau, who believed that warfare only existed among ancient mankind when they were hungry.[12] Russell Means and Marvin Wolf explain, "Before the whites came, our conflicts were brief and almost bloodless, resembling far more a professional football game than the lethal annihilations of European conquest."[13] This belief is erroneous and needs to be eliminated, as I show.

For example, there are many well-documented reports of this small-group warfare. Fossilized bones of a small group of people dating to around 10,000 years ago show that they were slaughtered by some other party. Of the twenty-seven skeletons found, there were at least eight women and six children. One of the women was in late pregnancy, as fetal bones were also discovered. Twelve of the skeletons were relatively complete, and, of these, ten showed signs of blunt-force trauma. There were also signs of trauma consistent with arrow wounds. The leader of this study, Dr. Marta Mirazón Lahr, expounds,

"These human remains record the intentional killing of a small band of for-agers with no deliberate burial, and provide unique evidence that warfare was part of the repertoire of inter-group relations among some prehistoric hunt-er-gatherers."[14] These strategies of plunder of material goods, destruction of such personal property as housing and crops, and the capture or murder of women and children are all features of total war among these small tribes.

While this incident may be the earliest record of warfare, it is by no means the only one.[15] According to Keeley, approximately 90 to 95 percent of all prehistoric societies engaged in some type of warfare.[16] The few that did not participate in these actions were invariably very small bands and tribes that were isolated by great distances of inhospitable land.[17] Alaskan Inuit, normally portrayed as "happy, peaceful, honest, smiling people who were friendly to-ward strangers" are actually described by their descendants as "quarrelsome and warlike."[18] Attacks against other groups of people were usually out of revenge. Because the areas had been inhabited for so many years, with little territorial movement, the possibility of long-term grudges with the concomitant desire for revenge was great.[19] Quest for land was limited, as any new land acquisition would necessitate the abandonment of the old land due to the lack of tribes-men to maintain the old land while populating the new land. Again, every able-bodied male was expected to participate in any action against rival clans, whether offensive or defensive.[20] Attacks against rival clans attempted to catch the group off-guard while they were in their lodges, setting the lodges on fire and either trapping them inside while they burned or killing the residents as they tried to leave. If they were successful in fleeing, they would typically re-turn to find their houses burned and food either taken or destroyed.[21] If instead the attacking party was successfully repelled, one or two survivors would be allowed to remain in order to return to their clan and warn them against at-tacking the victorious group again.[22]

Inuit oral tradition in the Saunaktuk region of the Canadian Northwest Territories tells of an encounter between the Inuit settlement there and the attacking Athabaskans (approximately 1370 AD). While the Inuit men were away whaling, the Athabaskan raiding party attacked the settlement, slaugh-tering old men, women, and children. The story of the Inuit being tortured, murdered, and mutilated is corroborated by archeological findings of bones showing clear signs of mutilation. J. J. K. Simon attributed these signs to "mor-tuary custom, which is not fully understood."[23] However, Jerry Melbye and Scott I. Fairgrieve refute this theory: Not only are there no other reports of such postmortem events anywhere in the Arctic involving a "mortuary custom of slashing or chopping the deceased, defleshing and dismembering the body, splitting the long bones, and scattering the remains in a random fashion about the site," but also the Inuit oral traditions corroborate the manner in which the human remains came to be located where they were, in the condition in which they were found.[24]

Canadian Inuit share a similar history. Long before the Europeans entered the picture, the Inuit attacked the Cree in the area of Hudson Bay:

> And this raid from the Inuit warriors attacked the group of Omush-
> kegok [Cree] people, mostly women and children and the elders,
> while the men were hunting. . . . The Inuit descended from their
> hiding place and attacked the camp, killing all the elders and the
> children and women. And also being so excited by killing, they
> begun to act very strangely and very savagely which led them to cut
> the breast of the women who were . . . nursing the children.[25]

Nicolas Jeremie in the early eighteenth century noted that the Inuit "make war on all their neighbours, and when they kill or capture any of their enemies, they eat them raw and drink their blood. They even make infants at the breast drink it, so as to instil [sic] in them the barbarism and ardour of war from the tenderest years."[26] This was one area where wars were fought for territory so that the conquering tribe could use the acquired land for seal, geese, and duck hunting.[27] When attacking the Cree tribes, the Inuits' preferred strategy was to arrive at the enemy's camp; hide until nightfall; then attack the sleeping, defenseless enemy.[28] The animosity between the Inuit and the Cree was such that when the Cree obtained efficient breech-loading and repeating firearms from the French, the Inuit were slaughtered and driven out of the southern areas they had inhabited: "There is such enmity between the Indians and Esquimaux, that the former have massacred all of the latter within their power."[29]

The weapons used by the various tribes during the prehistory era included not just the arrows and spears typically thought of when discussing ancient warfare but also poisons applied to the tips of the weapons. While it is difficult to document the existence of poisons, given their tendency to not survive in the soils in which the weapons were found, such examples exist. Some Nevada Shoshoneans drained the blood from the hearts of mountain sheep, placed it in a section of intestine, then buried it to allow it to rot before coating their arrowheads with it. These types of poison could only be useful in warfare: Allowing prey to develop infections over days or weeks before dying would serve no useful purpose, as the prey would have to be tracked during this time. However, allowing injured enemy warriors to develop this type of infection would kill just as effectively as a shot to the head or heart.[30] Other agents used as poisons included mud, orchid fibers, plant alkaloids (e.g., curare), snake venom, crushed red ants, scorpions, spiders, and poisonous plants (e.g., hemlock).[31] The one weapon that was not used to any appreciable extent was artillery. Weapons like catapults are only good against stationary targets, like buildings, or large bodies of combatants, like in the trenches of World War I. The primitive tribes had neither. The Sioux understood that "nobody with any brains would sit on his pony in front of it."[32] Not every weapon is suitable for every situation that the Indigenous peoples encountered.

Furthermore, the mortality rates of these primitive civilizations far exceeded those of any modern civilized population. The mortality rates of these primitive peoples ranged from around 10 percent to more than 40 percent.[33] Moreover, these rates seriously underestimate the true number of war deaths, as these statistics were compiled using only bodies that had projectiles embedded in their skeletons. Those who died of wounds that did not leave behind such objects can never be determined. In comparison, mortality rates among civilized states were merely a fraction of those percentages. France in the nineteenth century had a mortality rate of only about 3 percent—representing *all* the dead from warfare, regardless of cause (battle, disease, starvation, etc.).[34] The world would not see the approximation of such mortality rates as seen by ancient man until the Korean War in 1950.

Regarding these weapons and large death rates, M. Meggitt writes,

> We simply do not know how many infants and old people succumb to pneumonia in these [forced and sudden] flights, how many refugees are drowned when trying to cross boulder-strewn torrents, how many already sick and weak people die because food supplies are interrupted. These less obvious costs of war, I believe, accumulate significantly through time and . . . have played their part in effecting a relatively low rate of population growth in the past.[35]

And contrary to popular belief, the introduction of European warfare to the various tribal groups did not lead to an increase in violent trauma. The percentage of burials in coastal British Columbia showing signs of this type of trauma decreased from a range of 20 to 32 percent in prehistoric periods to 13 percent from 1774 to 1874.[36] As Keeley writes, "Primitive warfare is simply total war conducted with very limited means."[37]

It was not until the further refinement of civilization, with discrete divisions of labor, that the idea of separate military and civilian forces came into existence. With the rise of a money-based economy, cities and wealthy lords could hire soldiers who would work for them full-time instead of relying on their subjects to fight for them. This rise of the professional soldier allowed other noncombatants to remain less involved in war.[38] This was important not only for protecting the members of the society as a moral issue but also for allowing those who were not fighting to stay back and continue tending crops, fabricating clothing, and forging weapons for the soldiers.

As these cultures grew, the reasoning for such conflict advanced. Early battles evolved from small raids to obtain food from neighboring villages or avenge perceived injuries "into large-scale endeavors led by high-ranking individuals to achieve cultural prestige."[39] Such warfare continued into the Middle

Ages. One example of this is how the Mongols, especially under Genghis Khan, fought. A Mongol emissary would be sent to the ruler of a city with a very clear message: "Surrender and avoid a military confrontation."[40] Should the ruler decline to do so, the Mongols would annihilate the city by diverting rivers and constructing dams and walls to cut off the city from its surroundings and then set it on fire.[41]

Another example is Rome during the First Punic War (264–241 BC), who successfully defeated the Carthaginian navy at the Battle of Cape Ecnomus in 256 BC. After that, the Romans, under consul Atilius Regulus, invaded Africa and captured the town of Aspis. It was then that the Roman forces began pillaging the surrounding area in hopes of finishing the Carthaginians the following spring. They were so successful in this war of attrition that peace negotiations began between Rome and Carthage. While the question of who initiated these peace talks remains, the Roman plunder of the area of Apsis-Adys-Tunis was key to these talks. It was only because of Rome's greed with excessively punitive terms of surrender for Carthage that the Africans balked:

> According to [Cassius Dio], the consul asked the Carthaginians to abandon Sicily and Sardinia, to free the Roman prisoners without claiming for damages, to ransom the Carthaginian prisoners, to pay for the war expenses and to pay a yearly tribute in Rome's treasury. Also, the Africans were not allowed to make peace or declare war without Rome's consent. The consular peace project also stipulated the Carthaginians should give up their war fleet.[42]

The Carthaginians reorganized under the Spartan Xanthippus and crushed the Romans in the resulting battle. Of more than 15,000 men, only 2,000 survived. Regulus was captured and tortured to death, and the evacuation fleet was nearly destroyed by a storm, resulting in as many as 90,000 Romans drowning.[43] However, the Romans would have the last word. At their defeat, "Carthage was destroyed . . . [with approximately] 150,000 out of 200,000 Carthaginians massacred by the Romans: proverbially, the Romans covered the ruins of Carthage with salt to prevent anything from ever growing."[44] The remaining survivors, including women, were sold into slavery. This had all been at the urging of the Roman politician Cato the Elder: Roman historian Plutarch wrote, "The annihilation of Carthage . . . was primarily due to the advice and counsel of Cato."[45] Ben Kiernan calls this possibly the "first recorded incitement to genocide."[46]

Besides these more traditional manners in waging war, chemical, biological, and radiological (including, eventually, nuclear) weapons (CBRW) were frequently used. These methods, except for radiological and nuclear, which are less than one hundred years old, are as old as war itself. While it is very difficult to conclusively determine how these weapons were used or what they were, we can make some very educated guesses based on what little evidence we have. The Hittites in the fourteenth century BC are among the earliest evidence of the use of CBRW. They sent diseased rams, possibly infected with tularemia, to their enemies to weaken them, although this may have been primarily a means to rid themselves of disease with the infection of their enemies merely a bonus.[47] Furthermore, a Hittite manuscript specifically mentions the use of plague-infected arrows being used by the Luwian deity Yarri against the Hittites: "But then one draws a bow and puts an arrow in place. [The other] arrows he [Dandanku] scatters before him and says: 'God, keep shooting hence at the enemy land with arrows! But when you come into the land of Hatti, let the quiver remain closed for you, let the bow be loosened for you!'"[48]

Additionally, in one of the few ancient references to the specific use of poison on an enemy, the "Poem of Erra and Išum" ["Erra Epos"] tells of Erra, god of chaos and pestilence who, along with his helpers, used poison to defeat his foes.[49] Around 600 BC, the Athenian army used hellebore plants to poison the water supply of the city of Kirrha.[50] The siege against the city lasted from 595 to 591 BC, and when the city finally fell to the Athenians, it was razed to the ground, and the resulting plain was dedicated to Apollo.[51]

Herodotus in the fourth century BC wrote about Scythian archers poisoning their arrows with a mixture of decomposing cadavers of adders and human blood. This infection could have very well been *Clostridium perfringens* and *Clostridium tetani*.[52] Later during the Peloponnesian War, about 430 BC, the city of Plataea was initially besieged by the Peloponnesians. During the siege, every tactic was used to take the city and eliminate the Plataeans, including burning the city. Part of this burning involved sulfur and pitch to smoke them out of the city if the fire was not sufficient.[53] The use of these components was prescient, as Thucydides wrote:

> And then lighted the wood by setting fire to it with sulfur and pitch. The consequence was a fire greater than any one had ever yet seen produced by human agency, though it could not of course be compared to the spontaneous conflagrations sometimes known to occur through the wind rubbing the branches of a mountain forest together.
>
> And this fire was not only remarkable for its magnitude, but was also, at the end of so many perils, within an ace of proving fatal to the Plataeans; a great part of the town became entirely inaccessible, and had a wind blown upon it, in accordance with the hopes of the enemy, nothing could have saved them.
>
> As it was, there is also a story of heavy rain and thunder having come on by which the fire was put out and the danger averted.[54]

The ancient Greeks were well versed in the use of fire. In 672 AD, they introduced Greek fire, a major innovation in that it would burn on water. Invented
by a Christian named Kallinikos (Callinicus) of Heliopolis, it was composed
of a petroleum-based material or naphtha, to which sulfur, pitch, quicklime,
potassium nitrate, or other materials were added. It was employed with deadly
success, especially in naval engagements, as water would not extinguish the
fire—it required sand or vinegar to put out.[55]

In 1346, the Genoese city of Caffa was besieged by the Mongols. In doing
so, they reportedly threw plague-infected cadavers into the city. When the inhabitants fled the city, they supposedly carried the plague into the Mediterra-
nean basin, thereby unleashing the Black Death on Europe. Gabriel de Mussis
wrote,

> In 1346, in the countries of the East, countless numbers of Tar-
> tars and Saracens were struck down by a mysterious illness which
> brought sudden death. . . . The dying Tartars, stunned and stupefied
> by the immensity of the disaster brought about by the disease, and
> realizing that they had no hope of escape, lost interest in the siege.
> But they ordered corpses to be placed in catapults and lobbed into
> the city in the hope that the intolerable stench would kill everyone
> inside. . . . And soon the rotting corpses tainted the air and poisoned
> the water supply, and the stench was so overwhelming that hardly
> one in several thousand was in a position to flee the remains of the
> Tartar army. . . . Thus almost everyone who had been in the East, or
> in the regions to the south and north, fell victim to sudden death
> after contracting this pestilential disease, as if struck by a lethal
> arrow which raised a tumor on their bodies.[56]

Whether the story of the spread of plague to Europe is true or not, it is
nearly certain that the use of contaminated cadavers to force the inhabitants
of Caffa to flee did happen. However, one thing to remember about the use of
these biological weapons during these early days: It is very difficult to distinguish man-inflicted disease outbreaks from natural outbreaks that would frequently occur due to poor sanitation. Nonetheless, the use of CBRW in early
history was not unheard of (see table 1).

Sun Tzu, a preeminent strategist from 500 BC whose teachings are still
studied today, realized the need for quick warfare, stating that victory was the
main goal of war. If this goal is delayed, then morale suffers, and wars of long
duration drain the resources of the country.[57] He further postulated that troops,
while providing their own weapons, should live off the enemy's provisions:
"Hence the wise general sees to it that his troops feed on the enemy, for one
bushel of the enemy's provisions is equivalent to twenty of his."[58] Thus, while
wars were not to be long, drawn-out sieges, the troops were to travel lightly and

Table 1. Examples of Chemical and Biological Warfare Through History

Time/Year	Event
600 BC	Solon and the Athenian military use the purgative herb hellebore during the siege of Krissa (Kirrha).
479 BC	Peloponnesian forces use sulfur fumes against the town of Plataea.
1155	Emperor Barbarossa poisons water wells with human bodies in Tortona, Italy.
1346	Mongols catapult bodies of plague victims over the city walls of Caffa, Crimean Peninsula.
1495	The Spanish mix wine with blood of leprosy patients to sell to their French foes in Naples, Italy.
1650	The Polish fire saliva from rabid dogs toward their enemies.
1675	The Strasbourg Agreement, signed by Germany and France, outlaws the use of poison bullets, the first such international agreement.
1710	Russian troops catapult plague victims into Swedish cities.
1763	The British distribute blankets from smallpox patients to Native Americans.
1797	Napoleon floods the plains around Mantua, Italy, to enhance the spread of malaria.
1845	During the conquest of Algeria, French troops drive more than one thousand Berber tribesmen into a cave and then use smoke to kill them.
1863	Confederates sell clothing from yellow fever and smallpox patients to Union troops during the US Civil War.
World War I	German and French agents use glanders and anthrax as well as various chemical gases.
World War II	Japan uses plague, anthrax, and other diseases against the Chinese; several other countries experiment with and develop biological weapons programs.
1980–1988	Iraq uses mustard gas, sarin, and tabun against Iran and ethnic groups inside Iraq during the Persian Gulf War.
1995	Aum Shinrikyo uses sarin gas in the Tokyo subway system.

Adapted from Sarah Everts, "A Brief History of Chemical War," *Distillations Magazine*, May 12, 2015, https://www.sciencehistory.org/stories/magazine/a-brief-history-of-chemical-war/; Friedrich Frishknecht, "The History of Biological Warfare: Human Experimentation, Modern Nightmares and Lone Madmen in the Twentieth Century," *EMBO Reports* 4, no. S1 (June 1, 2003): S47, https://doi.org/10.1038/sj.embor.embor849; and Stefan Riedel, "Biological Warfare and Bioterrorism: A Historical Review," *Baylor University Medical Center Proceedings* 17, no. 4 (October 2004): 401, https://doi.org/10.1080/08998280.2004.11928002.

live off the enemy. This, by definition, entails forcing the civilian population of the enemy country to participate in the suffering of war. However, he also explicitly stated that the "best policy is to take a state intact; to ruin it is inferior to this."[59]

The Persians under Xerxes experienced the consequences of not being able to live off the land when they invaded Greece in 480 BC as part of the Greco-Persian Wars. The rugged Greek countryside provided very little vegetation upon which his troops and pack animals could survive. Nor was there much water, as Greece experiences a yearly eight-month-long dry period. The Greeks, knowing the land and its limitations, were able to turn these disadvantages into one of their major advantages in the defense of their land.[60] And while King Leonidas and his three hundred Spartans were eventually defeated at Thermopylae, it was only through the treason of a Greek named Ephialties.[61] This was not the only principle of Sun Tzu that the Greeks employed: From principles of leadership to those of positioning of troops, the Spartans provided real-life examples of Sun Tzu's teachings and other principles.[62] While we will probably never know if Leonidas and his Spartans could have continued to hold out against the Persians, it is obvious that these principles, when correctly applied, could allow a small force to successfully defend against a much larger one.

The modernization of civilization leads to the argument that war cannot be total because no modern society can devote its total resources to the war effort, and the more technologically advanced a society becomes, the less it can wage total war.[63] This clear division of labor also led later military to enact laws of war to protect noncombatants.

This is not to say that the rise of the professional soldier led to the protection of the various landowners as soldiers fought in battle. Quite the contrary, what passed for a rule of law in those days had been subjugated to the attacks of robber barons and allies who were engaged in numerous private wars they fought in an effort to obtain land. Because of the chaos that these private wars brought to the area of Aquitaine, in modern southwest France, in 1000 AD, a movement began in the Roman Catholic Church to try to curb these battles.[64] Known as the Peace of God followed by the Truce of God, these movements attempted to quell these clashes through the imposition of church-sanctioned punishments, up to and including excommunication. The Peace of God strove to limit the actual conflicts and impose a peace of sorts in an attempt to preserve property ownership by both the church and laypeople. The meeting set forth three offenses that would incur such a harsh measure as excommunication: taking ecclesiastical property by force, plundering agricultural resources from peasants, and attacking unarmed clerics.[65] The Truce of God set forth specific times of peace, fixed by the liturgical calendar. The movement was a tempered success, as there were multiple excommunications while it spread throughout France and into Spain and Italy.

1500s–1700s

From the 1500s to the 1700s, the thought process on the waging of war was slowly beginning to change. R. R. Palmer pointed out that the monarchies of the Enlightenment tried to spare their civilian populations the ravages of war for both reasons of humanity and as a means of protecting their sources of revenue.[66] The Thirty Years War (1618–1648) left Germany so devastated that it nearly descended into a post–Roman Empire Dark Age. Out of sheer necessity of preservation of civilization, European leaders realized that the way they waged war had to change.[67] It was this evolution in thinking that led to the rules of war forbidding the military involvement of nonmilitary personnel that has continued to our current day.

These rules were first published in Hugo Grotius's *De jure belli ac pacis* in 1625 and were published again by Emmerich de Vattel in *The Law of Nations* in 1758.[68] These two writers believed, contrary to the established customs of warfare, that women, children, the elderly, and the infirm should not be considered the enemy, and they disapproved of the use of poisonous weapons or the contamination of drinking water. Contributing to this line of thought was the further development of fortresses and the accompanying siege warfare that accelerated the separation of war from the civilian population. Limited resources by necessity limited the severity of the wars fought.[69] Slowly, the rules of war were changing, with an emphasis on protecting specific parts of the population.

Colonial America, 1700–1790

The American view of war differed greatly from that of the Europeans The differences between the new North American settlers and the Native Americans were great enough that the colonists had little trouble waging a more total type of war to achieve their goals.[70] Warfare on the frontier was more akin to the concept of total war than traditional European warfare during this time. Prisoners were frequently murdered, and civilians were often targeted to rid the areas of the European intruders or the Indigenous tribes. The story of the settlers of the New World exposing the Native Americans to smallpox and how that disease ravaged them is widely known. What may not be as well known is that this exposure was often done intentionally to deal with the Native Americans instead of fighting them. However, as pointed out earlier, distinguishing between the intentional introduction and the natural spread of a disease can be problematic. First, the disease had already been introduced to the Americas, so plausible deniability of this new infection is a possibility. Second, there is no control over what happens in combat, and people can be exposed to diseases that otherwise would not happen. Third, one cannot always prove intent, unlike such blatant acts as rape, pillage, and associated atrocities. Finally, even if the outbreak was intentional, the perpetrators may not have recorded the event, leading us to believe it was natural. What follows are a few of the many examples that lead us to believe that this method of warfare was more common than previously believed.[71]

In 1623, Dr. John Pott supposedly poisoned Native Americans in retaliation of a Powhatan attack in which 350 English died. Captain William Tucker left with twelve men to retrieve English prisoners. With him, he brought a drink that had been poisoned by Dr. Pott. The English met up with the Native Americans, whereupon they were offered the drink. The American Indians were smart enough to demand that the English drink first, but the English were prepared for such a demand and drank from a different container. After the meeting, some of the American Indians were walking with an English interpreter. At a prearranged signal, the interpreter fell to the ground, and the English unleashed a volley of shot into the group of American Indians. The English reported that about two hundred Native Americans died of poison and fifty died of gunshot wounds. The English were disappointed that the Native Americans' leader, Opechancanough, was not among the dead.[72]

In the early 1700s, according to Pierre-François-Xavier de Charlevoix, the Iroquois, during Queen Anne's War, employed these exact methods. He wrote that the English army "was encamped on the banks of a little river; the Iroquois, who spent almost all the time hunting, threw into it, just above the camp, all the skins of the animals they flayed, and the water was thus soon all corrupted."[73] British deaths from this act numbered more than one thousand.[74] Fenn claims that this account is noteworthy because this was seemingly the only such deed by Native Americans against the British and also because they did not use smallpox. This shows that Native Americans were already well aware of the deadly nature of the virus and believed that they would suffer just as much as, if not more than, the British would from the use of the virus.[75]

In 1757, during the Seven Years' War, the Potawatomis, who had helped the French, suffered a smallpox outbreak they attributed to the French. However, it also could have arisen from an attack by the tribe on unarmed prisoners who were sick with the disease and leaving Fort William Henry. Likewise, the Ottawa believed the disease came from their French allies in Montreal after this same conflict. Conversely, whether true or merely in an attempt to appease their allies, the French blamed the British for the outbreak.[76] And there is evidence that the British might have done so.

The British under General Jeffrey Amherst, commander in chief of British forces in North America, were told during this confrontation that

> Indians, Canadians and other painted savages of the Island, who will entertain them in their own way and preserve the Women & Children of the Army from their natural Barbarity. Indians spurr'd on by our Inveterate Enemys the French, are the only Bruts [sic] and Cowards in the Creation, who were ever known to exercise their Crueltyes on the Sex, and to Scalp and mangle the poor sick

soldiers, and Defenceless Women. When the light troops have by
Practice and experience, acquir'd as much Caution and Circum-
spection, as their spirits & activity these Howling Barbarians will
fly before them.[77]

It is clear from the language in this letter, especially the terms *bruts*, *cowards*,
and *barbarians*, that the British viewed the Native Americans as less than
human (a quality common throughout warfare, as I shall show) and would do
whatever it took to win. They very possibly could have planned and executed
such an attack. However, the French were also blamed when they paroled approximately three hundred British prisoners from the same Fort William Henry
clash. The British firmly believed that this was an attempt to introduce smallpox into Halifax, Nova Scotia, the destination of the vessels involved. Regardless, this outbreak failed to materialize.[78]

In 1763, Fort Pitt, built by the British between 1759 and 1761, where the
Monongahela and Allegheny Rivers join to form the Ohio River, in current-
day Pittsburgh, Pennsylvania, was attacked as part of Pontiac's Rebellion by
the local tribes. Both soldiers and civilians inside the fort had smallpox, and
the idea to use this "advantage" in the conflict arose. Two Delaware dignitar-
ies were permitted entry into the fort under the guise of a parlay. These two
Native Americans were given two blankets and a handkerchief laden with
the disease taken from the smallpox hospital in the hopes that it would spread
throughout the various tribes and decimate their ranks.[79]

Credit for this tactic is generally given to General Amherst, but it could
just as well have been conceived of by other commanding officers. Regardless,
Amherst quickly approved of the idea.[80] He had written a letter to one of his
commanders, Henry Bouquet, stating, "Could it not be contrived to Send the
Small Pox among those Disaffected Tribes of Indians? We must, on this occa-
sion, Use Every Stratagem in our power to Reduce them." Bouquet responded,
"I will try to inoculate the Indians by means of Blankets that may fall in their
hands, taking care however not to get the disease myself." Amherst replied,
"You will Do well to try to Innoculate the Indians by means of Blanketts, as
well as to try Every other method that can serve to Extirpate this Execreble
Race."[81] William Trent wrote, "Out of our regard to them, we gave them two
Blankets and an Handkerchief out of the Small Pox Hospital. I hope it will
have the desired effect."[82] It did. Shortly thereafter, the disease attacked the
American Indians in full force. Reports of the plague and its decimation per-
sisted into 1765.[83]

The American view of war differed greatly from that of the Europeans.
With the advent of muskets and cannons, the practical implementation of war
changed significantly. The drilling of troops to function as one cohesive unit;
flags and standards; and trumpets, bugles, and drums all became a part of warfare at this time.[84] Opposing troops faced each other in long lines to achieve

the maximum effect of these weapons and approached each other until they were close enough for their weapons to be useful. Then the discharge of weapons would ensue. Finally, they would fix bayonets and charge each other, assuming one side hadn't broken ranks and retreated.[85]

The colonists were considered traitors for not only seceding from Britain but also for breaking the rules of war. The colonists were smart enough to know that they didn't stand a chance fighting the British on their terms, especially after their defeat at New York in 1776. Realizing that his troops were dwindling, as were his chances of winning, George Washington knew something had to change. He then realized that he didn't necessarily have to *win* battles; he just needed *not to lose*. He changed tactics, adopting the Native Americans ways of hit and run. Also, compared to the Fabian strategy, this changed the focus from a decisive battle to multiple little skirmishes, leading to a war of attrition.[86] We know these tactics today as irregular warfare or guerrilla warfare. In this, the colonists had the advantage, as they were fighting on home soil, as the Greeks had against the Persians. Snipers were very effective against the British, as were the hit-and-run tactics.[87] Members of Congress were not happy with this decision because they viewed it as "unheroic." However, Washington knew that he had to keep his army at least somewhat intact in order to secure foreign help. Any impulsive decisions could lead to the destruction of the army and with it any hope of securing that help.[88]

Another tactic that differed was that many times, the colonists didn't aim to kill; they aimed to maim. The colonists would load their muskets with shrapnel and aim for the legs. They knew that in so doing, the legs of the British would be torn to pieces, but the redcoats would still be alive, thus necessitating prolonged care. This would result in not just one but at least two or three people unable to fight in the war, "to leave them as burdens on us, to exhaust our provisions and to engage our attention."[89]

Besides this change in tactics, Washington established spy networks for intelligence gathering. One of the biggest successes of these networks came in the winter of 1777, when the British were led to believe that the Continental Army consisted of about 12,000 troops instead of the actual 1,000. All of these changes allowed the colonists to endure seemingly endless hardships until the French joined their cause in 1779.[90]

The British began calling the colonists "savages" for employing the tactics of the "savage" Native Americans. They had very little problem, therefore, in using any means available to put down this rebellion: They implemented biological warfare by attempting to expose the colonists to smallpox. While the British were used to this disease, many of the colonists had never been exposed to it before, due to the very rural nature of the colonies, and the thought of being exposed to this horrific disease terrified them. While this tactic may have been morally questionable, the laws of war at this time apparently permitted it.[91] The works by Grotius and de Vattel as well as military tradition stated when these rules did not apply. Among these exceptions were rebellions, wars against enemies who broke the rules of war first, and wars against "savages" or "heathen" people.[92] In the eyes of the British, the American colonists fell into all three of these categories, similar to how the colonists viewed Native Americans. About the Native Americans, Amherst said, "their Total Extirpation is scarce sufficient Attonement for the Bloody and Inhuman deeds they have Committed."[93] The colonists deserved no less. As Machiavelli wrote, "When it is absolutely a question of the safety of one's country, there must be no consideration of just or unjust, of merciful or cruel, of praiseworthy or disgraceful; instead, setting aside every scruple, one must follow to the utmost any plan that will save her life and keep her liberty."[94] This maxim applied to both sides equally.

The use of smallpox as a weapon was not an isolated event. Charges of the introduction of smallpox came from Quebec, Boston, and Yorktown. In Boston, one British officer inoculated his entire family with smallpox without telling anyone and then let them continue their normal activities.[95] Once the Americans attacked Boston, the tables were turned, and the British and other residents of the city fell victim to the disease.[96] Washington initially refused to believe that the British were intentionally trying to infect the Continental Army. However, when a major outbreak occurred after refugees from the siege of Boston arrived in the company of the Continental Army, Washington changed his mind and decided that it was in fact occurring: "The information I received that the enemy intended Spreading the Small pox amongst us, I coud not Suppose them Capable of—I now must give Some Credit to it, as it has made its appearance on Severall of those who Last Came out of Boston."[97] The Americans were able to avoid an outbreak by quarantining those identified as being infected and disinfecting the patients and all of their belongings.[98] However, they failed to stop the rampant infections in Quebec. The army attempted to take the city on December 31, 1775, but could not and had to endure the winter of 1775–1776 there until British reinforcements came in May 1776 and the Americans were forced to retreat.

It was a common belief among the American troops that this epidemic was the direct result of British actions. However, British general Sir Guy Carleton ensured that captured Americans who were infected received proper care, which would seem at odds for a person who wanted the extermination of his enemy.[99] This is one aspect in which the Americans exceeded the efforts of the British—ensuring that the troops were inoculated against smallpox. Being a rural, agrarian society, many colonists were able to escape the ravages of smallpox. However, bringing so many of these individuals together in such confined quarters was an invitation for a major outbreak of that disease and many others. The inoculations helped prevent these outbreaks.

While more examples exist, as well as many more there is no record of, it is clear that the Revolutionary War was not fought with just muskets and cannons.[100] And as time progressed, so did man's ability to wage war and the propensity to do so. But to fight a war, one needs soldiers as well as support back home. During this time, people were being kept abreast of happenings at home and abroad with newspapers and pamphlets. Newspapers originally reported the happenings of the government, such as laws and proclamations, and foreign affairs. What was not a part of this was news from other colonies. It was not until the 1760s, starting with the Stamp Act, that they began reporting events from the other colonies.[101] Before this time, newspapers tried to remain impartial and merely reported what was happening. They relied on a network of contacts to supply them with the news, often copying stories directly from foreign newspapers. With the beginning of the colonial conflict, newspapers dropped this guise and took sides. Newspapers were known as Loyalist papers or Patriot papers, and the papers relied on specific sources that reflected their own beliefs. This polarization became so great that people associated with the papers had to fear for their lives. In Boston, for example, the supporters of two rival papers, the *Boston Chronicle* (Loyalist) and the *Boston Gazette* (Patriot), came to blows in the streets. People were injured, and guns were fired. It wasn't until the redcoats arrived with drawn bayonets that the scuffle broke up.[102]

Along with newspapers, pamphlets were common reading. These presented ideas to the people; John Bailyn stated that these pamphlets were the "most important and characteristic writing of the American Revolution."[103] They were inexpensive to produce and purchase and easy to carry. According to Bailyn, most of these were printed corresponding to three major events: the Stamp Act (1765–1766), the Townshend Duties and Boston Massacre (1767–1770), and the Boston Tea Party and Parliament's response to the Coercive Acts (1774).[104] Among these pamphlets were Thomas Paine's *Common Sense*, Thomas Jefferson's *A Summary View of the Rights of British Americans*, and John Adams's *Thoughts on Government*.[105] The colonists began publishing them in 1765, but the British didn't start until after the first meeting of the Continental Congress in September and October 1774. They dismissed the Patriot movement as inconsequential and not worth the time to counter the colonists' publications.[106] Thus, propaganda was present even in colonial times.

1790s–1860

While the rules of war were slowly codified, philosophers and thinkers continued to explore just what these rules might be. Niccolò Machiavelli realized that in war, extreme measures were sometimes necessary. Writing about the conquest of a new state, he stated that the best way to hold a newly acquired state was to destroy it so that no memories of their previous liberties remained to act as a catalyst for future rebellions:[107] "Men must be either pampered or crushed, because they can get revenge for small injuries but not for grievous ones."[108] He also recalled how the "Romans, in order to hold Capua, Carthage, and Numantia, destroyed them, and so never lost them."[109] Machiavelli determined that the defense of a state was the responsibility of the entire population and not just that of a select group.[110]

Jacques Antoine Hippolyte, Comte De Guibert, echoed these sentiments in his early work *Essai général de tactique*. He felt that because prisoners were no longer slaughtered in cold blood and that the consequences of conquest resulted in no greater hardship than experienced under the previous rulers, the people had grown soft and weak.[111] He also felt the need for troops to travel light and live off the countries they conquered.[112]

Napoleon

Napoleon Bonaparte is generally considered to be the father of modern total warfare through his use of all available resources in the pursuit of his military objectives. Before Napoleon, European battles were relatively small affairs. He introduced the concept of mass numbers in war. François-René de Chateaubriand wrote that the size of these battles, numbering in the hundreds of thousands of men, bore little resemblance to prior battles that left "peoples in place while a small number of soldiers do their duty."[113] Napoleon fought limited wars with limited means only when forced to do so.[114] While he did not write specifically about the concept of total war, he did allude to certain principles of the concept. He stated, "The country must supply you with provisions, clothes, remounts, and everything necessary for your army so that it will not cost me a sou," and "To know . . . how to draw supplies of all kinds from the country you occupy makes up a large part of the art of war."[115] He also declared that one "must order people to provide oats, and confiscate the oats from those who do not."[116] When faced with a shortage of provisions, he executed his prisoners in order not to have to feed or stand guard over them.[117] However, he also pointed out that the success of an army depends on order and discipline. Pillaging and wanton destruction destroy not only the occupied country but also the conquering army and turn the vanquished population into permanent enemies.[118]

How Napoleon changed warfare can be seen in the actual tactics he employed in his quest for European supremacy. His main objective was always the quick destruction of the enemy in battle.[119] This meant rapid movement of his forces. He achieved this through the use of mobile field guns instead of the heavier cannons used by the other armies. And instead of trying to move his entire army to meet the adversaries, he broke his army into smaller, more mobile units. We know these units today as divisions. He would use these divisions to encircle the enemy troops and attack in force. While he might be outnumbered with a smaller division, once the other two divisions joined forces with the one engaged in battle, he quickly had numerical superiority and would finish off the enemy.[120]

Other innovations were the elimination of supply trains, with his army living off the land. He also implemented conscription laws, giving him larger armies than ever before. These laws provided the troops needed to implement his tactics. With these innovations, Napoleon developed the modern warfare to which we are accustomed.

These "dishonorable" innovative tactics were first used while Napoleon was still a twenty-six-year-old general during the French attack on Italy in 1796. Outnumbered 37,600 to 52,000 Austrians and Piedmontese, Napoleon was at a distinct disadvantage, and, in fact, his forces were initially compelled to withdraw after the Austrians executed a surprise attack. However, in doing so, they left their forces isolated and ready for Napoleon to attack. When he did, he was able to successfully obtain numerical superiority, 9,000 to 6,000 Austrians in one battle and 25,000 to 20,000 in another. Soon thereafter, the Piedmontese were attacked in a battle and routed. A week after that loss, the Piedmontese officially surrendered, something the French had tried to accomplish for six years.[121]

Whether intentional or not, Napoleon was also involved in biological warfare. In the summer of 1809, the British made plans to invade the Dutch coast. The Austrians wanted help from the British to draw away some of the force that they, the Austrians, would be facing. This served double duty, as the British admiralty was issuing warnings that Napoleon was planning on invading Britain from the port of Antwerp. This port, along with the port of Flushing, which was further downriver from Antwerp, had been extensively modified to support the French ships of the line. The final plan's objectives were

1. Capture and destroy enemy whips at Antwerp, Flushing, and on the Scheldt River;
2. Destroy the arsenals and dockyards at Antwerp, Ternuese, and Flushing;
3. Capture Walcheren Island; and
4. Make the Scheldt unnavigable to ships of war.[122]

These plans were rendered unnecessary, as Napoleon had already defeated the Austrians in July 1809. The British decided to proceed with their plans anyway. However, despite the largest British expeditionary force ever assembled, they failed to involve the medical corps in any way. Had they done so, they would have been warned that the invasion area during this time of year was at high risk of contracting disease.[123]

The area of Walcheren was low-lying and covered with dikes, a topography of stagnant pools and swampland that provided the perfect breeding ground for the millions of mosquitos that inhabited the island. Combining the existing swampland with the heavy flooding from the previous year's rains caused even more water and created more swamps, and the hot and steamy August weather and heavy rains led to a perfect storm for disease to be especially present that summer. Additionally, in an attempt to slow the British, the French opened the dikes, which caused even more flooding and the associated disease. This act met with only limited success.[124]

Opening the dikes was done to slow the British advance, but whether the spread of disease was an intentional reason for doing so or a fortunate happenstance, the British were stopped. The attack began in early August, and at first, things went well for the British. They were able to take the towns of Middleburg and Veer and surround Flushing, later capturing it. However, things would slow down almost immediately. From August 6, when the British landed on Walcheren, until September 3, the number of "Walcheren fever" rose from fewer than seven hundred to more than eight thousand.[125] By August 27, 20 percent of the British forces were in the hospital, with more becoming ill by the hour.[126] By late October those stricken with the fever numbered 9,000, which was more troops than those who were healthy.[127] And when the expedition finally ended in February 1810, 60 officers and 3,900 soldiers were dead. Forty percent of the expedition came down with the disease. In comparison, only 100 were killed in actual battle.[128]

So, what exactly was the Walcheren fever? Evidence points to malaria being a major component. This should not come as a surprise, given the requirements for malaria of stagnant water and the right species of mosquito. However, it also appears that dysentery, typhoid, and typhus were present.[129] This appears to have been a particularly virulent and toxic combination of diseases. Diseases in general were of greater threat to the soldiers than was battle. Between 1793 and 1815, estimates put British losses at 240,000, with less than 30,000 of these from battle.[130]

Of course, any tactic works both ways, and biological warfare, even though unintentional, did exactly that. In 1802, Napoleon sent nearly 60,000 of his finest troops to Haiti to put down a slave uprising, thinking it would be easy to do so.[131] There is also evidence that he planned to use Haiti as a starting point for an invasion of the Mississippi Valley via New Orleans. However, while the Native people were immune to yellow fever, the French were not. The area around the ports of St. Domingue consisted of swamps, providing the perfect breeding grounds for the mosquitos that carry the disease. In contrast, the mountainous region of the island was relatively free from mosquitos and disease. Hence, the slaves stayed in the mountains, while the French stayed in the ports. Europeans in general knew of this safety, but the French General Charles Leclerc refused to station any troops in the mountains because of the Haitian guerrillas there. Moreover, before they left for Haiti, Leclerc abandoned most of his medical supplies. Additionally, there was a physician shortage.[132] This was a recipe for disaster.

As a result of this poor planning, the French were decimated. From 1802 to 1803, between 50,000 and 55,000 French troops were killed by the disease in the worst yellow fever outbreak in the New World from the late 1600s to the late 1800s.[133] As many as fifty soldiers per day were dying at one point.

Finally, even after receiving reinforcements, the French were forced to abandon the island in November 1803. Of all the troops sent to the island, only three thousand men made it back to France.[134] One benefit from all of this was that Napoleon gave up his ideas of moving into the Louisiana Territory and instead sold it to the United States for $15 million, effectively doubling the size of the United States.

This was neither the last nor only time Napoleon's ranks were decimated by disease. During the Egyptian campaign of 1798–1799, bubonic plague ravaged his troops. When the plague first struck, the mortality rate was 80 percent. At the height of the epidemic, upward of thirty men were dying each day.[135]

Antoine Henri de Jomini was an ardent advocate of the Napoleonic way of war—massing forces against a definitive weak area. He differentiated between national wars, those that involve the destruction of the enemy army and end with the occupation of the conquered country, and wars of opinion, in which dogmas and doctrines are propagated onto the opposing party. Such wars of opinion, he stated, should be avoided because they become vindictive, cruel, and deplorable.[136] Invading countries have only their army on which to rely, while the invaded country has not only their army but also the entire population to utilize in their defense. In foreshadowing the philosophies of guerrilla warfare as espoused by T. E. Lawrence and Mao Tse-Tung, Jomini specifically mentioned the 1823 war in Spain as an example of the use of peasants led by monks and priests in annihilating all but a single corporal of the invading army in one area.[137] In such cases, the amassment of an invading army would be pointless, as the entire population is the enemy, not a specific military body or location. To be effective, the invading army would have to be large enough to consist of both the attacking forces along with occupying forces left behind at every essential point in the country.[138] The alternative was to calm the people; wear them out over time; and be courteous, gentle, and just to them.[139]

However, Jomini also realized that invading armies needed sustenance and that the invaded country was the source of this nourishment. He credited Caesar with the opinion that war should support war, and while he supported living off the land and disdained the barbarian hordes of the fourth to thirteenth centuries, he acknowledged that this was not likely to change soon.[140] He also believed that the direction of war exclusively toward either organized armies or civilian areas was erroneous and that a middle road between these two extremes was the correct choice.[141]

Carl von Clausewitz was the major theorist of the concept of total war, or absolute war, as he referred to it. He espoused that "war is not merely an act of policy but a true political instrument, a continuation of political intercourse, carried on with other means. What remains peculiar to war is simply the peculiar nature of its means. . . . The political object is the goal, war is the means of reaching it, and means can never be considered in isolation from their purpose."[142] He recognized absolute war (how war should be waged) as differing from real war (how war was actually waged). Absolute war, in his opinion, was most closely achieved by Napoleon after the French Revolution, who waged war "without respite until the enemy succumbed."[143] After a review of the history of warfare, covering the ancient barbarity of the Tartar hordes up through the French Revolution, he, too, noted the progression of conflict from warfare involving entire populations up to the then-current practice of limited warfare among professional armies.[144] With the coming of Napoleon, warfare "again became the concern of the people as a whole . . . or rather closely approached its true character, its absolute perfection."[145] He stated that the fighting forces had to be defeated to the point that they could no longer fight, that the invaded country had to be occupied to prevent new troops from being raised, and that the enemy government and population had to be driven to the point of submission and asking for peace—their will had to be broken to truly be victorious in war.[146] He agreed with the principle of the invading army living off the land of the invaded country.[147] However, he also realized that there existed a gap between theory and actual implementation of the theoretical principles.[148] He felt, though, that with the political changes in the nineteenth century, the general population was more involved in the aspects of war, thus bringing war closer to its absolute form through a "degree of energy" that had previously been unimaginable.[149] He also noted the potential strength of the invaded people in the heart of their own country as they were forced to switch from offensive to defensive movements.[150] Clausewitz's writings were so significant that he is still studied and written about by military and nonmilitary writers today.

Nevertheless, during this time an important step toward the limitations of war was taken. Pierre-François Percy, Napoleon's surgeon in chief to the French army, conceived of the idea of bringing the hospital to the wounded and treating them while still on the battlefield rather than waiting until the battle was over and carrying them miles away to the hospital. Of course, this would mean that the attending medical personnel and the wounded needed to be afforded some kind of protection. In 1800, Percy wrote about military hospitals being off-limits to any kind of violence, becoming sanctuaries for those people, and being marked as such so that enemy troops would know not to attack there. This concept was addressed by the Red Cross upon its formation in 1863. Also in 1863, President Abraham Lincoln had Dr. Francis Lieber prepare a written document of the laws of land warfare. This document was issued as General Order 100, Instructions for the Government of Armies in the Field. Following this document, the Declaration of St. Petersburg of 1868 banned projectiles of less than four hundred grams. This was expanded in 1874 by the Brussels Declaration, which further banned the use of poisoned weapons and the murder of unarmed combatants. These documents would form the basis for the Hague Conferences of 1899 and 1907 and the Geneva Conventions, which codify the rules of warfare that are in use today.[151]

The American Indian Wars

In the States, American Indian wars had started shortly after the first European colonists came to North America and continued unabated from that early beginning to the 1800s. Some of the more notable confrontations occurring prior to the Civil War included the French and Indian War (1754–1763), in which France and Great Britain fought each other, with both sides utilizing American Indian allies, and ended in a victory and large land gains for Great Britain. Previously noted was Pontiac's Rebellion of 1763–1765. Then after the Revolutionary War occurred the Battle of Fallen Timbers on August 20, 1794, which is considered the final battle of the American Revolution and ended in a victory for the United States with the gain of territory that would become Illinois, Indiana, Michigan, Ohio, and Wisconsin.[152] Regardless of which side they fought with, the Native Americans were treated as nonentities. They were left out of peace talks and constantly lost land because of those talks. For example, during the War of 1812, after the Battle of Horseshoe Bend on March 27, 1814, the Creek relinquished approximately two million acres of land.[153]

Possibly the most famous of all these early conflicts was the forced removal of the Cherokee Nation from Georgia in 1838. Known as the Trail of Tears, the Cherokee were forced to walk from their homes in Georgia to their new homes in what is now Oklahoma. The march started in October of that year under the direction of President Martin Van Buren, who continued the policy set forth by President Andrew Jackson. Starting the march in the beginning of winter, they were forced to walk through bitter weather and the normal accompanying blizzards. Already weakened by being forced to live in stockades during the summer of 1838, with many dying of diseases like dysentery and then from exposure, malnutrition, and physical exhaustion, the march of about one thousand miles resulted in the death of more than 4,000 out of a population of more than 16,000 Cherokee.[154]

The Trail of Tears occurred as a direct result of Americans deeming that they were more entitled to the native lands than were the actual native inhabitants. Georgia was among the first to annul the nation's laws, annex the land, and begin distributing land plots by lottery, for the cultivation of cotton and prospecting for gold.[155] The Cherokee Nation took their issue to the Supreme Court with *Worcester v. Georgia* (1832). The Georgia State Legislature had passed a law banning missionaries from preaching to the Cherokee due to the tendency of the missionaries to side with the Cherokee instead of the state. When the missionaries refused to stop proselytizing, the governor tried to bribe, then intimidate, them, but did not succeed.[156] They were eventually arrested. The Supreme Court ruled that the laws of Georgia were secondary to the laws and treaties of the United States. However, the state had an ally in President Jackson. Having fought against the Creek and Seminole prior to his presidency and having forced the Creek from another 22 million acres of land in Georgia and Alabama, Jackson was no stranger to dealing with Native Americans.[157]

In 1814, Jackson massacred nine hundred Creek at the Battle of Horseshoe Bend in Alabama. The next "threat" to the United States were the American Indian tribes in Florida, specifically the Seminole. But seemingly the only act the Seminole did was to welcome escaped Black slaves as allies and allow them to farm the area as free men, requiring only a yearly tribute of a portion of their crops. While relatively few in number, with only about one thousand runaway slaves, the thought of free Blacks, including runaway slaves, was anathema to Jackson and the United States in general. On trumped-up charges, he invaded Florida, killed the Seminole and Black Seminole, and captured Pensacola while raising the US flag over the city, eventually annexing Florida. Rubenstein argues that this attack was justified for one simple reason—self-defense—with two parts: First, the presence of all the aforementioned Floridians threatened the very idea of the United States—free White men who were independent and disciplined, believing in a democratic society. Second, the presence of said foreigners threatened the very existence of the United States, whose people believed that if they didn't conquer and claim the land, then some other country would, putting a potential enemy on their doorsteps. Regardless, in order to achieve the security they wanted, Jackson and the United States murdered innocent people using false pretenses to do so.[158]

As president, Jackson had forced through Congress the Indian Removal Act, which allowed him to establish lands west of the Mississippi for tribal use. Now, as a follow-up to that act, he stated that the US government would not protect American Indians and declared that they should submit to state laws or move west of the Mississippi River.[159] In his message to Congress on December 6, President Jackson stated,

> It gives me pleasure to announce to Congress that the benevolent policy of the Government, steadily pursued for nearly thirty years, in relation to the removal of the Indians beyond the white settlements is approaching to a happy consummation. Two important tribes have accepted the provision made for their removal at the last session of Congress, and it is believed that their example will induce the remaining tribes also to seek the same obvious advantages.
>
> The consequences of a speedy removal will be important to the United States, to individual States, and to the Indians themselves. The pecuniary advantages which it promises to the Government are the least of its recommendations. It puts an end to all possible danger of collision between the authorities of the General and State Governments on account of the Indians. It will place a dense and civilized population in large tracts of country now occupied by a few savage hunters. By opening the whole territory between Tennessee

on the north and Louisiana on the south to the settlement of the
whites it will incalculably strengthen the southwestern frontier and
render the adjacent States strong enough to repel future invasions
without remote aid. It will relieve the whole State of Mississippi
and the western part of Alabama of Indian occupancy, and enable
those States to advance rapidly in population, wealth, and power. It
will separate the Indians from immediate contact with settlements
of whites; free them from the power of the States; enable them to
pursue happiness in their own way and under their own rude institutions; will retard the progress of decay, which is lessening their
numbers, and perhaps cause them gradually, under the protection
of the Government and through the influence of good counsels, to
cast off their savage habits and become an interesting, civilized, and
Christian community.[160]

After the *Worcester* decision was rendered, Georgia continued to flaunt its authority over federal laws and treaties and delayed releasing the missionaries,
while President Jackson said that he would not intervene. This was not the first
time Georgia brandished its authority at the federal government. In 1828, the
state legislature and the governor resolved that they would no longer recognize federal law or its treaties.[161] Additionally, when a Cherokee member was
accused of murdering another tribesman, Georgia, not the Cherokee Nation,
arrested and tried the accused. When the Supreme Court demanded a post-
ponement of the execution, state authorities hanged the accused anyway.[162] It
was in these circumstances that Jackson supposedly said, "John Marshall has
made his decision, now let him enforce it."[163] The US government had un-
officially declared war on the American Indians and would not rest until it
had taken all of their lands and either placed the inhabitants on reservations
or murdered them. In the following years, the United States and American
Indians fought regularly, often by way of massacres of either American soldiers
and homesteaders or, more often the case, the American Indians. Such events
as the Sand Creek Massacre, the Battle of Little Bighorn, and the massacre at
Wounded Knee were common during this time.[164] This is covered more in the
next chapter.

    During this time, the United States continued its westward expansion in
its attempt to fulfill its "manifest destiny," the belief that the country would
encompass all of North America. This was the beginning of American imperialism, which saw the United States claim territory after territory at the expense
of the Native Americans living there. Beginning with the Louisiana Purchase,
the United States repeatedly acquired land, often by warfare. Such purchases
were needed, as the US population grew from around 5 million people in 1800
to more than 23 million people in 1850.

The expansion westward continued through the end of the century. In the 1840s, the United States expanded into the West Coast of California and Oregon, the Oregon Territory was established in 1846, and California was bought from Mexico in 1848. The Mexican-American War of 1846–1848 ended with the Treaty of Guadalupe Hidalgo and the sale for $15 million of the land that would become New Mexico, Utah, Nevada, Arizona, California, Texas, and western Colorado.[165] Texas was annexed in 1845, and the Gadsen Purchase in 1853 gave the United States the remainder of Arizona and New Mexico.

To protect US citizens from threats, either real or perceived, the army established a system of forts from which they would launch efforts against the Native Americans. In the 1850s, the army fought a series of battles against tribes in the Pacific Northwest, in Nebraska down through Kansas, into the future Oklahoma, and further into the Southwest. These battles followed a pattern that varied little: Settlers encroached into the Native American lands, the respective tribes then attacked the intruders to drive them out of their homelands, and the army, with superior forces, counterattacked, leading to the decimation of the American Indians.

The American Indian wars in the years immediately preceding and following the Civil War were a vastly different type of warfare compared to what the army was used to in fighting the Eastern American Indians. These Western Native Americans knew how to ride and utilize horses in guerrilla-type warfare. What the army had collectively learned in previous years of American Indian warfare helped very little in these wars:

> Those who approached their new opponent with respect and learned his ways became the best Indian-fighters and in some cases the most helpful in promoting a solution to the Indian problem. Some who had little respect for the "savages" and placed too much store in Civil War methods and achievements paid the penalty on the battlefield. Capt. William J. Fetterman would be one of the first to fall as the final chapter of the American Indian Wars opened in 1866.[166]

The Utah War

The American Indians were not the only group with whom the US Army contended. The Church of Jesus Christ of Latter-Day Saints (Mormons) was organized in 1830 in Upstate New York.[167] However, they did not stay there long. Persecution drove them to Ohio (Kirtland), then onto Illinois (Nauvoo) and Missouri (Far West), where Governor Lilburn W. Boggs issued an extermination order against them. Their founder, Joseph Smith, was lynched in 1844, so in 1847, they headed west to the Utah Territory, then a part of Mexico, and the area of the Great Salt Lake, hoping to get away from the persecutions and from the threat of the US government. However, soon after their arrival, Mexico ceded this land to the United States, leading to renewed conflict between the US and Mormon governments. This animosity reached its climax in 1857.

Non-LDS officials had been sent to Utah earlier in the year to supervise the new territorial government, composed of LDS members. Conflict soon arose between these two sets of officials, with the non-LDS officials levying charges of intimidation and destruction of government records against the LDS officials, namely Governor Brigham Young (who had been appointed by President Millard Fillmore), who was also the president of the LDS Church. Upon his inauguration as president of the United States, James Buchanan received these reports and was led to believe that the Saints were in open rebellion against the federal government. He therefore declared the territory in rebellion and sent 2,500 soldiers and Alfred Cumming to serve as the new governor (the Utah Expedition) to ensure that the "Mormon rebels" would cooperate with the transition and with government overview. And while there were no major battles, there were, as can be expected, numerous skirmishes and deaths from guerrilla warfare.[168] Commonly referred to as a "bloodless war," the resulting Utah War (1857–1858) was anything but. Fatalities on both sides of the altercation numbered approximately the same as during the altercations in what became known as "Bloody Kansas."[169]

Upon hearing that the United States had sent troops to Utah to put down a rebellion, President Young, acting in his role as governor but not knowing that he had been relieved of that position, decreed martial law and sent their militia, the Nauvoo Legion, to delay their arrival. Given their past experiences with the government, the Saints fully expected the army to continue the history of persecution, possibly to the point of death. In addition, one of the church's apostles, Parley P. Pratt, had been murdered in Arkansas on May 13, 1857. Along with the prior murder of Joseph Smith, it is of little surprise that the Saints and the church would feel the need to prepare for the worst. With the order issued by Young, the legion burned three supply trains and herded hundreds of government cattle back to the Salt Lake Valley.[170] They caused enough disruption that the main army force had to winter near burned-out Fort Bridger.

Young, in an attempt to prevent the Utah Expedition from being able to use anything belonging to the Saints, ordered them to be ready to burn their homes and crops should the need arise. He also ordered the majority of the Saints to move south to Provo and other towns in central and southern Utah, away from any potential conflict, resulting in the exodus of approximately 30,000 people.[171] Upon entering Salt Lake City, the federal troops found a near–ghost town, with the remaining citizens armed with torches to burn whatever was necessary. Finally, a peaceful resolution was reached on June 12, 1858, as President Young relinquished his role as governor in exchange for a pardon, issued by President Buchannan, for all the Saints.[172]

However, perhaps the most significant event of this episode happened in southwest Utah (see figure 1). Known as the Mountain Meadows Massacre, the event resulted in the death of 120 men, women, and children, leaving only 17 small children alive, as they were deemed too young to remember or explain what had happened.[173] A small group of emigrants from Arkansas, where Pratt had been murdered, was trying to move to California. The Baker, Dunlap, Fancher, Miller, and Tackitt families were some of the groups trying to emigrate to California from Arkansas. They met up in Salt Lake City, formed one large wagon train known as the Baker-Fancher Party, and headed into southern Utah. As fate would have it, they arrived in August 1857 as the church was preparing for war. The group encountered less-than-friendly Saints, who argued with them over grasslands and supplies.[174] While the Saints were partially to blame for these disputes, the emigrants were also responsible for abusing the local Paiute tribes and taunting the Saints about how they had been chased out of Missouri.[175] These disagreements came to a head in September of that year.

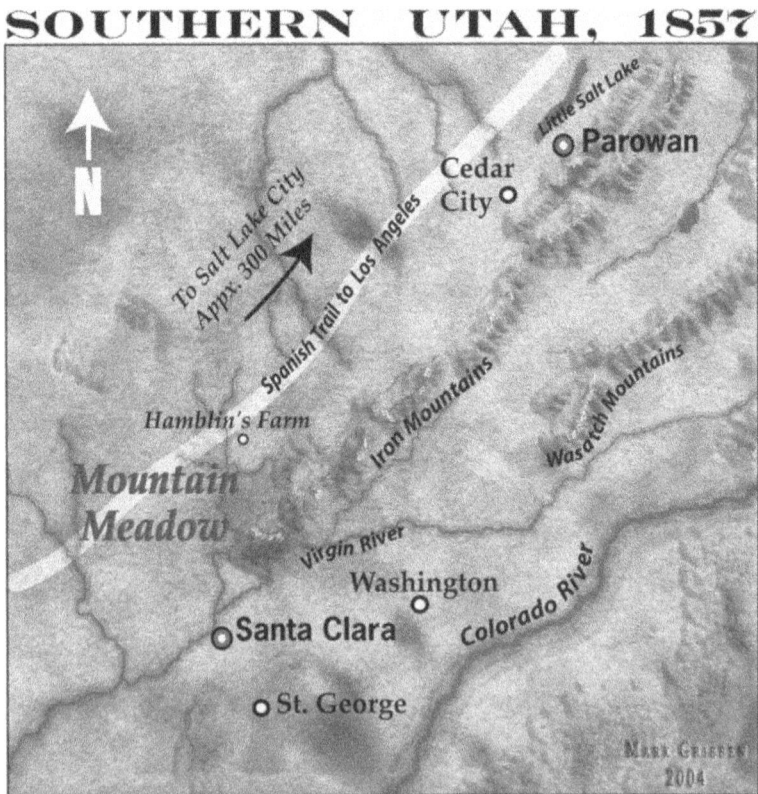

Figure 1. Southwest Utah, 1857, with the locations of Cedar City and Mountain Meadows. Note that the map has labeled Mountain Meadow incorrectly. It is Mountain Meadows. SOURCE: MARK LAMAGNE, HTTPS://WCHSUTAH.ORG/MAPS/WASHCO1857.JPG.

Upon arriving at Cedar City, Utah, the emigrants encountered a renegade group of LDS guerrillas posing as guardians to protect the wagon train from the Paiutes. However, they had in fact joined with the Paiutes in a plan to eliminate the emigrants. They told the Paiutes that the approaching wagon train was a threat to both them and the Saints, with rumors abounding that the train might poison the water and cattle as they passed.[176] On September 7, after the wagon train arrived at Mountain Meadows, they were attacked by these American Indians. The emigrants were able to defend themselves for several days and in doing so saw that these renegade Saints had been involved, disproving their later claims that the attacks had been the act of only the American Indians. In direct contravention of the area church leaders, on September 11, the renegades acted.

"In regard to emigration trains passing through our settlements," Young stated,

> "we must not interfere with them untill [sic] they are first notified to keep away. You must not meddle with them. The Indians we expect will do as they please but you should try and preserve good feelings with them. There are no other trains going south that I know of[.] If those who are there will leave let them go in peace. While we should be on the alert, on hand and always ready we should also possess ourselves in patience, preserving ourselves and property ever remembering that God rules.[177]

Posing again as guardians of the emigrants, the renegades convinced them that if they surrendered their arms, the renegades would escort them out of the area. Once disarmed, the Saints opened fire on the men, while the Paiutes murdered the women.[178] The massacre was covered up, but finally, twenty years later, the leader of the renegades, John D. Lee, was identified as the leader, tried, found guilty, and executed at Mountain Meadows at the site where he and his accomplices carried out the heinous crime.[179] This massacre was the "greatest incident of organized mass murder of civilians in the nation's history until the Oklahoma City bombing of 1995."[180]

The whole Utah War was a major stain on Buchannan, to the point where the whole fiasco was known as "Buchannan's Blunder."[181] This incident, along with his inept handling of the secession crisis of 1860–1861, led to him being labeled as one of the worst presidents in US history: A 2024 survey of experts on American presidents found him in next-to-last place.[182] And while the Mountain Meadows Massacre was just one part of the Utah War, the entire confrontation demonstrated the purposeful targeting of its own civilians, not just Native Americans, during war.

CHAPTER 2

CIVIL WAR, 1860–1914

"War is the ultimate tool of politics."

—R. Buckminster Fuller

THE AMERICAN CIVIL WAR, 1861–1865

The difference between European and American thought and practice became most apparent during the Civil War with the strategies and tactics of Generals Ulysses S. Grant and William T. Sherman. In the early years of the Civil War, battles were waged in traditional manners. General George B. McClellan stated, "I have not come here to wage war upon the defenseless, upon non-combatants, upon private property, nor upon the domestic institutions of the land."[1] Sherman seemed to believe similarly during these early years. In August 1861, he wrote to his wife, "Our soldiers are the most destructive men that I have ever known. It may be that other volunteers are just as bad— indeed the complaint is universal and I see no alternative but to let it take its course. . . . My only hope now is that a common sense of decency may be inspired into the minds of this soldiery to respect life and property."[2] While he was in Kentucky in the winter of 1861–1862, he issued orders prohibiting his soldiers from taking fresh food, sleeping in vacant houses, and using fence rails for firewood. Because of this, the Thirty-Third Indiana Regiment ended up with more than half its men in the hospital and sixty-two deaths in one month due to exposure and inadequate rations.[3]

However, as the war progressed, both Grant and Sherman came to realize that the Confederacy was so resolved to win that the Union's tactics had to change, or else that would indeed happen. Sherman wrote, "The Government of the United States may now safely proceed on the proper rule that all in the South *are* enemies of all in the North; and not only are they unfriendly, but all who can procure arms now bear them as organized regiments, or as guerrillas."[4]

Both Grant and Sherman determined that the only way to win and stop the war was to take the war to not just the Confederate Army but also the entire population of the South and make them demand peace. Grant stated,

> I gave up all idea of saving the Union except by complete conquest. Up to that time it had been the policy of our army, certainly of that portion commanded by me, to protect the property of the citizens whose territory was invaded, . . . whether Union or Secession. After this, however, I regarded as humane to both sides to protect the persons of those found at their homes, but to consume everything that could be used to support or supply armies.[5]

Sherman had been pondering the situation and arrived at the same conclusion and was more than willing to implement the plan. He realized that "man has two supreme loyalties—to country and to family . . . but even the bonds of patriotism, discipline, and comradeship are loosened when the family is itself menaced."[6] As to how he would deal with the people of the area, he wrote, "If the people raise a howl against my barbarity and cruelty, I will answer that war is war, and not popularity-seeking. If they want peace, they and their relatives must stop the war."[7] In doing this, he then realized that the quickest way to defeat the Confederates was to get them to lose interest in fighting and return to their homes and families. To do that, he had to threaten those civilians in some manner.

As he further refined his thoughts on defeating the South, Sherman realized, as Machiavelli had earlier, that "they cannot be made to love us, but may be made to fear us, and dread the passage of troops through their country."[8] The ideas of the Mongol hordes had returned but for different reasons. Considering my original definition of *total war*, Sherman's actions necessitate an additional clause: "[the deliberate use of] military force against the civilian population of the enemy *that affects both their psyche and economic support.*"[9] It was the South that had started a total war against themselves when they armed their citizens. The inhabitants of Georgia and South Carolina would feel the full wrath of an army intent on implementing the three facets of Sherman's philosophy of total war: the destruction of civilian property to shorten the war by depriving the Confederate army of supplies needed to wage war; the destruction of civilian possessions, which would also lead to a loss of desire for war and succession and result in their demand for peace; and the collective responsibility of the entire South for causing the war of succession that necessitated their collective defeat.[10] Anything that could be done to shorten the war and increase the probability of a Union victory would be done. This strategy was in direct contravention of the Articles of War dated April 10, 1806, which stated in article 1, paragraph 52,

Any officer or soldier who shall misbehave himself before the enemy, run away, or shamefully abandon any fort, post, or guard which he or they may be commanded to defend, or speak words inducing others to do the like, or shall cast away his arms or ammunition, or who shall quit his post or colors to plunder and pillage, every such offended, being duly convicted thereof, shall suffer death, or such other punishment as shall be ordered by the sentence of a general court martial.[11]

By the latter part of 1862, Sherman had had enough and simply ignored these rules:

I would not let our men burn fence rails for fire or gather fruit or vegetables though hungry. . . . We at that time were restrained, tied by a deep-seated reverence for law and property. The rebels first introduced terror as a part of their system. . . . Buell had to move at a snail's pace with his vast wagon trains. . . . Bragg moved rapidly, living on the country. No military mind could endure this long, and we were forced in self-defense to imitate their example.[12]

Sherman, upon the capture of Atlanta, wrote to General Hood of the Confederate Army, offering the inhabitants of the city the chance to leave, stating that Atlanta was "no place for families or non-combatants."[13] Hood replied that they would fight to the death and that it would be "better to die a thousand deaths than submit to live under you or your Government and your negro allies," and James M. Calhoun, mayor of Atlanta, wrote to Sherman asking him to reconsider his orders.[14] Sherman replied that he would not revoke them:

We must have peace, not only at Atlanta, but in all America. To secure this, we must stop the war that now desolates our once happy and favored country. . . . War is cruelty, and you cannot refine it; and those who brought war into our country deserve all the curses and maledictions a people can pour out. . . . You cannot have peace and a division of our country. If the United States submits to a division now, it will not stop, but will go on until we reap the fate of Mexico, which is eternal war.[15]

Desperate times required desperate means. As Adam Badeau stated, "It was not victory that either side was playing for, but existence. If the rebels won, they destroyed a nation; if the government succeeded, it annihilated a rebellion."[16] Sherman, therefore, would not directly target the civilians but would target everything the civilians used. He would make life as difficult for them as possible, thereby leaving nothing for them to live on or with which to supply the Confederate army. The soldiers would feel the effects of this strategy, and fighting would be more difficult due to the lack of supplies. This, along with concern for their families, would hopefully get the Confederates to desert in great enough numbers to finally end the war. With fewer Confederate soldiers fighting, more soldiers could make it home without being maimed or otherwise becoming another casualty of war.

This strategy was first implemented by General Philip Sheridan when he took command of the Army of the Shenandoah on August 6, 1864. He had previously said,

> Those who rest at home in peace and plenty see but little of the horrors . . . [of war] and even grow indifferent to them as the struggle goes on, contenting themselves with encouraging all who are able-bodied to enlist in the cause. . . . It is another matter, however, when deprivation and suffering are brought to their own doors. Then the case appears much graver, for the loss of property weights [sic] heavy with the most of mankind; heavier often, than the sacrifices made on the field of battle. Death is popularly considered the maximum of punishment in war, but it is not; reduction to poverty brings prayers for peace more surely and more quickly then [sic] does the destruction of human life.[17]

After taking command, Grant told him, "Give the enemy no rest. . . . Do all the damage to railroads and crops you can. Carry off stock of all descriptions, and negroes, so as to prevent further planting. If the war is to last another year, we want the Shenandoah Valley to remain a barren waste."[18] When Grant told Sheridan that he was to follow the railways east and destroy General Lee's supply lines, Sheridan replied, "My judgment is that it would be best to terminate this campaign by the destruction of the crops, &c., in this valley, and the transfer of troops to the army operating against Richmond," to which Grant replied, "You may take up such position in the Valley as you think can and ought to be held, and send all the force not required for this immediately here. Leave nothing for the subsistence of an army on any ground you abandon to the enemy."[19] For thirteen days, Sheridan's army destroyed anything useful to the Confederates, except for houses, empty barns, property of widows, single women, and orphans. However, the army was told there would be no looting.[20] As Lance Janda points out, in the Articles of War of 1806, any soldier convicted of plunder and pillage should suffer death as a punishment.[21]

The effects of this strategy were apparent. Sheridan reported to Grant, "I have destroyed over 2,000 barns filled with wheat, hay and farming implements; over 70 mills, filled with flour and wheat; have driven in front of the army over 4,000 head of stock, and have killed and issued to the troops not less than 3,000 sheep." He also said, "The people here are getting sick of war."[22]

The new strategy was working as would become even more apparent during the battle and occupation of Atlanta.

By this point in the war, most of the South had given up the expectation of winning the war in combat. Instead, they focused on a policy of winning by not losing. They wished to drag the war out until November, when the Northerners would hopefully tire of the war and elect the Democratic candidate who would, with any luck, seek an armistice with the Confederates.[23] Facing off against General Joseph E. Johnston, Sherman was confident of success because, among other reasons, Johnston was a defensive, not offensive, general to the point of being timid in attacking opposing forces.[24] Finally, when Sherman was five miles from Atlanta, President Jefferson Davis replaced Johnston with the very aggressive General John B. Hood. Following three unsuccessful attacks by Hood, Sherman began cannonading the city and its few remaining residents while also cutting the railroad lines that supplied Hood's army. On September 1, Hood was forced to abandon Atlanta, while the Union troops entered the city the next day.[25]

After two months, as Sherman prepared his march to Savannah (see figure 2), he ordered the destruction of anything that might be useful to the Confederates: "Make preliminary preparations for the absolute destruction in Atlanta of the railroad track, depots, car and store houses, shops, and indeed everything that might be used to our disadvantage by an enemy."[26] Rumors had circulated among the soldiers that Sherman was going to burn Atlanta, so they were ready and eager to do so, whether official orders came down or not. On the night of November 11, Union soldiers began torching private buildings and residences, with fires ensuing each subsequent night. The burning of public facilities and private homes had occurred earlier in the year at Jackson, Brandon, and other locations, giving the soldiers the impression that such actions in Atlanta would go unpunished. They were correct: Sherman stated, "I never ordered burning of any dwelling—didn't order this, but can't be helped."[27]

The engineers proceeded with their duties, heating the railroad rails and twisting them until they were useless and destroying brick and granite buildings with battering rams and black powder. Finally, on November 15, the last detonations of brick buildings by the engineers began. The regular soldiers decided to "help" by looting, pillaging, and burning businesses and residences. The fires continued to grow throughout the day and into the night. This unsanctioned arson resulted in a fiery glow over much of the city. One witness reported "immense and raging fires lighting up whole heavens . . . huge waves of fire roll up into the sky; presently the skeleton of great warehouses stand out in relief against sheets of roaring, blazing, furious flames."[28] While reports vary, Atlanta residents returning to the city found the destruction to be not as bad as anticipated. Using these accounts by various residents, Dr. Gordon Jones of the Atlanta History Center determined that "only forty percent of Atlanta was left in ruins."[29]

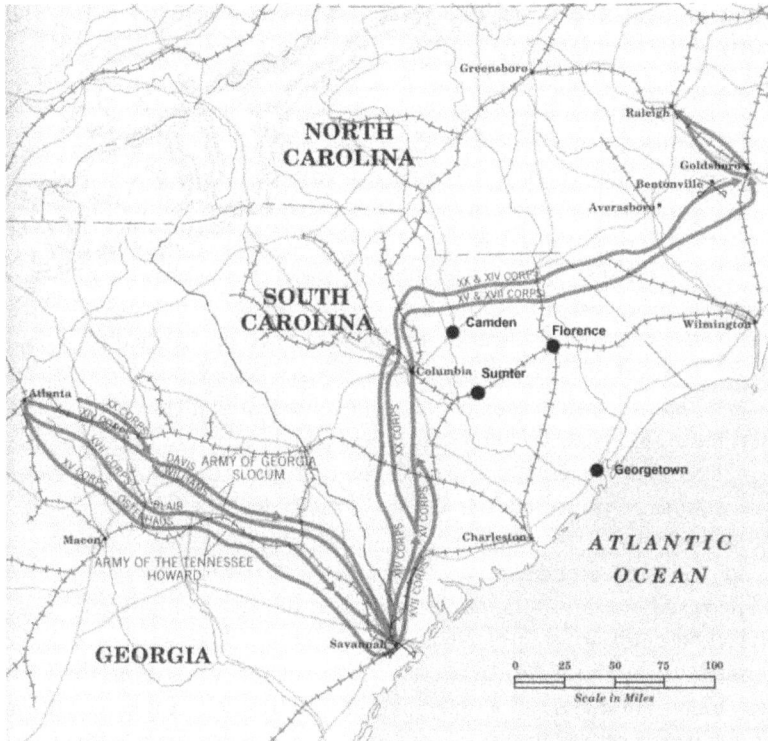

Figure 2. Savannah Campaign (Sherman's March to the Sea) and South Carolina
Campaign. Source: Hannah Matthews, "Civil War in South Carolina," Sutori. February 19, 2020,
https://www.sutori.com/en/story/civil-war-in-south-carolina--bkFDdpEfY1BXt4YbpbpJnV5W.

The next morning, Sherman took four corps of 62,000 men and headed for
Savannah, cutting a path up to 60 miles wide and 250 miles long. Upon leaving Atlanta, they cut the remaining telegraph lines, and for just over a month,
his army was cut off from all communication with the rest of the Union. Everything that the Union army could use, it did, and everything that it could not
use was destroyed. To achieve these results, certain adaptations had to be made.
To travel light and move fast, he eliminated some of the artillery pieces they
normally used. And in living off the land, the army still had to carry supplies
in case foraging did not provide enough food, so he dispensed with a supply
base in Atlanta. This not only allowed him to travel faster, but it also allowed
him to take with him the men he normally would have had to leave behind to
guard and man the supply base.

In keeping with his desire to prevent civilian casualties, he adopted a
"live-and-let-live" doctrine: As long as the civilian populace left the Union
soldiers alone, the soldiers would leave them alone. However, should anybody
attempt guerrilla warfare or directly fight them, or should they attempt to slow
the army by indirect means, such as burning bridges or other methods, then the

Union army had direct orders from Sherman that they were free to destroy any property of the offending people that they wanted.[30] Sherman ordered,

> In districts and neighborhoods where the army is unmolested, no destruction of such property should be permitted; but should guerrillas or bush whackers molest our march, or should the inhabitants burn bridges, obstruct roads, or otherwise manifest local hostility, then army commanders should order and enforce a devastation more or less relentless, according to the measure of such hostility.[31]

In one instance, a young officer on his horse stepped on a buried torpedo, killing the horse and blowing off one of the officer's legs. When Sherman found out, he ordered Confederate prisoners to march along the road, acting as human mine sweepers.[32] There was, however, one group of Southerners who were not to receive mercy of any type: the rich Southern elite. The common people were to be protected, but the rich were to feel the full force of punishment for supporting the rebel cause. It was the rich who had voted for secession, not the common person.[33]

The results were exactly as Sherman had hoped. The only food that his army lacked upon arriving at Savannah was bread. They started with 2,500 wagons drawn by six mules each, which had not recovered from the Chattanooga ordeal. These were shot and replaced with fresh mules taken by the army along the way, resulting in the loss of some 15,000 mules. They left with 5,000 head of cattle and arrived with more than 10,000. They lived off the turkeys, chickens, sheep, hogs, and cattle of the land, along with sweet potatoes, corn, apples, and other crops that had recently been harvested. The army procured food by means of foragers sent out before dawn each morning. They would return with wagons full of "bacon, corn-meal, turkeys, chickens, ducks, and every thing that could be used as food or forage."[34] Sherman recounted,

> The custom was for each Brigade to send out daily a foraging party of about fifty men, on foot, who invariably returned mounted, with several wagons loaded with poultry, potatoes, &c, and as the army is composed of about forty Brigades, you can estimate approximately the quantity of horses collected. Great numbers of these were shot by my orders, because of the disorganizing effect on our Infantry of having too many idlers mounted.[35]

General Sherman further said,

> We have consumed the corn and fodder in the region of country thirty miles on either side of a line from Atlanta to Savannah as also the sweet potatoes, cattle, hogs, sheep and poultry, and have carried away more than 10,000 horses and mules as well as a countless

number of their slaves. I estimate the damage done to the State of Georgia and its military resources at $100,000,000 [$1.99 billion in 2024]; at least $20,000,000 of which has inured to our advantage and the remainder is simple waste and destruction.[36]

Though the Union army took great pleasure in taking or destroying everything allowed, there were still rules of discipline that were enforced. While foraging was encouraged, the troops were not to pillage any residences. While rapes inevitably occurred, they were kept to a minimum. And aside from a few battles during the march, there were no murders reported.

Savannah fell on December 21, 1864, surrendered by Mayor R. D. Arnold. Sherman telegraphed President Lincoln the joyous news, which was received on Christmas Eve: "I beg to present you as a Christmas-gift the City of Savannah, with one hundred and fifty heavy guns and plenty of ammunition, also about twenty-five thousand bales of cotton."[37] In a letter to chief of staff Major General Henry Halleck, Sherman wrote,

We are not only fighting hostile armies, but a hostile people, and must make old and young, rich and poor, feel the hard hand of war, as well as their organized armies. I know that this recent movement of mine through Georgia has had a wonderful effect in this respect. Thousands who had been deceived by their lying papers into the belief that we were being whipped all the time, realized the truth, and have no appetite for repetition of the same experience. To be sure, Jeff. Davis has his people under a pretty good state of discipline, but I think faith in him is much shaken in Georgia; and I think before we are done, South Carolina will not be quite so tempestuous.[38]

In covering the effects of Sherman's march on the civilian population, one must remember that the ultimate goal was ending the war. Sherman knew that as long as General Robert E. Lee still commanded an intact Army of Northern Virginia, the war would not end. This army was a symbol of the Confederacy and secession and needed to be eliminated from the fight. As bad as the devastation was in Virginia, Mississippi, and Georgia with Sherman's March to the Sea, the destruction heaped on South Carolina was the worst, in that the soldiers held that state directly accountable for the instigation of the war and secession from the Union and would have no difficulty in expressing their enmity toward that state in no uncertain terms:

The truth is the whole army is burning with an insatiable desire to wreak vengeance upon South Carolina. I almost tremble at her fate, but feel that she deserves all that seems in store for her. Many and many a person in Georgia asked me why we did not go to South Carolina, and when I answered that I was *en route* for that State, the

invariable reply was, "Well, if you will make those people feel the se-
verities of war, we will pardon you for your desolation of Georgia."[39]

After conferring with Grant, Sherman resupplied his men and headed to South
Carolina and on to Goldsboro, North Carolina. Grant had wanted Sherman's
armies to go, by way of the Union navy, to the James River in Virginia to avoid
the inclement weather and flooded terrain that they would have had to march
through. He also wanted him there in order to be positioned south of Richmond and to put more pressure on the Army of Northern Virginia. In the end,
Grant relented and approved Sherman's march to North Carolina. This allowed Sherman to continue to isolate the Confederate army and prevent them
from receiving any supplies. And Goldsboro was the perfect location to resup-
ply his men as well as ready them to act as Grant needed. His first destination,
though, was Columbia, South Carolina.

The results were exactly as Grant, Sherman, and Sheridan had hoped—
the elimination of the Confederate ability and will to wage war. As one Confederate soldier wrote, "i hev conkluded that the dam fulishness uv tryin to lick
shurmin Had better be stoped. we hav bin gettin nuthin but hell & lots uv it
ever sinse we saw the dam yankys & I am tirde uv it. . . . Thair thicker an lise
on a hen and a dam site ornraier."[40] Columbia would receive more hell than
could have been imagined.

Upon entering South Carolina, Sherman's army resumed destroying railways with more zeal than in Georgia. Columbia had already had most of its
inhabitants flee the city to avoid any contact with Sherman and his troops,
having heard of his work in Georgia. The few Confederate defenders who had
been there had also fled the city. When they reached Columbia, Sherman's
three wings met up. Fifteenth Corps in the right wing under General Howard
moved through the city unopposed and encamped outside on Camden Road.
The mayor of the city had already surrendered in hopes of showing the Union
soldiers that they would not oppose them and also letting them know that they
had opposed secession. However, in a writing that proved to be quite prescient,
Sherman stated, "I look upon Columbia as quite as bad as Charleston, and I
doubt if we shall spare the public buildings there, as we did at Milledgeville."[41]

On the night of February 17–18, 1865, Columbia was engulfed in flames
(see figure 3). The impact of this inferno is evident in writings by those who
were there. In her diary, Emma LeConte wrote,

> By the red glare we could watch the wretches walking—
> generally staggering—back and forth from the camp to the
> town—shouting—hurrahing—cursing South Carolina—
> swearing—blaspheming—singing ribald songs and using [such]
> obscene language that we were forced to go indoors. The fire on
> Main Street was now raging, and we anxiously watched its progress

Figure 3. The burning of Columbia, South Carolina, February 17, 1865, sketched by William Waud on April 8, 1865. Source: Library of Congress Prints and Photographs Division, https://www.loc.gov/item/2003668338/.

from the upper front windows. In a little while however the flames broke forth in every direction. The drunken devils roamed about setting fire to every house the flames seemed likely to spare. . . . The wind blew a fearful gale, wafting the flames from house to house with frightful rapidity. By midnight the whole town (except the outskirts) was wrapped in one huge blaze. . . . Imagine night turned into noonday, only with a blazing, scorching glare that was horrible—a copper colored sky across which swept columns of black rolling smoke glittering with sparks and flying embers, while all around us were falling thickly showers of burning flakes. Everywhere the palpitating blaze walling the streets with solid masses of flames as far as the eye could reach—filling the air with its horrible roar. On every side the crackling and devouring fire, while every instant came the crashing of timbers and the thunder of falling buildings. A quivering molten ocean seemed to fill the air and sky.[42]

There is still debate about who started the inferno, with both sides blaming the other. Union Major General Henry W. Slocum wrote, "A drunken soldier with a musket in one hand and a match in the other is not a pleasant visitor to have about the house on a dark, windy night, particularly when for a series of years you have urged him to come, so that you might have an opportunity of performing a surgical operation on him."[43] And there were plenty of drunk Union soldiers around. Thomas Osborn, a Union officer, recalled, "When the brigade occupied the town the citizens and negroes brought out whiskey in buckets, bottles and in every conceivable manner treated the men to all they would drink."[44] Sherman claimed that he never gave any orders to do anything like what happened. Likewise, the majority of the Union soldiers placed the blame on the Confederates for having piled bales of cotton in the streets to be burned before they retreated. Many of the city's residents, however, claimed to have seen Union soldiers purposely setting fire to buildings by using torches. Sherman possibly bears some of the blame for not keeping so much alcohol out of the hands of the Union soldiers. And the wind only exacerbated the problem. By around 4:00 a.m., the fire finally died down. What remained was about one-third of Columbia having been destroyed, resulting in approximately 30,000 homeless. Regardless, not all the Union soldiers were drunk, and many of them tried to help the city's residents:

> The block of buildings directly opposite the burning cotton of that morning was on fire, and that it was spreading; but he had found General Woods on the ground, with plenty of men trying to put the fire out, or, at least, to prevent its extension. The fire continued to increase and the whole heavens became lurid. I dispatched messenger after messenger to Generals Howard, Logan, and Woods, and received from them repeated assurances that all was being done that could be done, but that the high wind was spreading the flames beyond all control.[45]

While the burning of Columbia appears to be accidental, it could have happened for numerous reasons. Regardless, this destruction is a major part of total war—the indiscriminate targeting of guiltless civilians. However, this campaign lacked one very important aspect of total war: the actual death of innocent civilians. While their property was destroyed, the population was left alive and able to rebuild once the war had ceased. Despite all of this, some humanity managed to shine through:

> Marster told father and mother they could have the house free and wood free, and he would help them feed the children, but mother said, "No, I am goin' to leave. I have never been free and I am goin' to try it. I am goin' away and by my work and the help of the Lord I will live somehow." Marster then said, "Well stay as long as you wish, and leave when you get ready, but wait until you find a place to go, and leave like folks." Marster allowed her to take all her things with her when she left. The white folks told her goodbye.[46]

The Civil War is often referred to as the first "modern" war. Many new inventions made their debut during this clash. For example, iron-clad ships first appeared at this time in the form of the *Monitor* and the *Merrimac* (captured and rechristened the *Virginia* by the Confederate navy). The battle between these two vessels on March 9, 1862, at Hampton Roads ushered in a completely new era of naval warfare. The day before, March 8, 1862, the *Virginia*, along with several other Confederate ships, attacked a fleet of Union wooden warships at Newport News, Virginia, and decimated the fleet, destroying the sloop *Cumberland* and the fifty-gun frigate *Congress*. This was just one discovery that arose during the Civil War.

Some other advancements included repeating rifles that could fire more than one bullet before needing to be reloaded and were more accurate at a longer range. This gave the users a huge advantage over other soldiers. While the railroad was already in use, it conferred a significant advantage in carrying supplies and troops to needed destinations. The telegraph allowed much faster communication between parties and allowed for real-time strategy meetings and the issuing of orders. President Lincoln was the first president able to communicate directly with his officers on the battlefield. The Gatling gun was the precursor to the modern machine gun. Its inventor, Richard Gatling, invented it in the hopes that seeing the destruction that it could wreak would convince men to stop killing each other. It didn't. These, along with many others, allowed for more efficiency in warfare, which meant more dead, which continued to advance warfare closer to the ideal total-war confrontation.

PRISONERS OF WAR

So far, the focus has been on the conduct of combatants against traditional civilians. Now I touch on the treatment of another important class of noncombatant: prisoners of war. During antiquity, the concept of prisoners of war was foreign. This is because the losing side in a battle was usually annihilated. Holding onto enemies was only done if they were kept as slaves, if women could be married or used for breeding, or if children could be raised as one of the victors. In other words, keeping those from the losing side was a matter of expediency.[47] Keeping any others required great expense in food that could otherwise feed the victorious tribe and in the men required to guard the prisoners from escaping or attempting revenge on the captors. As noted in the previous chapter, the Mongols offered a besieged city an option—surrender or die. Surrender meant life but as slaves supporting the Mongol army.

With the advent of money, it became profitable to exchange prisoners for ransom. Therefore, it became expedient to keep prisoners alive. It wasn't necessary to keep them in perfect health, so not much consideration was given to housing or feeding prisoners. The rise of organized states led to the view that the soldiers employed were property of the conscripting nation. It was in their best interest to have their property returned. The main objective was preventing the return of the prisoner to battle. Therefore, if a prisoner gave his solemn oath that he would lay down his weapon and not resume battle until returned to his own country, he would be released. This custom was seen in the Civil War.[48]

During the American Revolution, colonial soldiers were viewed as traitors, not prisoners of war, in England's eyes, as the Revolution was a rebellion,
not a war. Therefore, captured soldiers and sailors were treated as criminals.
This treatment included confinement in the dank holds of British vessels an-
chored in such large cities as New York.[49] The British made no attempt to
hide the treatment of captured colonists—it was a warning of what might happen to any who were considering joining the rebel cause.[50] The Americans,
hoping that their example of proper prisoner treatment would lead the British to properly care for theirs, gave their British POWs very good care. But
this conduct did not change the British treatment of American prisoners—
American Revolutionary prisoners (i.e., "traitors") of war had a mortality rate
of 72 percent (18,000 dead out of 25,000 captured).[51] There were, however,
limited exchanges of officers between the two sides. The War of 1812 saw similar treatment of prisoners on both sides of the conflict.[52]

And just as during the Revolutionary War and the War of 1812, prison
camps in the Civil War tended to be hell holes: "As we entered the place, a
spectacle met our eyes that almost froze our blood with horror. . . . Before us
were forms that had once been active and erect—stalwart men, now nothing
but mere walking skeletons, covered with filth and vermin. . . . Many of our
men exclaimed with earnestness, 'Can this be hell?'"[53]

As would happen again in World War I, both sides in this conflict expected a short war, and neither was prepared to deal with the logistics of large
numbers of prisoners. By war's end, some 194,000 Union and 214,000 Confederate soldiers were prisoners of war.[54] Of these, nearly 30,000 soldiers from
either side died while imprisoned.[55] Ultimately, more than 150 prisons were
established during the war.[56] The story was the same at nearly every prison—
inadequate supplies, overcrowding, lack of sanitation with the concomitant
propagation of disease, filth, starvation, and death.[57] Gary Flavion cites several
examples of camps that started out as tolerable situations but, due to lack of
planning and prison population explosion, turned into the aforementioned hell
holes. Salisbury Prison in North Carolina started as a relatively tolerable camp
in which to be interred. The death rate in 1861, with a low population count,
was about 2 percent. By 1865, with the population having exploded, the death
rate exceeded 28 percent.[58] Elmira Prison in New York had a similar death rate
of 25 percent, but this was due more to the exploits of commissary general of
prisoners Colonel William Hoffman. As Flavion notes,

> Col. Hoffman forced Confederate prisoners to sleep outside in the
> open while furnishing them with little to no shelter. Prisoners relied
> upon their own ingenuity for constructing drafty and largely inade-
> quate shelters consisting of sticks, blankets, and logs. As a result, the
> Rebels spent their winters shivering in biting cold and their sum-
> mers in sweltering, pathogen-laden heat.[59]

Camp Douglas in Illinois was noted for the "amount of standing water, of un-policed grounds, of foul sinks, of general disorder, of soil reeking with miasmic accretions, of rotten bones and emptying of camp kettles . . . was enough to drive a sanitarian mad."[60] Belle Isle in Virginia was the same:

> In a semi-state of nudity . . . laboring under such diseases as chronic diarrhea, scurvy, frost bites, general debility, caused by starvation, neglect and exposure, many of them had partially lost their reason, forgetting even the date of their capture, and everything connected with their antecedent history. They resemble, in many respects, pa-tients laboring under cretinism. They were filthy in the extreme, covered in vermin . . . nearly all were extremely emaciated; so much so that they had to be cared for even like infants.[61]

As bad as these were, the worst of these, whose name became nearly synony-mous with the horror that was Civil War prison camps, was Andersonville, Georgia. In only fourteen months, 45,000 prisoners came to the camp, and nearly 13,000 died.[62] This was the highest mortality rate of any Civil War pris-on.[63] Besides the usual dysentery from relieving themselves into the stream where they retrieved their drinking water and scurvy from lack of vitamin C, there were roving gangs who beat and murdered the other prisoners for food, supplies, and booty.[64] Although the guards tried to stop some of these gangs, they were largely ineffective. The prisoners took it upon themselves to deal with the gangs and hanged six of the leaders on July 11, 1864. After this, a new police force comprised of prisoners tried to enforce discipline, such as im-plementing sanitation practices and stopping the robberies. Prisoners felt that there had been no change between the old gangs and the methods of the new police force.[65] The camp's commandant, Captain Henry Wirz, was found guilty of war crimes and hanged on November 10, 1865. He was one of the very few people to be punished in such a manner.[66]

SAND CREEK MASSACRE
On November 29, 1864, in the midst of the Civil War, more than 600 Col-orado Territory Militia attacked the winter encampment of more than 500 Cheyenne and Arapaho and killed approximately 163 to 230 while suffering only a small number of casualties (estimates vary on casualties for both sides—the military suffered in the area of 24 dead and 52 wounded).[67] The previous summer, a family had been murdered about twenty-five miles southeast of Den-ver, Colorado (not yet a state). Their bodies were brought to Denver and pub-licly displayed, stoking fears that the Cheyenne or Arapaho had caused their death. The citizens of Denver also feared that the Native Americans might be agents of the Confederacy, sent to keep the White settlers out of the area. Territorial Governor John Evans ordered that all Native Americans in the area

were to place themselves under the protection of several of the military or risk being shot on sight. Evans also ordered the White settlers to arm themselves and actively pursue those Native Americans who did not surrender themselves.

By November, there were about 750 Native Americans camping by the Big Sandy Creek, flying white and American flags from their camps, expecting the military to protect them. However, the militia, the Third Colorado Cavalry, who had been taunted by the residents of Denver during their unenthusiastic training and who had seen no action, demanded blood.

On November 29, the cavalry rode in from Fort Lyon and opened fire. The commanding officer, Colonel John M. Chivington, had instructed the men to take no prisoners. The Native Americans were shot point-blank, and howitzers were fired on the few defenses the Native Americans were able to cobble together. This mob action went on for seven hours, after which the dead bodies were mutilated, and their jewelry stolen. More than half of the casualties were women and children—of the total that had camped there, only one hundred were of fighting age. Of the chiefs who were murdered, most had called for peace with the White pioneers and the US government. The troops left with six hundred horses from the Native Americans and rode through Denver proudly displaying the scalps and other body parts as trophies. The US Army would stop at nothing to rid themselves of the Native Americans.

CHEMICAL WARFARE

Chemical and biological warfare techniques were envisioned during this time. While they weren't implemented to any appreciable degree during the war, the ideas came to fruition some fifty years later. Though the government typically deals with the research and development of these types of weapons, a surprising number of suggestions came from ordinary citizens. These agents were typically easily found in daily activity. Many involved the use of cayenne pepper to render the enemy temporarily incapacitated through violent sneezing.[68] The delivery method most often suggested was artillery shells loaded with these substances that would break open upon explosion of the shell, thereby dispersing the material among enemy forces.[69] One person suggested using such substances to bombard the Confederate ship *Virginia*.[70] Chloroform was another frequently suggested chemical agent. Other suggested substances included chlorine, hydrogen cyanide, arsenic and sulfur compounds, and acids.[71]

Most of these ideas focused on their effects against the enemy, with little if any regard to those who had to actually handle the material. Nor did the innovators consider if the agents could be delivered in suitable quantities to have the intended effects.[72] As such, most of these ideas were rejected as impractical. And morals played a big role in determining the use of these agents, as the chief ordnance officer of the US Army flat out stated that he would not allow the use of chloroform in artillery shells.[73] In this sense, these types of weapons were considered unethical and were barred by the US War Department.[74] This brings the moral quandary of whether war can be fought with rules and what these rules should be. Many proponents of these weapons had no qualms with using them to put down an open rebellion or to shorten the war, thereby saving countless lives.[75]

Black writes of three developments of significant importance during the US Civil War and the German Wars of Unification: the development of trench warfare, the targeting of civilian property, and guerrilla/insurrectionary and counterinsurrectionary warfare.[76] Trench warfare did not originate during the Civil War, but it closely resembled the warfare that became infamous during the coming war. For example, barbed wire was first used in May 1864 at Drewry's Bluff.[77] Sherman was not the only US general to employ scorched-earth techniques. John Pope, commander of the Union's Army of Virginia, destroyed large amounts of property in 1862, as did Sheridan in the Great Valley of Virginia in 1864.[78] A similar situation arose in the Franco-Prussian War with the German siege of Paris from September 19, 1870, to January 28, 1871.[79] Finally, guerrilla warfare became common during this time. The Confederates did what they could to sabotage the Union forces and had some success. The German situation is covered shortly.

JAPAN

Until now, I have focused on the European/American concept and implementation of total war. Let us now turn our attention to the Far East. The Japanese movement toward total war began in the late 1800s. When Commodore Matthew Perry came to Japan in 1853, the Japanese realized just how far behind the rest of the world they were. The ships that accompanied Perry showed them what they needed to do to compete in the global theater. In 1868, the Meiji Restoration brough Japan out of isolation and set it on a course of modernization that caused the world to take notice of this quiet giant.

An important aspect of this modernization involved bringing the military up to the standards of the militaries of the Great Powers. This was demonstrated by, among others, the motto "Enrich the country, strengthen the armed forces!" Another part of modernizing involved establishing territories and colonies apart from the Japanese home islands. Given their proximity to Japan, China and Korea were the logical choices for expansion.

Of course, all this activity caused other countries to take notice. In an especially prophetic statement, naval officer V. Golovnin wrote in 1816,

> If a sovereign similar to our Peter the Great reigns over this numerous, smart, subtle, resourceful, patient, hardworking and capable nation, then with the benefits and treasures that Japan has in its bowels, in a few years he will bring it to a position to rule over the entire eastern ocean. And what would happen then to the coastal regions in eastern Asia and western America, remote from the countries that should protect them?[80]

Little did he know that in just over one hundred years from the date of his
writing, this exact thing would happen.

During the First Sino-Japanese War of 1894–1895, the Japanese easily defeated the Chinese and gained the island of Taiwan for its territories.[81] However, the other countries did what they could to prevent the ascension of Japan
to the status of a great power. At the end of this war, the Great Powers humiliated Japan by forcing it to give back the territories that it had obtained during
the war. Japan further realized that the only way to deal with these countries as
equals was to have a strong military. At the end of the Russo-Japanese War in
1905, the Japanese emerged victorious. For the first time in modern history, an
Asian country was able to defeat a more powerful European country, resulting
in a thorough shake-up of the global balance of power. These ramifications were
important enough that the Russo-Japanese War is often referred to as "World
War Zero."[82] The war was very unpopular in Russia and led to the Revolution of
1905 and contributed to the Russian Revolution of 1917. Japan retained control
over Korea, which it annexed in 1910. After Russia left Manchuria and gave
Port Arthur to the Japanese, Japan was in control of much of eastern Asia.[83]
The Treaty of Portsmouth, negotiated by President Theodore Roosevelt, was a
balancing act by the United States to keep either Russia or Japan from becoming too dominant in the Orient, at a detriment to US interests there.[84]

While initially favoring the Japanese, as did the US public, Roosevelt
eventually sided with Russia in not forcing the payment of an indemnity to
Japan. Neither Russia nor Japan was able to negotiate the treaty from a position
of strength, resulting in this compromise. However, the Japanese public was in-
censed at the treaty, feeling they had soundly defeated Russia. Anti-American
rioting broke out in Tokyo and lasted three days. Moreover, the treaty fueled
a rising imperialism not just of the military but also of the ultranationalistic
public, both of which continued to view the involvement of both Europe and
the United States as unfavorable to Japan. In fact, Steinberg posits that Japan's
"losing the peace" was what led that country into World War II.[85] This victory
was an important step in helping the military enhance its position in the national government. In fact, the government could no longer exist without the
express consent of the heads of the army and navy. Despite these attempts to
"put Japan in its place," the Japanese were still able to annex Manchuria in the
northeast portion of China in 1931.

THE UNITED STATES
Back in the United States in 1865 to 1890, the US Army battled the Native Americans of the western United States. Captain Fetterman was one of
the first to fall in this new round of warfare. On December 21, 1866, after
boasting that, with eighty men, he could ride through the entire Sioux Nation, he proceeded to do just that. Against the direct orders of Colonel Henry
B. Carrington, Fetterman and his men followed Sioux decoys directly into an

ambush. The Sioux wiped out the entire force, despite having inferior weap-ons.[86] Such battles continued until the stalemate was broken and the Native Americans were placed on reservations. General Sheridan decided to change tactics and attack the Native American's shelter, food, and livestock, thereby destroying their entire way of life. Such tactics were, at best, morally question-able for many of his soldiers.[87] Regardless, the lessons learned during the Civil War were extensively used in the American Indian Wars of 1865 to 1890.

The sight of wagon trains was common as the westward migration of peo-ple continued from 1845 to 1870. The expansion into the West originally em-ployed these wagons over trails, but the completion of the first transcontinental railroad in the United States made life easier for the new settlers and aided in the migration. Barring Alaska, which was purchased in 1867 from Russia, the only place left for the United States to expand was overseas.

The Native Americans no longer had land to be forced onto, so reser-vations were created. Some Native Americans continued to hold out against the army and even had some success in fighting them. In December 1875, the US Indian Bureau notified the Sioux and Northern Cheyenne tribes that they were to vacate their winter quarters and return to their reservation. Because it was winter, there was no easy way for them to do this. They chose to stay where they were. The army attempted to remove them in March 1876 but were unable to do so. Skirmishes continued until June, when another major campaign was started. Plans were made for two forces to trap the Native Amer-icans in the Little Bighorn area. Lieutenant Colonel George Armstrong Custer and his Seventh Cavalry led one of these forces. He didn't know, however, that he was heading directly into six thousand to seven thousand Sioux and North-ern Cheyenne, two thousand warriors, and leaders like Crazy Horse and Sitting Bull. Custer and his Seventh Cavalry were no match for the overwhelming number of Native Americans. While this was a great victory for the Sioux and Northern Cheyenne, it led to an invasion by the US Army that permanently took down the Native Americans.[88]

Native Americans were not the only group to feel the wrath of the Ameri-can people. In the aftermath of the Mexican-American War, those of Mexican origin were frequently targeted for lynchings. From 1848 to 1879, there were 473 lynchings per 100,000 Mexican people, compared to 52.8 Black victims per 100,000. In total, 597 Mexicans were lynched between 1848 and 1928.[89] These lynchings are not considered war per se, but given the circumstances surrounding them, they were seemingly a continuation of the war against Mex-ico. The lynchings were not committed just by ordinary citizens but also often involved the local law enforcement and judiciary. In February 1857, a justice of the peace forced a group of Mexicans to watch as he decapitated a person and then repeatedly stabbed the corpse.[90]

Such treatment was not applied to all Mexicans equally. The native ruling class was seen as racially superior to the working class, to such an extent that members of the ruling class were sometimes invited to participate in vigilance committees and were often treated as White. Meanwhile, those of the working class were not usually considered White and were subjected to most of the ill treatment that was wrought on Mexicans. As one person said in the late 1920s, "They are an inferior race. I would not think of classing Mexicans as whites."[91] Clearly, White European settlers had little tolerance for those they considered to be unlike them.

1890–1914

By the end of the 1800s, the American worldview, while still very isolationist, was slowly turning outward, as opposed to an American-centric view. In 1890, the Census Bureau deemed that the frontier no longer existed. With the termination of inward expansion, the nation looked externally for growth. In 1878, the United States acquired the rights to develop a coaling station in Samoa, and in 1889, they recognized Samoan independence.[92]

The Industrial Revolution gave birth to new ways of manufacturing equipment as well as new ideas. However, it was during the so-called Second Industrial Revolution, beginning in the 1870s, when economic development finally caught up to technological development, allowing for such developments as improved and increased steel production, electrical grids, and mass production of machine tools. During this time, the United States was becoming a manufacturing giant. In 1890, it surpassed Great Britain as the world's leading producer of pig iron and steel, as well as coal. With this, custom fabrication of everything used in daily life was gone. Instead, everything was now standardized, and equipment parts were now interchangeable. Such things as clothing and farm machinery could be produced in bulk, thereby facilitating the ease of purchasing as well as the repairing of equipment.

Of course, military arms were a part of this standardization. With the ability to mass-produce weapons, countries were all too eager to have the newest and best equipment for their militaries. Artillery increased their range, while explosive shells increased the killing power of artillery. The Gatling gun had been present in the US Civil War, but the first true machine gun was developed in 1881. Innovations made it lighter, more accurate, more reliable, and faster. Additionally, small arms followed suit, and the range of rifles improved from around four hundred yards to more than two thousand yards.[93] All these developments led to greater and greater ability to kill. Sir Edward Grey, British foreign minister, said in 1914,

> A great European war under modern conditions would be a catastrophe for which previous wars afforded no precedent. In old days, nations could collect only portions of their men and resources at a time and dribble them out by degrees. Under modern conditions, whole nations could be mobilised at once and their whole life blood

and resources poured out in a torrent. Instead of a few hundreds of thousands of men meeting each other in war, millions would now meet—and modern weapons would multiply manifold the power of destruction. The financial strain and the expenditure of wealth would be incredible.[94]

These weapons were frequently used in the years leading up to World War I. As to naval advances, the HMS *Dreadnought* was launched in 1906. It was a state-of-the-art battleship under the British flag. Its four propellers were powered by steam turbines, which gave it a top speed of twenty-one knots, unparalleled at the time. Other advances incorporated into its design rendered all other battleships at the time obsolete. This one ship launched the first arms race of the twentieth century. Other nations quickly built their own *Dreadnoughts*, making the HMS *Dreadnought* obsolete by the end of World War I. Land weapons followed suit, making that war the bloodiest and most deadly the world had seen to that point.

Having conquered as much land as they could in North America, the United States was now setting its sights on territory elsewhere. Whereas to this point, as previously mentioned, the United States had been vehemently isolationist in all its dealings, it now embarked on the course of imperialism that the other nations had already started. In another display of contempt toward Indigenous peoples, the United States overthrew the government of Hawaii. In 1893, after Hawaii threatened to terminate their concessions to the United States, including a site for a naval base at Pearl Harbor, the United States tried to annex the islands. Wanting to keep European powers out of the islands and the sugar that those islands produce, the United States did all they could to claim the islands, including having the marines invade. The issue stayed in limbo until 1898, with the outbreak of the Spanish-American War and the need for a way station on the way to the Philippines. Again, land was taken from the Indigenous peoples without their consent:

> Joint Resolution To provide for annexing the
> Hawaiian Islands to the United States.
>
> Whereas, the Government of the Republic of Hawaii having, in due form, signified its consent, in the manner provided by its constitution, to cede absolutely and without reserve to the United States of America, all rights of sovereignty of whatsoever kind in and over the Hawaiian Islands and their dependencies, and also to cede and transfer to the United States, the absolute fee and ownership of all public, Government, or Crown lands, public buildings or edifices, ports, harbors, military equipment, and all other public property of every kind and description belonging to the Government of

the Hawaiian Islands, together with every right and appurtenance thereunto appertaining: Therefore,

> *Resolved by the Senate and House of Representatives of the United States of America in Congress assembled*, That said cession is accepted, ratified, and confirmed, and that the said Hawaiian Islands and their dependencies be, and they are hereby, annexed as a part of the territory of the United States and are subject to the sovereign dominion thereof, and that all and singular the property and rights hereinbefore mentioned are vested in the United States of America.[95]

While in the end any resistance was futile, the queen of Hawaii did lodge a formal complaint with the House of Representatives:

> I, Liliuokalani of Hawaii, named heir apparent on the 10th day of April, 1877, and proclaimed Queen of the Hawaiian Islands on the 29th day of January, 1891, do hereby earnestly and respectfully protest against the assertion of ownership by the United States of America of the so-called Hawaiian Crown Islands amounting to about one million acres and which are my property, and I especially protest against such assertion of ownership as a taking of property without due process of law and without just or other compensation.
>
> Therefore, supplementing my protest of June 17, 1897, I call upon the President and the National Legislature and the People of the United States to do justice in this matter and to restore to me this property, the enjoyment of which is being withheld from me by your Government under what must be a misapprehension of my right and title.[96]

The annexation of Hawaii was completed in 1898, over the objections of the queen. During this time, both Presidents Grover Cleveland and William McKinley tried to keep the United States out of other nations' conflicts, specifically the affairs with Spain and Cuba. However, the destruction of the US battleship *Maine* in the harbor of Havana, Cuba, ended all hopes for this, and on April 25, 1898, the United States declared war on Spain to help Cuba gain their independence.

Alfred Thayer Mahan of the US Navy had written extensively about the need for a blue-water navy to protect the nation's interests, both militarily and commercially. His seminal work, *The Influence of Sea Power upon History, 1660–1783* (1890), was world renowned for its insights, to the point that Kaiser Wilhelm II supposedly ordered a copy of the book placed on every German warship.[97] In this book, he lists six essential factors of sea power: "I. Geographical Position. II. Physical Conformation, including, as connected therewith, natural productions and climate. III. Extent of Territory. IV. Number of Population. V. Character of the People. VI. Character of the Government, including therein the national institutions."[98] Analyzing these factors, he correctly predicted the United States becoming the geopolitical heir to Great Britain.[99] Because of his writings, the US Navy was basically prepared for this conflict. The army was not. However, they were, for the most part, able to overcome their initial chaos, recruit the needed manpower, develop a semicoherent strategy, and fulfill their mission.[100] Fortunately, the navy carried the brunt of the important action in the war.[101]

The result was the end of Spanish rule in the Americas and the United States gaining control of Cuba, Puerto Rico, the Philippines, and Guam. Overnight, the United States became a world power, with a stake in world politics that continues to this day. While the United States withdrew its troops from Cuba in 1902, the Platt Amendment, which was incorporated into the Cuban Constitution, guaranteed a permanent US presence in Cuba in the form of a naval base at Guantanamo Bay, still occupied by the United States today. However, the Philippines, under Emilio Aguinaldo, who had been the leader of the government of the insurrection against Spain, revolted against their new "owners" and declared independence and then declared war on the United States. The insurrectionists were only concerned with independence and switched the focus of their attacks from Spain to the United States. The United States in turn wanted to annex the islands for themselves. Having purchased the Philippines from Spain in February 1899, the United States treated this uprising as an insurrection.[102] While they achieved multiple victories against Aguinaldo and his insurgents, they were unable to capture Aguinaldo and end the uprising.

Realizing that he could never defeat the United States using conventional warfare techniques, Aguinaldo switched to guerrilla tactics. He was able to effectively use nationalism, paternalism, propaganda, and terror to maintain control over the Filipino population. Part of the terror campaign included the assassination of pro-American Filipinos. His goal was to tie up the US forces long enough to make Americans tired of the war and hopefully elect a president, William Jennings Bryan, who was vehemently anti-imperialism.[103] To combat this, the Americans established a series of posts throughout the country that served three purposes: "They helped protect the population from guerrilla intimidation; they interfered with the ability of the population to provide food and recruits to the guerrillas; and they served as launching pads for innumerable small-unit patrols and raids into the bush in search of the guerrillas and their bases."[104] This strategy included such incentives as amnesty for those who surrendered and rewards for turning in their weapons. Then, in a carrot-and-stick move, those of the insurrection who refused to surrender faced imprisonment, deportation, and execution. Those supporting the rebellion without actually engaging in fighting still faced the prospect of the confiscation and/or destruction of personal property, thereby punishing the guerrillas.[105] When this didn't work, the army began destroying buildings and crops, imposing population concentration measures, and torturing people to gain information for their fight against the insurgents.[106] Here were additional uses of war against civilians to obtain the desired results.

These atrocities were well documented in letters from the conflict. C. R. Coulter, in a letter to the *San Francisco Call* on August 1, 1899, wrote, "The Americans immediately upon entering a captured village would proceed to ransack every house, church and even hold up the natives and procure everything of value, also that all natives discovered coming toward the lines with a flag of truce were shot down."[107] Another soldier wrote,

> The town of Titatia was surrendered to us a few days ago, and two companies occupy the same. Last night one of our boys was found shot and his stomach cut open. Immediately orders were received from General Wheaton to burn the town and kill every native in sight; which was done to a finish. About 1,000 men, women and children were reported killed. I am probably growing hard-hearted, for I am in my glory when I can sight my gun on some dark skin and pull the trigger.[108]

These are just a sample of the letters written; many more were documented in the 1899 Anti-Imperialist League pamphlet entitled *Soldiers' Letters: Being Materials for a History of a War of Criminal Aggression*.[109] Additionally, one general was retired as a result of his "Kill and Burn" order.[110] In 1946, the United States finally recognized the independence and sovereignty of the Republic of the Philippines, but even then, they retained their military bases there until they finally left in 1992.

During this war, disease, not battle, proved to be the major killer. Of the three thousand Americans who perished during the war, only a small fraction of these were killed in battle. The vast majority were claimed by yellow fever and typhoid. The Spanish forces had already been reduced from 230,000 troops to only 55,000 because of illness.[111] As long as disease remained a scourge, battle was not needed to decimate the ranks involved in the conflict. One soldier wrote home that he didn't believe there was a Spanish bullet that would kill him, but it was disease that he was most afraid of.[112]

Propaganda reached new heights during this time as yellow journalism came into being. Named for the comic strip *The Yellow Kid* (Mickey Dugan; see figure 4), many newspapers embraced yellow journalism and switched from traditional print layout of bland narrow columns and few illustrations to a more tabloid style, with headlines and illustrations to fuel the emotions of the readers, leading to more copies sold. William Randolph Hearst and Joseph Pulitzer engaged in a battle to sell more newspapers and embraced the sensationalism of the new newspaper layout. The Spanish-American War provided the perfect chance to grab events and turn them from a boring exposition of the account

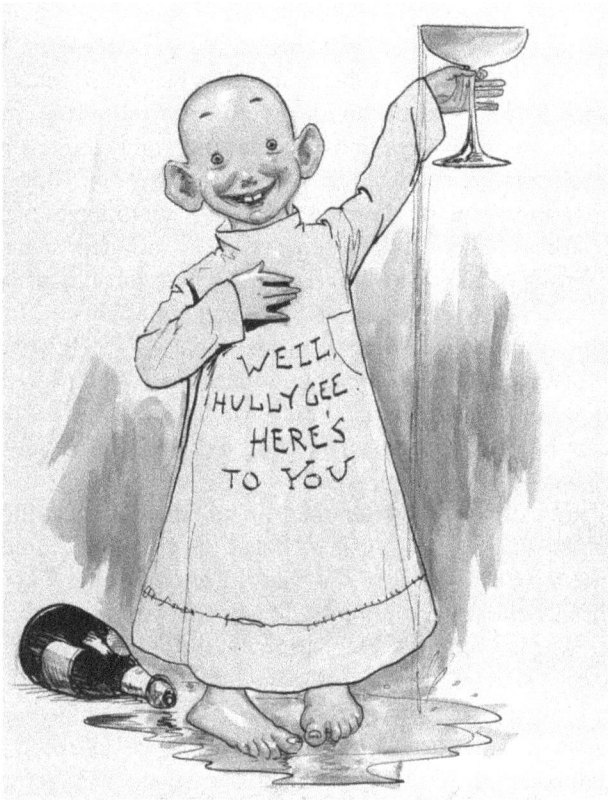

Figure 4. *The Yellow Kid*, by Richard F. Outcault, 1895. SOURCE: WIKIPEDIA, "THE YELLOW KID," ACCESSED MAY 17, 2025, HTTPS://EN.WIKIPEDIA.ORG/ WIKI/THE_YELLOW_KID#/MEDIA/FILE:YELLOWKID.JPEG.

to a more sensationalized story. The stories often became little more than propaganda, as facts were embellished, often to a point that the story was demonstrably false. However, as long as these stories sold papers, the publishers cared little about things as trivial as facts.[113] They utilized such tropes as depicting Spaniards as subhuman, while Cuba was shown as an innocent White girl being attacked by the Spanish monster.[114] These motifs appeared frequently in the years to come.

In 1915, the United States occupied Haiti. They had contemplated annexing the island of Hispaniola, comprised of Haiti and the Dominican Republic, as early as 1868 under President Andrew Johnson.[115] Haiti had been unstable since 1902, when it suffered through a civil war, where the government had prevailed against the rebels. Even then, the government was not stable, as seven presidents were overthrown or assassinated between 1911 and 1915.[116] In 1915, President Jean Vilbrun Guillaume Sam was assassinated, leading to more instability in the country. Under his rule, armed supporters broke into the national penitentiary and assassinated 167 political prisoners.[117]

With the pretext of preventing the country from descending into anarchy, President Woodrow Wilson sent the US Marines. In reality, Wilson was
protecting US interests and preventing invasion by foreign countries, specifically Germany.[118] The presence of US forces did little to stop the atrocities
rendered against the native Haitians. Several thousand civilians involved in
the Caco Rebellion were murdered by US troops. Many more were tortured by
such means as hanging by the genitalia and forced absorption of liquids.[119] According to historian Roger Gaillard, other methods included "summary executions, rapes, setting houses on fire after gathering their inhabitants inside them,
lynchings, . . . torching civilians alive," and being buried alive.[120] In foretelling
Japanese actions during World War II, an additional 5,500 civilians died during
forced-labor operations, such as road building. Those who slowed down the
pace of work were killed with machetes.[121] Moreover, US planes bombed and
shot at the civilians of Thomazeau without verifying what groups they were
bombing.[122]

Other countries were experiencing similar situations. During this same
time, the British were dealing with the Boxer Rebellion in China. It was the
first time since the Revolutionary War that the United States had joined with
other nations in a military operation.[123] The British were also embroiled in the
Second Boer War (October 11, 1899–May 13, 1902) and were similarly dealing with guerrilla warfare, for which they were unprepared. As did the Americans in the Philippines, the British countered with a scorched-earth policy,
in which the Boer farms, as well as entire towns, were destroyed and people
were placed into camps. While these camps started out as refugee camps, they
quickly devolved into concentration camps, where between 22,000 and 28,000
Boer (Dutch settlers) women and children and at least 20,000 Africans died of
disease and malnutrition.[124] One report stated,

> This removal took place in the most uncivilised and barbarous man-
> ner, while such action is . . . in conflict with all the up to the present
> acknowledged rules of civilised warfare. The families were put out of
> their houses under compulsion, and in many instances by means of
> force. . . . [The houses] were destroyed and burnt with everything
> in them, . . . and these families among them were many aged ones,
> pregnant women, and children of very tender years, were removed
> in open trolleys [exposed] for weeks to rain, severe cold wind and
> terrible heat, privations to which they were not accustomed, with
> the result that many of them became very ill, and some of them died
> shortly after their arrival in the women's camps.[125]

And as had happened in the Philippines, while no doubt many participated in
such heinous acts against the locals, only a noteworthy few were disciplined.
After the death of his commander, Australian Lieutenant Harry "Breaker"

Morant and Lieutenant Peter Handcock went on a killing rampage, during which at least one Boer prisoner of war and eight civilians, along with possibly a missionary, Reverend Carl August Daniel Heese, were murdered. While those involved were acquitted of the murder of Reverend Heese, they were convicted of the murder of the others and executed by firing squad on February 27, 1902. In fact, prior to being executed, the accused had left written confessions in their jail cells.[126]

Germany was also dealing with its own issues with the Wars for Unification from 1864 to 1871. Existing as a loose confederation of thirty-nine sovereign and independent states, including Austria and Prussia, from the defeat and expulsion of Napoleon in 1814, German nationalists increasingly insisted on a unified German state. Such efforts were in vain, as the varying viewpoints continued to argue among themselves about how this unification should occur, how the final product would appear, and who would be strong enough to accomplish this act. The military aspects of the Wars of German Unification are not of concern here. It is the mobilization of the German peoples that is of interest. In the early years of the 1800s, after the defeat of Napoleon, the unification was of great concern in an attempt to rid German lands of the French. Ernst Moritz Arndt stated, "Let the unanimity of your hearts be your church, let hatred of the French be your religion, let Freedom and Fatherland be your saints, to whom you pray!"[127] As Hamza Elshakankiri writes, "The new enthusiasm of German people took on the nature of a Crusade; the goal was Germany, and the antichrist was Napoleon."[128] Led by Frederick William III of Prussia, the fight was no longer confined to just the miliary. Civilians of the entire population were eager to join in the fight for unification. However, this populism was short lived, as rulers realized that any revolutionary group willing to amass against a foreign power like France could amass against any power, including their own conservative rulers.

It wasn't until Bismark came to power that the dreams of unification again came to life. Conservative and aristocratic, he knew that to achieve such a goal required working with the liberals. In 1866, the kingdom of Prussia withdrew from the German Confederation. Beginning with the defeat of Austria, Prussia, under the command of Helmuth von Moltke the Elder, chief of staff of the army, was able to utilize such modern equipment as railroads for the rapid deployment of troops, as well as the rifling of gun barrels and artillery pieces and faster breech-loading rifles. Prussia was the only country to implement these advances at this time.[129]

After this victory, Bismark realized that the unification was still not a sure thing. Needing an outside source against which to rally, he set his sights on France. By provoking France into declaring war on Prussia, Bismark was able to keep the other European nations out of the fight and created such an anti-Franco sentiment throughout Germany that for the first time in history, every other German state joined in the fight against France.[130] Bismark was

able to do what no other German leader in history had been able to do: mobilize the sentiment of the entirety of the German states into a unified effort
against a common enemy, resulting in a unified country (see figure 5).

This is not to say that the Germans had an easy time. The third of the
unification wars, the Franco-Prussian War, was longer than the first two wars.
Bismark was only fighting enough to win the war. He wanted a quick, relatively
bloodless affair and had no desire to destroy the enemy. The first two wars,
the Second Schleswig War and the Austro-Prussian War, were exactly that.
The Franco-Prussian War was longer and more difficult because there was more
territory to cover. The deeper into France Bismark got, the longer his supply
lines became, and the more difficult it was to maintain them. Moreover, this
deep incursion into France led to more encounters with the noncombatants,
the farmers and others not involved in the war. At least they weren't involved
until the Prussian army came through. It was then that Black's third point, previously touched on, came into play.

Figure 5. Map of the Unification of Germany, 1815–1871. Source: Thaddeus Chang, "European Union Economics," SlideServe, March 15, 2019, https://www.slideserve.com/thaddeus-chang/
european-union-economics-powerpoint-ppt-presentation.

Though there weren't many and they didn't do much damage, the insurgents/ guerrillas became a thorn in Bismark's side. Smallpox played a bigger role in casualties, with 180,000 German civilians dying from infection by French prisoners of war in a nearby camp.[131] The insurgency movement and its potential to escalate were something for which the Prussians were not operationally or psychologically prepared.[132] The number of these insurgents, the damage they did, and the extent of the Prussian reprisals remain contested by various scholars, but the issue of guerrilla/insurgent/counterinsurgent warfare remains valid.[133]

In closing, there have been many comparisons of the German Wars of Unification to the US Civil War. However, the one major difference between the two is that the Germans were fighting to defeat an enemy, while the United States was fighting to exterminate their enemy, a hallmark of total war. This was touched on previously, but another quote seems appropriate. Jefferson Davis, president of the Confederacy, in reply to President Lincoln's amnesty proclamation of December 1863, stated,

> The North was mad and blind; it would not let us govern ourselves; and so the war came, and now it must go on till the last man of this generation falls in his tracks, and his children seize his musket and fight his battle, unless you acknowledge our right to self-government. We are not fighting for slavery. We are fighting for Independence,— and that, or extermination, we will have. . . . Amnesty, Sir, applies to criminals. We have committed no crime. You may "emancipate" every negro in the Confederacy, but we will be free! We will govern ourselves. We will do it, if we have to see every Southern plantation sacked, and every Southern city in flames.[134]

CHAPTER 3

WORLD WAR I, 1914–1918

"Laws are silent in time of war."

—Marcus Tullius Cicero

While the Civil War saw the introduction of such technological advances as the railroad and rifled weapons, World War I was the first truly mechanized war. This mechanization led to the increasing involvement of civilians in order to supply the materials necessary to wage war. When the war first broke out, it was fought in relatively conventional terms. By the time the war ended, traditional strategies had ceased to exist, replaced by a strategy of escalation: that is, continuing to elevate the military involvement of a country despite the lack of ability to sustain such elevation. This was war based on ideology rather than on rationality and mobilized entire countries for total, unlimited war. Germany, with its dwindling resources toward the end of the war, should have scaled back its goals. Instead, it threw everything it had into the final battles of 1918 and, as Michael Geyer notes, began the passage into apocalyptic war.[1] This strategy was in direct contradiction to the writings of Sun Tzu, who stated, "Invincibility lies in the defense," and "One defends when his strength is inadequate; he attacks when it is abundant."[2]

THEORY
The invention of the airplane in the twentieth century revolutionized the theory of total war. Giulio Douhet realized that battle was no longer limited to the front lines of the opposing armies, with the civilian population residing safely and quietly behind those lines. Instead, the "battlefield will be limited only by the boundaries of the nations at war, and all of their citizens will become combatants, since all of them will be exposed to the aerial offensives of the enemy. There will be no distinction any longer between soldiers and civilians."[3] He recognized that Germany lost the war not because of military defeat but because of moral defeat caused by the attrition inflicted during the war.

He envisioned bombers flying over strategic targets and completely destroying
them in one attack, thereby eliminating the physical as well as the moral ability to wage war. He proposed the use of explosive, incendiary, and poison gas
bombs to inflict this damage: explosives to destroy the buildings, incendiaries
to burn anything combustible that remained, and poison gas to prevent the
population from extinguishing the fires.[4] He rationalized this devastation in
the same way that Grant and Sherman did: By ending the war as quickly as
possible, the minimum number of lives would be lost. Soldiers' lives were no
less precious than civilian lives simply because they were soldiers; all life was
precious.[5] Therefore, it made no difference who was killed, as long as it was
done as quickly and efficiently as possible in order to minimize the total loss—
fast war would result in fewer lives lost than long war.

William Mitchell echoed these beliefs about the importance of airpower in
waging total war:

> To gain a lasting victory in war, the hostile nation's power to make
> war must be destroyed—this means the manufactories, the means of
> communication, the food products, even the farms, the fuel and oil
> and the places where people live and carry on their daily lives. Not
> only must these things be rendered incapable of supplying armed
> forces but the people's desire to renew the combat at a later date
> must be discouraged.[6]

He viewed the threat of such total warfare as a deterrent to any warfare: "The
menace will be so great that either a state will hesitate to go to war, or, having engaged in war, will make the contest much sharper, more decisive, and
more quickly finished. This will result in a diminished loss of life and treasure
and will thus be a distinct benefit to civilization."[7] These two viewpoints fore-
shadow the doctrine of mutually assured destruction in the nuclear age: The
potential of two countries launching nuclear weapons at each other, thereby
possibly annihilating the entire human race, keeps either side from using such
weapons.

The teachings of Alfred Thayer Mahan demonstrate his understanding that the United States is geopolitically an island, like Great Britain, and
the defense of each of those powers from Continental Europe depended on a
strong navy. He also accurately predicted the political battles that would arise
in World Wars I and II, stating, "A German navy, supreme by the fall of Great
Britain, with a supreme German army able to spare readily a large expeditionary force for over-sea operations, is one of the possibilities of the future. The
rivalry between Germany and Great Britain to-day, is the danger point, not
only of European politics but of world politics as well."[8]

He also foresaw the Cold War with the Soviet Union, noting the "vast,
uninterrupted mass of the Russian Empire, stretching without a break . . . from

the meridian of western Asia Minor, until to the eastward it overpasses that of Japan" and realizing that it would take an alliance of the United States, Great Britain, France, Germany, and Japan to contain Russia from overexpanding into other countries, which is what occurred from 1945 to 1991.[9] Additionally, he predicted the rise of China as a problem that would need to be addressed. In 1893, he recommended the annexation of Hawaii for the United States to control the Pacific Ocean and prevent the overexpansion of China, which could threaten numerous other nations. Following Mahan's foundations of sea power, Francis Sempa notes, "China is situated in the heart of east-central Asia and has a lengthy sea-coast, a huge population, a growing economy, growing military and naval power, and, at least for now, a stable government," all aspects that led to the United States becoming a world power.[10] Additional works reference the writings of Mahan and the rise of the Chinese threat and note that China has fully adopted his teachings.[11] These issues are covered further in chapter 6. With the teachings and writings of Mahan, the United States would be prepared for the events that culminated in the First World War

WORLD WAR I

On June 28, 1914, Archduke Franz Ferdinand, heir to the Austro-Hungarian throne, and his wife, Sophie, Duchess of Hohenberg, were assassinated by Bosnian Serb student Gavrilo Princip while being driven through Sarajevo. Austria-Hungary had been worried about the Serbian threat to its empire and suspected Serbia of backing Princip and the assassination. Using this incident as a pretext, Austria-Hungary declared war on Serbia, hoping to crush the country and its threat. As a result of a series of multinational treaties among the various European nations, once Austria-Hungary declared war on Serbia, the rest of the countries were obliged to fulfill their treaty obligations and declare war on each other. Russia supported Serbia against Austria-Hungary. As Russia was allied with France and Great Britain, Austria-Hungary wanted assurances from Germany that it would come to its aid should Serbia, Russia, France, and Great Britain all align themselves together. In this way Russia, France, and Great Britain came to be at war with Austria-Hungary and Germany (see figure 6).

Initially predicted to be a quick war, it shortly stagnated into the trench warfare for which it became so well known. Germany crossed through Belgium to attack France, with France successfully counterattacking and driving the Germans back. While both sides tried to outmaneuver each other via flanking maneuvers toward the North Sea, neither side succeeded. Failing this, they eventually ran out of land to use to outflank each other. They were forced to dig in and move from a war of annihilation to a war of attrition. Once the trenches had been established, it became virtually impossible to gain any appreciable territory, as troops leaving their trenches entered the area between known as no-man's-land. Here they were quickly mowed down by machine gun

Figure 6. Europe pre– and post–World War I. SOURCE: RomanItalianEuropean, "Europe Before and After World War I," Reddit, posted 2019, https://www.reddit.com/r/europe/comments/crbnqa/europe_before_and_after_world_war_i/..

fire from the opposing side. These trench lines eventually extended from the Swiss border in the south to the North Sea coast of Belgium. As close to the ocean as they were, the trenches were constantly wet and cold. If the soldiers weren't actively engaged in battle, then they were focused on keeping warm and dry. As Jeremy Black phrases it, "This was a world that was very different from civilian society, at once both disorienting and very frightening."[12]

Trench warfare was a terrible experience. The two main items present in the trenches were mud and rats. The mud could be knee deep, while the rats could be as big as cats. Additionally, overflowing latrines, lice, trench foot, and trench fever were constant companions of the soldiers who lived in them.[13] While the German trenches were generally superior to those of the Entente, they were still trenches with the same general issues. The longer the war lasted, the more developed the trenches became, to the point that they nearly became small cities. By 1918, some German trenches were so developed that they ran for up to fourteen unbroken miles (see figure 7).

Besides the machine gun, artillery was the other constant mechanized companion of trench warfare. In the years leading up to the First World War, artillery technology and lethality had also advanced. Better sights, better propellants, steel-coated shells, high explosive contents, and new systems to allow

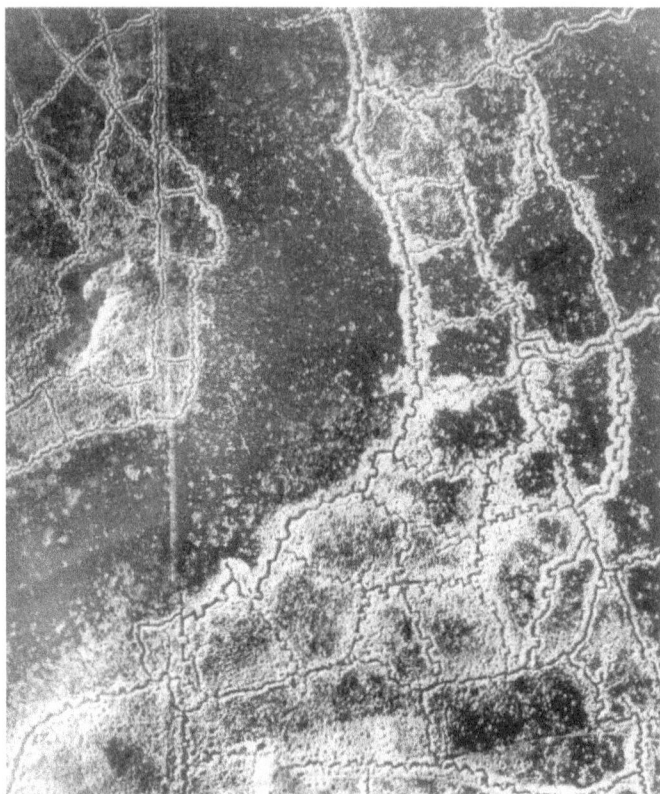

Figure 7. An aerial reconnaissance photograph of the opposing trenches and no-man's-land between Loos-en-Gohelle and Hulluch in Artois, France, taken at 7:15 p.m., July 22, 1917. German trenches are at the bottom right; British trenches are at the top left. The vertical line to the left of center indicates the course of a prewar road or track. The location is hill 70, attacked and taken between August 15 and 25, 1917, by the Canadian Corps. SOURCE: WIKIPEDIA, "AERIAL VIEW LOOS-HULLUCH TRENCH SYSTEM JULY 1917," ACCESSED MAY 21, 2025, HTTPS://EN.WIKIPEDIA.ORG/WIKI/ FILE:AERIAL_VIEW_LOOS-HULLUCH_TRENCH_SYSTEM_JULY_1917.JPG.

the artillery mechanism to recoil without the entire apparatus moving and to return automatically to its original firing position all served to turn artillery into a deadly instrument requiring some kind of counterforce.[14] Combining artillery and machine guns with the trench systems on both sides led to the trenches becoming a near-definitive defense system. The defender's inner lines led to the ability to quickly amass forces to repulse nearly any attack. When the attackers did manage to punch a hole in the enemy lines, they had usually suffered so many casualties that they were unable to exploit the opening. When they were able to attempt exploitation, the artillery that was needed to help them advance had an extremely difficult time maintaining the pace with the infantry, as the shell-cratered area of exploitation made it nearly impossible for the artillery to move together with the infantry. Moreover, those breakthroughs that occurred prior to late 1917 suffered from poor wireless technology, resulting in the inability to quickly apprise headquarters of the breakthrough. This rapid communication and response would have allowed more attacking forces to be rushed to aid in the attack. Finally, the defenders, as they fell back, could quickly bring in reinforcements via railway.[15]

These vast improvements in artillery caused most of the casualties of the war. In fact, Black states that artillery combined with mortars accounted for up to 60 percent of all casualties.[16] The stability of the trenches made artillery of prime importance. The artillery could be brought up to firing distance, then rain their high explosive shells down on the enemy trenches. While they may not have been of much use in the follow-up to the initial breakthrough, the initial barrage was devastating. However, even this artillery had its limitations. For example, in the Battle of Verdun, approximately two hundred artillery rounds were required for one casualty.[17] Clearly, the effectiveness of artillery was dependent on the volume of shells fired and not its accuracy: Given enough shells, there was no need for pinpoint accuracy.

Technology had made life easier in one sense, but when applied to war, it led to astronomical casualties. World War I saw the large-scale use of submarines, airplanes, machine guns, and flamethrowers. It also saw the invention and first use of tanks and chemical warfare. The result of these new inventions and new uses of old weapons, combined with obsolete tactics left over from earlier wars, resulted in casualties unheard of previously (see tables 2 and 3). There were several battles in which there were more than one million casualties. To illustrate, on July 1, 1916, the first day of the Somme Offensive, Britain attacked with 120,000 troops. On that day alone, it suffered 57,470 dead or wounded, or nearly 50 percent of its troops.[18] Generals were fighting as they had in previous wars, when armies attacked in formation. Using these tactics sent troops into the proverbial meat grinder. Either new tactics, new inventions, or a combination of the two were needed to allay the carnage.

The advent of the tank was one such invention. While some ascribe the demise of the German army in 1918 to the introduction of the tank, this is not

Table 2. Mobilized Strength and Total Casualties by Country in World War I

Nation	Mobilized	Dead	Wounded	Prisoners or Missing	Total Casualties
Allies					
United States	4,272,521	67,813	192,483	14,363	274,659
British Empire	7,500,000	692,065	2,037,325	360,367	3,089,757
France	7,500,000	1,385,300	2,675,000	446,300	4,506,600
Italy	5,500,000	460,000	947,000	1,393,000	2,800,000
Belgium	267,000	20,000	60,000	10,000	90,000
Russia	12,000,000	1,700,000	4,950,000	2,500,000	9,150,000
Japan	800,000	300	907	3	1,210
Rumania	750,000	200,000	120,000	80,000	400,000
Serbia	707,343	322,000	28,000	100,000	450,000
Montenegro	50,000	3,000	10,000	7,000	20,000
Greece	230,000	15,000	40,000	45,000	100,000
Portugal	100,000	4,000	15,000	200	19,200
Total	39,676,864	4,869,478	11,075,715	4,956,233	20,882,226
Central Powers					
Germany	11,000,000	1,611,104	3,683,143	772,522	6,066,769
Austria-Hungary	6,500,000	800,000	3,200,000	1,211,000	5,211,000
Bulgaria	400,000	201,224	152,399	10,825	364,448
Turkey	1,600,000	300,000	570,000	130,000	1,000,000
Total	19,500,000	2,912,328	7,605,542	2,124,347	12,542,217
Grand Total	59,176,864	7,781,806	18,681,257	7,080,580	33,434,443

Source: Library of Congress, "Mobilized Strength and Casualty Losses," accessed May 19, 2025, https://www.loc.gov/collections/world-war-i-rotogravures/articles-and-essays/events-and-statistics/mobilized-strength-and-casualty-losses/.

Table 3. Casualties by Battle, from Greatest to Least

Battle	Location	Dates	Number of Casualties
Brusilov Offensive	Carpathian Mountains, Galicia, Austria-Hungary	June 4–September 20, 1916	2,317,800
Hundred Days Offensive	Amiens, France, to Mons, Belgium	August 8–November 11, 1918	1,855,369
Kaiserschlacht (Kaiser's Battle)	Northern France/West Flanders, Belgium	March 21–July 18, 1918	1,539,715
Battle of Somme	Somme River, France	July 1–November 18, 1916	1,219,201
Gorlice-Tarnów Offensive	Galicia, Austria-Hungary	May 2–July 13, 1915	1,087,000
Battle of Verdun	Verdun-sur-Meuse, France	February 21–December 20, 1916	976,000
Battle of Passchendaele	Passchendaele, Belgium	July 31–November 10, 1917	848,614
Battle of the Lys and Second Battle of the Somme	Flanders, Belgium, to Northeast France and Somme River, France	April 2–29, 1918, and August 21–September 3, 1918	804,100
Battle of Galicia (Lemberg)	Lemberg, Galicia, Austria-Hungary	August 23–September 11, 1914	655,000
First Battle of the Marne	Marne River near Paris, France	September 5–12, 1914	483,000
Battle of Gallipoli	Gallipoli Peninsula, Turkey	April 25, 1915–January 9, 1916	473,000
Battle of Kolubara	Kolubara River, Serbia	November 16–December 16, 1914	405,000
Battle of Arras	Arras, France	April 9–May 16, 1917	278,000
Battle of the Argonne Forest	Argonne Forest, France	September 26–November 11, 1918	187,000
Battle of Tannenberg	Near Allenstein, East Prussia	August 26–30, 1914	182,000
Ninth Battle of the Isonzo	Soča Valley, Slovenia	October 31–November 4, 1916	138,000

Adapted from Alex Browne, "15 Bloodiest Battles of World War One by Casualty Figures," History Hit, November 9, 2018, https://www.historyhit.com/biggest-battles-world-war-one/, and Infoplease Staff, "World War I Battles with the Most Casualties," Infoplease, updated August 5, 2020, https://www.infoplease.com/us/military/world-war-i-battles-most-casualties.

the case.[19] Regardless, the tank was an important invention. Its strengths lay in its ability to accompany soldiers as they attacked the enemy lines, thereby giving the attackers (the Allies) protection from the opposing trenches. Additionally, the fact that they could even accompany the soldiers through the craters of no-man's-land was a significant advantage.

The tank was first used on September 12, 1916, in the Battle of Flers-Courcelette, with mixed results.[20] On August 8, 1918, 430 British tanks broke through the German lines. The British took 12,000 prisoners, and the Germans were unable to stop the Allied advance.[21] As with any new technology, the first tanks had numerous problems to overcome, such as dependability, speed, and communications.[22] Many would break down before even reaching their deployment points. However, the potential was there, and with improved mechanical ability came improved results. And while effective tactics needed to be defined and then utilized, the tank played an important role in warfare from this point on.

CRIMES AGAINST CIVILIANS

As covered in chapter 2, Germany, in the Franco-Prussian War, believed the French to be engaging in a major partisan warfare effort against their country. They believed the same possibility existed in the current war. While such civilian resistance was to be expected and was actually allowed under the Hague Convention, the Germans made it very clear that they would not tolerate it.[23] This was true on all fronts: the Western Front as they advanced through Belgium into France, the Southern Front as troops invaded Serbia, and the Eastern Front where the Russians invaded Prussia. While there was no evidence of such guerrilla warfare, the Germans imagined it was occurring and acted accordingly. General actions, such as hostage taking, human shields, and mass reprisals with mass executions were used in all three areas.[24] Rumors and press coverage caused even more retributions against the enemy civilians, with military losses leading to increased reprisals. On all these fronts, mass exoduses began, as civilians tried to flee the invading forces, displacing tens of thousands of refugees.[25]

The Rape of Belgium

In early August 1914, German forces entered Belgium, and the atrocity known as the Rape of Belgium began. On August 19, 1914, German troops entered the village of Andenne on the Meuse River, unopposed by the civilians left there by the retreating Belgian forces. At 4:30 the next afternoon, shots rang out, which were countered by the Germans. At that point, the Germans began the slaughter of the village inhabitants. For the next two hours, the villagers were gunned down by machine gun fire. When the initial round of carnage ended around 7:00 p.m., the village was torched. The following morning, more villagers were driven into the town square, where three men were shot and a fourth was bayonetted. More were taken to the banks of the Meuse and shot. In total, about four hundred villagers were murdered.[26] On August 23, in the Belgian town of Dinant, 674 civilians were murdered, shot in revenge for perceived actions against the Germans. A report compiled the following year by the Committee on Alleged German Outrages detailed many more of these crimes.

The Belgians had been ordered by provincial civil governors and town burgomasters to "offer no provocation to the invader" and to surrender their firearms to the local officials.[27] However, this did little to stop the ensuing bloodbath. Multiple villages were torched, and when the inhabitants, including children, attempted to escape, they were shot.[28] Numerous priests were accused of firing on the German soldiers as a pretext to round up hostages and execute them, but the shots were usually from Belgian troops.[29] Another reason the report gave for the slaughter was the "army as a whole wreaked its vengeance on the civil population and the buildings of the city in revenge for the setback which the Belgian arms had inflicted on them."[30] The specific incident to which this statement refers occurred in and around the village of Leuven. The Germans initially occupied the village without incident. The soldiers were well behaved except for the occasional looting of empty houses. However, after six days of occupation, on August 25, 1914, the carnage began after the Germans had been repelled by the Belgian troops. Once the orgy of violence began, it followed the same pattern as before. The university library and its 300,000 volumes, along with the church of St. Peter and numerous houses, were burned to the ground.[31] Corpses, many of them burned beyond recognition, were left on the sides of the roads and in the streets. The murders weren't limited to the citizens of Leuven. Large numbers of people were brought to Leuven from the surrounding areas, where they were subjected to the same actions as had the people of Leuven.[32] In a refrain that echoed twenty to thirty years later, several German soldiers, NCOs, and officers offered the excuse that they were "acting under orders and executing them with great unwillingness" and that failure to comply with the orders would result in the disobedient soldier being himself shot.[33] In total, the report provides thirty pages of documented war crimes committed by the Germans against the Belgian civilians.[34]

The report points out that most of these systematic atrocities occurred in an area forming an irregular Y from the Belgian frontier to Liège onward to Charleroi and from Liège to Malines (see figure 8), from August 4 to 30. The barbarities began as soon as the Germans crossed the frontier. However, from August 4 to 18, these acts were confined mainly to the area around Liège. From August 19 to 30, the brutality spread toward Charleroi and Malines. Coincident with these acts were issues the German army was having in the area: Either the Belgians put up unexpected resistance, or the Germans needed quick passage through Belgium at any cost.[35] Isolated incidents of heinous acts occur in any war, as I have shown in this book, but the emphasis is on isolated. The actions in Belgium can only be labeled as organized and methodical and, as

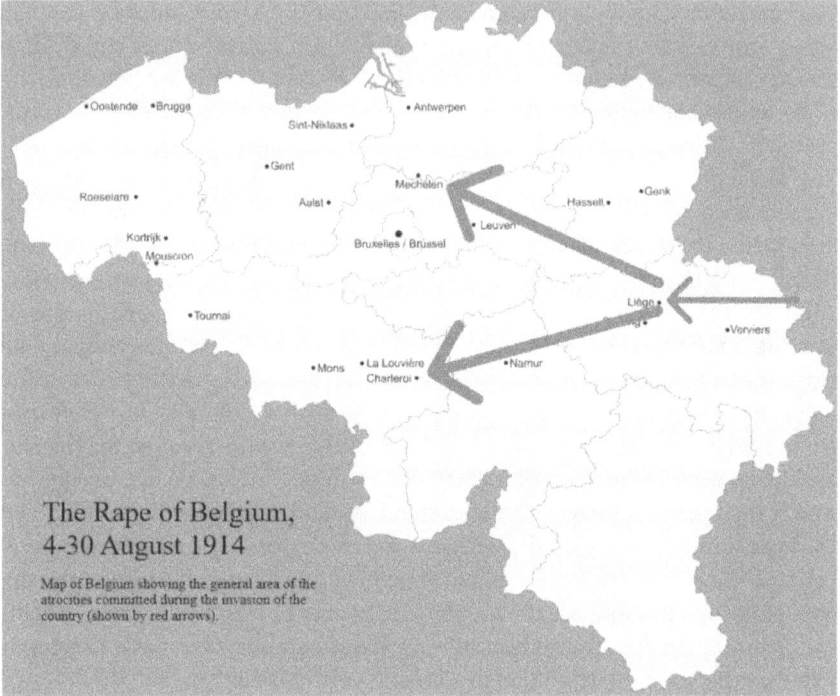

The Rape of Belgium,
4-30 August 1914

Map of Belgium showing the general area of the
atrocities committed during the invasion of the
country (shown by red arrows).

Figure 8. Map of Belgium showing the general areas (arrows) of the atrocities com-
mitted during the Rape of Belgium, August 4–30, 1914. EDITED; ORIGINAL IMAGE FROM
GUIDE OF THE WORLD, "MAP OF BELGIUM'S MAIN CITIES," ACCESSED MAY 19, 2025, HTTPS://WWW
.GUIDEOFTHEWORLD.COM/WP-CONTENT/UPLOADS/MAP/BELGIUM_MAIN_CITIES_MAP.JPG.

such, can hardly be attributed to the isolated, infrequent acts of individual soldiers. As covered in chapter 6, such actions were justified as militarily necessitated or in direct retribution for civilians firing on German troops.[36]

Still, organized and methodical war was not the only cause of death of
civilians. Being displaced typically meant being without food, clothing, and
shelter. During the winter of 1914–1915, around 100,000 Serbian civilians died
as a direct result of the war leaving them without their necessities.[37]

Armenian Genocide

In a preview of the genocides perpetrated from 1937 to 1945, the Armenian
people were nearly exterminated by the Turkish government. The Armenian
people had inhabited the area of eastern Turkey and beyond the eastern border
of the Ottoman Empire and into Russa since antiquity, and for nearly as long,
they had been persecuted.[38] Christian Armenians faced numerous incursions
from outside forces. In the fifteenth and sixteenth centuries, they were over-
taken by the Ottoman Empire.[39] The Armenians were able to remain a semiautonomous entity with control over their religious, social, and legal structures,

but they were often heavily taxed.[40] However, the Kurdish nomads that inhabited the same area treated them cruelly. And while they were able to maintain some control over their legal system, they were ultimately under the control of the Kurds. Because the local courts and judges typically backed the Muslims, the Armenians had little legal recourse.[41] And in spite of being supported by Russia, the persecution continued, especially after Russia defeated the Ottomans in the Russo-Turkish War of 1877–1878.[42]

However, in the late 1800s, this changed. By this time the Ottoman Empire was beginning to fall apart with revolts by the Christian subjects. In 1895, the Armenians in the Sasun region refused to pay a crushing tax. Ottoman troops, along with the Kurdish tribesmen of the area, proceeded to kill thousands of Armenians in the region.[43] In 1895, a demonstration by Armenians in Istanbul turned into a slaughter, while from 1894 to 1896, hundreds of thousands of Armenians were murdered in the Hamidian massacres.[44] And in urban riots in Adana and Hadjin in 1909, some 20,000 Armenians were killed.[45]

Later, in 1908, a group of aspiring, disgruntled junior army officers known as the Committee of Union and Progress (CUP, or Young Turks, as they were popularly known) seized control of the government. While initially friendly to the Armenians, the Young Turks gradually came to believe that the Armenians were working with foreign countries.[46] When the Ottomans were soundly defeated in the First Balkan War of 1912–1913, the Armenians were able to force the Turks to be overseen by European powers. The Young Turks, in turn, took that arrangement as evidence of the Armenian complicity with Europe.[47] When World War I started, the Young Turks wanted not only the Ottoman Armenians but also the Russian Armenians to fight for the Ottoman Empire. When told that those two factions would fight for their respective empires, the Young Turks saw another act of betrayal.[48]

The CUP utilized the available media to stoke the fires of genocide against the Armenians. A speech given to members of the CUP in February 1915 included the following: "It is absolutely necessary to eliminate the Armenian people in its entirety so that there is no further Armenian on the earth and the very concept of Armenia is extinguished."[49] Their message was also printed in numerous newspapers throughout the country. The Turks portrayed the Armenians as less than human, as animals, as infestations, and as fifth-column agents of Russia attempting to destroy Turkey from within, and they were able to whip the population into a fever pitch. Such messages as "Turkey could only be revitalized if it rid itself of its non-Muslim elements"; portraying them as an "invasive infection in Muslim Turkish society"; and stating rhetorically, "Isn't it the duty of a doctor to destroy these microbes?" had the desired effects.[50] Widespread grassroots efforts to indoctrinate the people included Muslim clerics preaching these ideas during worship services, as well as town criers, who would use their platforms to exacerbate the already-present anti-Armenian sentiment in the nation.[51]

This all came to a head in January 1915, when the Ottoman Empire suffered their worst defeat of the war fighting against the Russians at the Battle of
Sarikemish. Rather than place the blame where it belonged, on poor leadership
and severe conditions, they blamed the Armenians. The government enacted
laws to deport anyone they deemed a security threat to confiscate abandoned
Armenian property. All their guns were seized, and the people were forced to
cross the Syrian desert without food, water, or shelter to concentration camps.
Many died during these marches, and many more were executed into mass
graves.[52] Armenian adult males were ordered to report for military duty. The
Armenians believed this was merely an excuse to ensure that they were all
killed. Subsequently, on April 19, 1915, the Armenians staged an armed uprising in the city of Van.[53] Finally, on April 24, 1915, several hundred thousand
Armenians were rounded up and executed. The leadership of the Armenian
people was removed from the city. Very few survived.[54] These acts continued
into 1917. When the war started, there were 2,133,190 Armenians. By the
time the war was over, there were 387,800.[55] Some Young Turks said that they
wanted the Armenians destroyed, and they very nearly succeeded.[56] These
crimes were well documented by foreign diplomats, journalists, military officers,
and missionaries, leading to widespread outrage against the Young Turks. Even
Germany was later found to have expressed horror at what was happening.[57]

Russia, however, was not as benevolent to other races as they were to the
Armenians. After defeating the Hungarians in the Carpathians in the spring
of 1915 and capturing the city of Przemyśl, they immediately began persecut-
ing the Jews there. On April 17, 1915, the Russians began a full-scale pogrom
against the Jews. They were assaulted, robbed, and driven out of the city. Besides being innately anti-Semitic, the Russians believed the Jews were actively
aiding Russia's enemies.[58] I show later that this is a common trait among people regarding especially Jews.

PRISONERS OF WAR

One very important part of both the Hague Conferences and the Geneva Convention was the treatment of prisoners of war. The Hague Convention of 1907,
Annex to the Convention, section 1, chapter 2 sets forth the manner in which
prisoners of war are to be treated. Articles 4 through 20 contain detailed requirements for their treatment. Article 4 states, "They must be humanely
treated," and article 7 states, "Prisoners of war shall be treated as regards board,
lodging, and clothing on the same footing as the troops of the Government who
captured them."[59] However, just how closely this was followed is up for question.

Prisoners held by France and Great Britain tended to be treated better
than those held by the Central Powers due to the effects of the British naval
blockade. Russia had their own problems, in that even the soldiers couldn't
always get basic necessities, which was only exacerbated by the Revolution of
1917. This was due in large part to the fact that no one expected the war to last

for any appreciable length of time and certainly not into winter. There was no place to serve as camps to put the prisoners, along with insufficient clothing, food, and housing.[60] In total, some nine million people were taken prisoner at one point or another.[61] The vast majority of these prisoners were taken on the Eastern Front, where battle was a much more fluid affair, compared to the Western Front and its static trench warfare. It wasn't until the failed German offensive of 1918 and the resulting Allied breakthrough that large numbers of Germans were taken prisoner, approximately 340,000 between July 18, 1918, and the armistice on November 11.[62]

Treatment in the camps largely depended on class and rank. Officers were almost always treated better than enlisted men. Overall, the higher the rank, the better treated the prisoners were. Nearly all were forced to work, whether as mundane as making beds or as grueling as helping construct the Murmansk railway in Russia, where, of 70,000 prisoners used in the construction, 25,000 died. Romanian prisoners in Germany had a mortality rate of 29 percent.[63] Forced labor was an effective way to keep the prisoners occupied. While the Hague Convention mandated that prisoners not be used to work on the captor's war effort, both sides largely ignored this. As Heather Jones notes, "In a total war, with whole economies geared towards military production, this stipulation was rapidly abandoned."[64]

Some of the purported mistreatments were laughable. In 1914, British officers reported mistreatment when they were transported in second-class carriages. In 1915, the British and French governments complained that their respective captives were being held with Russian prisoners rather than being separated.[65] However, there were some very serious cases of prisoner mistreatment. In one particularly egregious case, a British parliamentary paper in 1918 documented that a British prisoner of war, Able Seaman John P. Genower, was burned to death in March of that year. He, along with other prisoners (in total, eleven Russians, two French, and one English, Able Seaman Genower) had been confined to one building designated for "rebellious" prisoners. A fire broke out in the building, but the guard ignored the calls for help. When Genower tried to escape the building, the guard sank his bayonet into him, forcing him back into the burning building. Of the fourteen prisoners, only four (three Russian, one French) were able to escape. The other ten burned to death.[66] In another instance, medical personnel and wounded soldiers were murdered after a hospital had already surrendered.[67] Additionally, there were multiple episodes of a tit-for-tat mentality among countries.[68]

Nevertheless, some of those tit-for-tats were well deserved. In spring 1917, the Germans announced that all newly captured, uninjured prisoners would be kept behind the lines. This was in retaliation for the French use of German prisoners to labor in the Verdun battlefields under very harsh and dangerous conditions as well as the British decision to use German rank prisoners to work in labor companies on the Western Front.[69]

Attitudes of the involved countries played a role in prisoner treatment. Italians viewed captured Italians as cowards and traitors.[70] And international animosities played a role. Russians favored Slavic prisoners over German, while Ottoman captors favored Muslim captives over Hindu, Sikh, and Christians. The favored captives were given the best food; were sent to the best camps; and received the best lodging, the best work details, and more lenient punishments.[71]

Besides mistreatment, disease was a constant threat. In 1915, there was a major typhus epidemic in both the German and Russian prisoner-of-war camps. Again, this was partly due to the overcrowded conditions of the camps as a result of the lack of planning for the number of prisoners. In the Totskoe prisoner camp in Russia, the typhus outbreak was so severe that of the 25,000 prisoners present, at least 10,000 died.[72] The outbreaks in the German camps were used as propaganda by the Allies to demonstrate the poor conditions under which the prisoners were subjected. These outbreaks forced the camps to improve their living conditions. Delousing upon arrival at the camps became standard operating procedure.

Typhus was not the only disease to run rampant in the camps. The Ottoman Empire at one point reported 771,844 army casualties, with disease accounting for 466,759.[73] Another report based on historical analysis of Ottoman records suggested the following casualties from disease: "dysentery (40,000), typhus (26,000), malaria (23,351), recurrent fever (4,000), or syphilis (150)."[74] Typhoid fever, diphtheria, tuberculosis, cholera, malaria, smallpox, and pneumonia were all common in prisoner camps.[75]

Prisoners changing sides was not uncommon. Prisoners with ethnic backgrounds similar to those of the captors were often confronted with the opportunity to change sides. Germans tried to recruit Muslims on behalf of the Ottoman Empire. The French tried to recruit those from the area of Alsace-Lorraine. One notable example was the Czech Legion, formed by those captives of Austro-Hungarian heritage. At its largest size, it comprised some 60,000 men fighting for Russia in a highly trained and effective unit. During the Russian Civil War, they stayed on the side of the Allies and ran the trans-Siberian railway until they were repatriated.[76]

Repatriation was a difficult process. The Allies demanded release of their prisoners as soon as the armistice was signed. However, getting those of the Central Powers took a great deal more time than did those of the Allies. German prisoners held captive by Britain were repatriated in late 1919. The French used its prisoners to clear the battlefields of remaining munitions, which led to a number of casualties. They were finally released in 1920. Those captured by the Russians held the distinct possibility of having adopted Russian Bolshevik beliefs; Russians held by Germany were used in the German revolution as pawns of the left-wing groups. Prisoners in Japan and Siberia weren't released until 1922. Surprisingly, given the lack of support that exists today, most of the prisoners seem to have adjusted to civilian life relatively well.[77] The experiences of the prisoners of war bore results some years later at the Geneva Convention, as extensive rules regarding prisoner care were codified and are still in force today.

CHEMICAL WARFARE

On April 22, 1915, at about 5:00 p.m., during what became the Second Battle of Ypres, on a four-mile (6.5 km) front just north of Ypres, between the hamlets of Langemark and Gravenstafe, Allied soldiers in the trenches saw something unusual happening—a greenish-yellow cloud coming toward them:

> Just at dawn they opened a very heavy fire, especially machine-gun fire, and the idea of that was apparently to make you get down. And then the next thing we heard was this sizzling—you know, I mean you could hear this damn stuff coming on—and then saw this awful cloud coming over. A great yellow, greenish-yellow, cloud. It wasn't very high; about I would say it wasn't more than 20 feet up. Nobody knew what to think. But immediately it got there we knew what to think, I mean we knew what it was. Well then of course you immediately began to choke, then word came: whatever you do don't go down. You see if you got to the bottom of the trench you got the full blast of it because it was heavy stuff, it went down.[78]

Another survivor recalled, "I witnessed from the air the first gas attack when the Germans used chlorine gas in the Ypres Salient. Suddenly we saw to the north of us in the salient this yellow wall moving quite slowly towards our lines. We hadn't any idea what it was. We reported it of course when we landed. And an hour or so later the smell of chlorine actually reached our aerodrome."[79]

None of the Allies knew what this was, nor did the Germans really understand what to expect. Even though they had effectively opened the way for a complete German advance, they were unprepared to exploit the situation, and the Allies held most of their positions. However, the results of this gas quickly became apparent as the soldiers who had breathed in the gas were soon dead. Approximately 160 tons of chlorine gas had just been released by Germany against the Allies, ushering in the formal beginning of the age of chemical warfare.[80] By the time the conflict was over, an estimated 113,000 tons of chemical weapons had been used, with 92,000 dead and more than 1 million injured.[81]

Fritz Haber of Germany was a chemist who had devised a way to extract nitrogen from the atmosphere, which could then be used as fertilizer for farming. It was to be one of history's greatest agricultural achievements.[82] This discovery was recognized with the Nobel Prize in Chemistry for 1918.[83] However, Haber became infamous for his role in the development of chemical cylinders that released toxic clouds of chlorine and other gases in World War I. This led to him being acknowledged as the "father of modern chemical warfare."

Haber was a staunch German patriot who felt it his duty to help Germany win the war in any manner feasible. Because he was a chemist, it was only natural that he would turn to chemistry in the quest. His first attempts to convince the German military of the practicality of chlorine gas in war were rejected. However, as the war turned into a stalemate, the military turned to Haber for any advantage possible. Upon seeing the effects of the chlorine gas, the military promoted Haber, who started this endeavor as a sergeant, to captain.[84] With the successful implementation of gas warfare, continued research under Haber (as well as the further use of gas on the Eastern Front under his direction) led to the development of phosgene gas and mustard gas.[85]

While Germany may be recognized as the first country to use poison gas in war, the French had used suffocating grenades prior to this incident. When blamed for having started the chemical gas war, Germany cited a French note dated February 21, 1915, describing that the French had already used such agents and were planning on using them again.[86] Clearly, both sides were culpable in the initiation of gas warfare, and said warfare was going to happen regardless of which side instigated it.

The very next day following the initial gas attack, the French began addressing the German initiative by identifying the gas and looking for a defense against it.[87] While they declared the abhorrence of the act, the French fully embraced the idea that retaliation in kind was an absolute necessity.[88] The Allied reaction was summed up by French General Maxine Weygand: "The Germans took the initiative to use inhuman means of warfare that had been banned by international treaties. But for us, it was not about procedures but about preparing as fast as we could the means to protect ourselves and retaliate in kind to these attacks. . . . A new step toward *total warfare* has been taken by our enemies."[89] And by June, the concept of using airplanes to drop these gas bombs was considered.[90]

The initial use of chlorine gas resulted in the death of five thousand Allied troops, although that number is subject to debate.[91] The Germans followed this initial gas attack with another at Ypres on April 24, 1915, and four times in May. The Allies hurriedly determined the best ways to combat the use of gas via gas masks and protective clothing, but each new chemical used required new protection. Additional research on both sides led to the development of diphosgene, chloropicrin, lewisite, hydrogen cyanide, cyanogen chloride, and mustard gas. Of these, phosgene and mustard were the most feared.

Phosgene is highly toxic and resulted in approximately 85 percent of all chemical weapon–related death.[92] Phosgene is heavier than air at room temperature. It has industrial uses in the manufacture of plastics and pesticides. It works as a pulmonary irritant, converting to hydrochloric acid in the lungs, damaging the pulmonary alveoli, and leading to pulmonary edema. Due to its low water solubility, its effects may not be known for several hours after exposure.[93]

Mustard gas causes severe burning and blistering of the skin, along with damaging the mucous membranes of the respiratory system. Additionally, it is very long lasting, remaining in the soil for days. It also had many imaginary effects, dreamed up by the paranoid, such as causing sterility and making arms and legs fall off a person's body.[94] Mustard gas was first used on July 12, 1917. It, too, took the Allies by surprise. Technically not a gas but a viscous brown liquid that gives off vapors, it was not what the Allies were expecting. Other than a slight irritation of the eyes and throat, the liquid apparently did nothing, and the soldiers went back to their normal routines.[95] But several hours later, the symptoms began.

Initially the soldiers complained of pain in the eyes, like sand or grit had gotten in them. Violent vomiting followed, along with blister formation. The vapors could easily pass through clothing, making it even more dangerous. As bad as the blisters were, it was the destruction of the respiratory system through the stripping of the mucous membranes that usually led to death. However, exposure was usually not fatal; it killed fewer than 5 percent of those who received medical care after exposure.[96]

In September 1915 came the first large-scale Allied chemical gas attack at Loos, Belgium.[97] Phosgene was subsequently used in the summer of 1916. It was a mustard gas attack by the Allies in October 1918 that injured a young German corporal by the name of Adolf Hitler.[98] This injury played a significant role two decades later. In spite of the horrific effects of these gases, Haber remained a staunch proponent of the use of gas in warfare, believing gas to be a more humane weapon than modern artillery.[99]

Meanwhile, back in the United States, not only were the Americans late in joining the war, but they were also just as late in starting their own chemical research program. Without any dedicated research facilities or staff, they were forced to implement a major recruitment effort among private chemists and locate NGO research facilities, one of those being the campus of American University, founded twenty-five years prior but still so incompletely built as to barely qualify as a skeleton of a university.[100]

Everything about the organization was built from scratch. The new director of gas service, Amos A. Fries, wrote, "It's a big job and a vital one to the army, and no one seems to have wanted to go into the thing so the dearth of information is startling."[101] When Fries started, "He had only two officers, no gas masks, no gas, no literature, and no organization."[102] The American public response to these plans for chemical warfare were mixed, with headlines reading "U.S. Prepared for Barbarity" and "Uncle Sam to 'Fight Devil with Fire.'"[103]

What did gas warfare accomplish during all of this? Strategically, nothing. No significant ground was lost or captured as a result of the use of gas. And while there were significant numbers of casualties, Haber concluded in 1921, "Poison gas caused fewer deaths than bullets."[104] The main advantage in the use of gas was the element of surprise. Guns and bullets are loud, overt weapons. When used, everybody around is immediately alerted. Gas is a quiet, covert weapon, able to inflict its damage before the intended victims even know they've been targeted. This element of surprise accounts for its repeated use later in history.

SUBMARINE AND NAVAL BLOCKADE

Britain, the world's leading naval power, was able to start a blockade of Germany from the outset of the war. The British hoped to turn this blockade into a two-way street: Besides keeping supplies from reaching Germany, they hoped to keep German products from being exported, thereby cutting off income to that nation.[105] While this was only partially effective due to Germany being able to borrow money to finance the war, it did reveal a critical shortage for Germany: nitrates for the manufacture of explosives. The only thing that saved the German war effort was the fact that Fritz Haber shortly before the war had invented a way to capture nitrogen from the air.[106]

Likewise, Britain was an island nation that could produce neither enough food to feed its citizens nor enough building supplies to keep its military armed. They relied on the United States for the matériel to keep the nation armed and fed. As such, they had to keep the sea lanes open for US cargo ships. The Germans knew this and by 1915 had begun a campaign of unrestricted submarine warfare, targeting all shipping. Originally bound by the cruiser rules, which required submarines to warn merchant ships before attacking them, the Germans realized these rules were impractical and ceased to recognize them.[107]

While these early submarines were primitive, they were highly effective. Antisubmarine measures in these early years of the war were nearly nil and allowed the submarines to hunt as they wished. However, after sinking of the HMS *Lusitania* on May 7, 1915, and the SS *Arabic* on August 19, 1915, and fearing US reprisal for the loss of 128 Americans onboard the *Lusitania*, unrestricted warfare was terminated. In fact, Admiral Tirpitz was upset about this restriction to the point that he removed all U-boats from the Atlantic.[108] And as effective as their submarine warfare had been, they had neither the submarines nor the trained crews to man them to continue such tactics.[109]

Besides submarines, the Germans did utilize their surface ships against the British, at least in the early phases of the war. In December 1914, their battle cruisers bombarded the towns of Scarborough and Hartlepool. The British were livid over the attack against civilians and quickly responded by sinking the German battle cruiser *Blücher* on January 24, 1915.[110]

On January 31, 1917, Germany again instituted unconditional submarine action against all surface vessels. Whereas in 1915 Germany feared alienating the United States and causing their entry into the war, in 1917 they made the conscious decision to risk the US entry into the war. They felt that their submarine fleet was such that a quick, decisive victory could be achieved, even though the statistics to justify the effectiveness of the U-boats were seriously skewed.[111] They anticipated that by August 1, Britain would sue for peace. Besides, they felt that the United States was already helping Britain so much that for all practical purposes, they were in the war, officially or not.

Partially in response to this new strategy, the United States declared war on Germany on April 6, 1917.[112] As would be the case in World War II, implementing the convoy was used to stop the effectiveness of the German U-boats.[113] Once the convoys began on May 10, 1917, Allied shipping losses plunged dramatically: Only 393 out of 95,000 ships lost were in a convoy. This had the added benefit of being able to transport more than two million US troops to Europe, with the loss of only three transports.[114] This influx of men finally gave the Allies numerical superiority over the Germans. This was even with the transfer of sixty-two German divisions from the Eastern Front after the capitulation of the Russians.[115]

While the U-boat blockade may not have been as effective as hoped, the blockade in general was. The estimated number of German deaths as a result of the blockade was 424,000. Many more who did not die were severely affected: Malnutrition struck many, to the point that in Freiburg, Germany, ten-year-old boys in 1918 were on average two centimeters shorter than they had been in 1914.[116] The blockade continued after the armistice was signed in an effort to force Germany to sign the peace treaty. The estimated death toll as a result of the postarmistice blockade was 100,000.[117]

Submarines in Britain were initially viewed as "underhand, unfair, and damned un-English" as well as inhumane, dishonorable, and in contravention to the rules of warfare among gentlemen.[118] Nevertheless, some in Britain saw the submarine for what it could do and helped procure funding for research and construction of them, diverting money from their traditional shipbuilding budget.[119]

As far as surface naval action, there was very little other than what was previously noted. The only battle worthy of mention was the Battle of Jutland, off the coast of Denmark, from May 31 to June 1, 1916. While Britain lost more ships and men than Germany in this battle, they had a much larger navy and were able to so badly damage the German fleet and the confidence of the German commanders that they won the encounter.[120] In a bit of irony, in the arms race and development of the *Dreadnought* class of battleships at the turn of the century, the countries involved spent so much money on these warships that they were afraid to use them for fear they would be damaged or sunk.

AIRPOWER
Airpower played a small but important role in the conflict. In the early days of the war, reconnaissance was the airplane's main role. This was in accordance with the views of the Madrid Session of the Institute of International Law in 1911. One member of this institute was quoted as saying, "I regret very much . . . that the progress of science has made aviation possible."[121] As Spaight then pointed out, there have been naysayers about every technological advance

made throughout history. This institute sought to limit the use of aircraft to reconnaissance only. In fact, the findings were deemed to be governed by article 25 of the Hague Règlement, which stated, "The attack or bombardment, by any means whatever, of undefended towns, villages, dwellings, or buildings, is forbidden," although semantics rendered this argument questionable.[122] As such, the associated airplanes lacked any kind of weaponry.

Originally, they flew over their assigned territory and reported back what they had seen. In efforts to improve efficiency, modifications were made, such as making planes with two seats so a dedicated observer could accompany the pilot, using cameras to take actual photos of the sites, and installing wireless telegraph equipment to allow messages to be sent to the ground while still in the air. There was a sense of camaraderie between the pilots from various countries, who would wave at each other as they flew by. However, they soon realized that those pilots were performing the same missions that they were and that they should be stopped if possible. The crew began to carry pistols to shoot at the enemy aircraft. The ground crews then began to install machine guns on the planes and then on dedicated pursuit planes designed to take down the enemy's slow two-seated reconnaissance planes. These reconnaissance planes and crews performed valuable services to their respective armies, and the smaller, faster pursuit planes introduced the world to a completely new manner of warfare. As with all new technologies, there was a learning curve for how to best avoid mistakes. One German pilot reported that a British unit was running around in complete chaos and panic. They were playing a game of soccer.[123]

With the advent of the reconnaissance and fighter planes came the bomber planes. In a precursor to Giulio Douhet's *Command of the Air* (1921) and Billy Mitchell's *Winged Defense: The Development and Possibilities of Modern Air Power—Economic and Military* (1925), Spaight correctly predicted the use of airplanes in the role of bombers: "Bombarding seems bound to become in time as usual and important a part of the duties of the military airman as reconnaissance is to-day."[124] Spaight then proceeded to accurately predict how bombers would be used. He pointed out that H. G. Wells (accurately) "painted a vivid picture of the wholesale destruction of capital cities by aeroplanes, driven by atomic engines and discharging atomic bombs." And while Spaight found it "unlikely that civilised nations will ever wreck one another's purely residential and commercial cities," he was wise enough to realize that that was what probably would and did happen.[125]

In January 1915, zeppelins began their raids over Britain. The kaiser initially authorized their use against military targets as well as docks along the English coast. He did this in an attempt to follow the laws of warfare practiced at that time, plus to avoid the British royal family, to whom he was directly related.[126] The Germans hoped that the zeppelins would have a profound strategic effect, but they were too slow, too vulnerable to attack by enemy airplanes, and carried too few bombs to have any real effect. By September and October

1916, Germany had lost three of its newest zeppelins. And rather than cause panic among the British population, it caused primarily anger, especially that the Germans would target defenseless women and children.[127] Though the zeppelin attacks continued to sporadically target Britain, it was time to try something new.[128] However, the seeds had been sewn for what would happen to Britain in a short twenty-five years.

On May 25, 1917, twenty-one German Gotha bombers took off from Belgian airfields and conducted their first strategic bombing run, the first systematic and sustained bombing campaign of the war.[129] Able to carry more than a half-ton of bombs, the Gotha was unlike anything people had seen to this point in the war. Ninety-five people died, and 260 were wounded on this raid. The Germans had decided to target civilians not just to win the war but also to terrorize the civilians. Kaiser Wilhelm changed his mind about targeting civilians and felt that they had become too soft from the industrial age, a recurring belief. Plus, such raids would force Britain to keep necessary resources from reaching the front.[130] The raid was supposed to bomb London, but at first, there were too many clouds for the aircraft to see the city. They got their chance on June 13. Clear weather over London allowed the aircraft to drop their bombs. In total, 132 people were killed, and 432 were injured, including 20 students from the Upper North Street School killed with an additional 30 injured.[131] Initially starting with daylight sorties, the Germans continued with night raids through the early winter.[132]

The successor to the Gotha was the Giant, a four-engine plane with a wingspan just three feet shorter than the World War II US bomber B-29 Superfortress. The crew consisted of the airplane commander, two pilots, a wireless operator, two flight mechanics, and a fuel attendant, with the capability to add two gunners as needed. The bomb load was nearly two tons.[133] Additional raids were met with varying success. However, the British were able to learn important lessons from these raids that would prove invaluable during the 1940 German air attacks.[134]

THE HOME FRONT
Propaganda
World War I was the first great war in history that was fought not just on the battlefront using arms but also at home using mass media. While newspapers had been around for more than three hundred years, as had fliers, pamphlets, and wall signs, print media were finally reaching their full potential in their ability to sway public opinion. Additionally, for the first time, motion pictures could be used to help influence the masses. Propaganda can be and is a powerful tool to help persuade people to do what one wants. Propaganda is just advertising with a different name, and advertising works, as evidenced by an estimated nearly $1 *trillion* spent globally on advertising in 2024.[135]

Propaganda could be used to arouse hatred of the foe, warn of the consequences of defeat, and idealize one's own war aims in order to mobilize a nation, maintain its morale, and make it fight to the end. It could explain setbacks by blaming scapegoats, such as war profiteers, hoarders, defeatists, dissenters, pacifists, left-wing socialists, spies, shirkers, strikers, and sometimes enemy aliens so that the public would not question the war itself or the existing social and political system. Propaganda could also demoralize the enemy, incite soldiers to desert, and stir up unrest among the enemy state's civilian population.[136]

Propaganda, when used properly, is a powerful tool. As such, it is difficult to talk about the events at home without discussing how propaganda affected activities of the people residing there. I touch on this topic again in chapter 6. What follows is a brief discussion of specific acts pertaining to World War I.

Prior to World War I, propaganda was synonymous with publicity and did not receive much attention. As Randall L. Bytwerk points out, propaganda came into the English lexicon through the Catholic Church and its *sacra congregatio christiano propagando*, or Sacred Congregation of Propaganda.[137] Propaganda was neither good nor bad; it was merely meant to persuade. By the end of the war, every country in the war was involved in propaganda. With the war changing from army fighting army to nation fighting nation, people saw that much of the propaganda had been based on, if not lies, then gross exaggerations, and the word began to morph into its current meaning:

> the spreading of ideas, information, or rumor for the purpose of helping or injuring an institution, a cause, or a person
>
> ideas, facts, or allegations spread deliberately to further one's cause or to damage an opposing cause.[138]

And while the buzzword du jour since the US presidential election of 2016 has been *misinformation*, the actual implementation of misinformation has been acknowledged since at least 1965. It was then that Jacques Ellul wrote, "Propaganda will always triumph over information. . . . Wherever there is propaganda, information, if it is to survive, must utilize the same weapons. It must engage in a struggle against the inaccuracy of the facts proclaimed by propaganda."[139]

In the beginning of the war, not much was needed by way of formal persuasion, as nearly everyone anticipated this being a short war: Men volunteered by the scores, and most citizens rallied around their government. What propaganda existed was generally of the "feel-good" type, portraying positive aspects of the war. University professors gave public lectures, preachers sermonized their congregations, and schoolteachers indoctrinated their students by such means as calculating enemy losses.[140]

As time went on and it became apparent that it was going to be a long, bloody affair, more work was required to keep the citizens voluntarily supporting the governments. Each of the governments had official offices that handled these matters, and each was involved to varying degrees. As so often happens when dealing with government bodies, these offices started small but soon swelled to near-bloated entities. The German department "employed hundreds of officers and countless writers, painters, caricaturists, photographers, and technicians."[141]

The actual trench warfare and trench conditions were presented to the German public as having all the comforts of home.[142] And the trenches received their share of propaganda, as literature was placed in "propaganda grenades" and launched by mortar or dropped by airplane into the opposing trenches. Messages were played by loudspeaker toward the enemy lines.[143]

The German propagandists cultivated a sense of *kultur* among the people; they compared their honesty with the hypocrisy of the English and French of waging a war of defense when they really wanted to expand their empires. For the Germans, the war was portrayed as a defense of such values as "devotion to community," "order," "performance of duty," and "discipline."[144]

The British used this same *kultur* to represent German lawless brutality, devoid of morality.[145] The British propagandists were more than simply parroting German propaganda modified for British use. The movie *The Battle of the Somme* was sent to theaters while the actual battle was still being waged. A silent documentary that followed the British during the first day of the battle, the movie was quite successful, with some estimating that half of Britain watched it.[146] Germany immediately responded with their own film version of the battle, *Bei unseren Helden an der Somme* [*With Our Heroes at the Somme*], in January 1917.[147] However, in Britain it seemingly sent a mixed message. Some viewers saw the movie as strongly antiwar due to its graphic battle scenes. Most, however, saw it as an exclamation that while war was hard and death was inevitable, so, too, was the inevitable British victory. However, the movie conveniently left out any scenes of unburied corpses or the injured awaiting treatment.[148] As time progressed, film played less of a role in all countries except the United States, where Charlie Chaplin and Mary Pickford were box office hits and had a big influence on helping sway public opinion.[149]

There were other challenges at home. Wars are expensive to wage. As a result, taxes increased, as did inflation. As a result of the British blockade, the Germans were getting just a fraction of the food and other supplies that the country needed, seriously affecting German economy and living standards and negatively affecting German morale.[150] The winter of 1916–1917 came to be known in Germany as the "turnip winter" due to the lack of potatoes.[151] By September 1916, Vienna citizens were standing in line daily as early as 3:00 a.m. for such foodstuffs as bread, potatoes, milk, and eggs.[152] However, a side benefit to the Central Powers losing battles was that the more casualties there

Figure 9. Perhaps one of the most famous US posters and images from the First World War, *I Want You for U.S. Army*, by J. M. Flagg, 1917. Source: Wikimedia Commons, "J. M. Flagg, I Want You for U.S. Army Poster (1917)," accessed May 20, 2025, https://commons.wikimedia.org/wiki/File:J._M._Flagg,_I_Want_You__for_U.S._Army_poster_(1917).jpg.

were in battle, the fewer soldiers there were to have to feed, thereby allowing more food for civilians.[153] Such private industries as railway, coal mines, and flour mills were taken over by the British government. Propaganda, ever present during war time, changed from invoking the patriotic duty of citizens in 1914 (1917 in the United States; see figure 9) to portraying the enemy as depraved beasts in 1917–1918 (see figure 10).

In the summer of 1916, General Ludendorff attempted to increase supplies to the military and match the output of the Entente by more fully mobilizing German industry. To do this, Hindenburg said that the output of shells should be doubled and the number of machine guns manufactured should be tripled, regardless of cost. And while doing so was in line with what industry wanted to do, it also led to price gouging and profiteering.[154] However, as is the usual case in economics, especially when the government is involved, the law of unintended consequences came into effect, leading to reduced wages; decreased availability of consumer goods, including food; inflation and increased cost of goods; and the inevitable increase in black market activity.[155] Meanwhile, the British imposed strict, militaristic constraints on its factory workers. Workers were banned from striking and were required to obtain permission to change jobs.

World War I was not only revolutionary in terms of military advances, but the home front was also radically changed. The mobilization of young men to

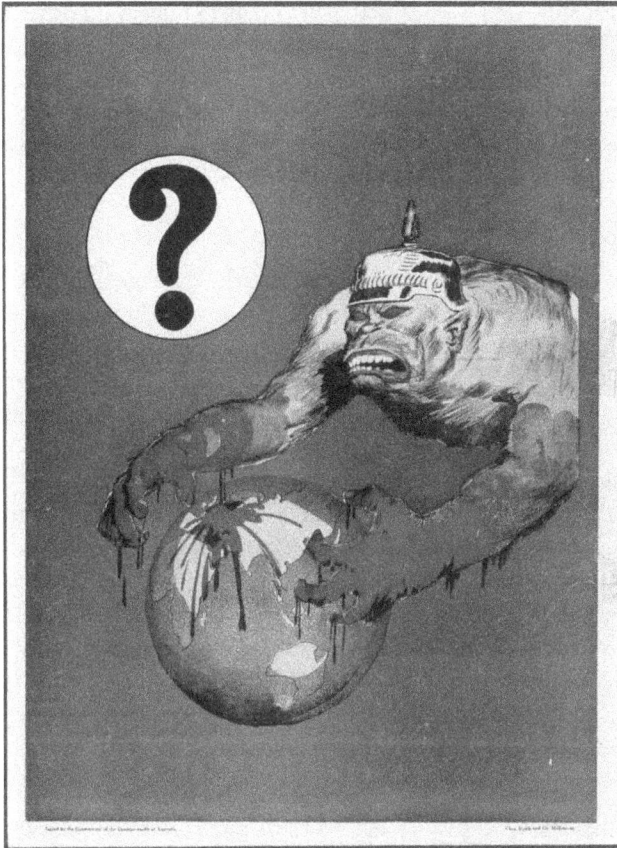

Figure 10. Australian World War I poster from 1918 by Norman Lindsay depicting an ape monster in a German helmet, with bloody hands reaching around the globe. SOURCE: WIKIMEDIA COMMONS, "WORKS BY NORMAN LINDSAY," ACCESSED MAY 20, 2025, HTTPS://COMMONS.WIKIME DIA.ORG/WIKI/CATEGORY:WORKS_BY_NORMAN_LINDSAY#/MEDIA/FILE:?_(QUESTION _MARK)_AUSTRALIAN_WWI_POSTER_-_NORMAN_LINDSAY.JPG.

fight the war is obvious. In Britain, by 1917–1918, one in three of the male labor force was serving in the military.[156] However, faced with the daunting task of supplying these troops, by necessity entire nations were mobilized to meet this challenge. Women, in particular, were called on to help. Many of these women were recruited to work in munitions factories. They were primarily married, working-class women going back to work or young women leaving the textile mills.[157] Another benefit to women entering the workforce was the improvement of working conditions in the factories.[158]

Military nursing numbers also swelled. In Britain in 1898, there had been seventy-two army sisters working in military hospitals. From 1914 to 1918, 32,000 women served as military nurses. These women were part of the command structure and were able to give orders to male ward orderlies.[159] They had much more to accomplish as far as women's rights: They were regarded as temporary workers, expected to return home after the war, and they were paid less than men doing the same job.[160] However, they had made great strides toward equality, and women in Britain were given the right to vote immediately after the war as a direct result of their effectiveness during the war.

United States

While the United States did not experience the carnage and resulting loss of life that the other countries experienced, there were incidents related to the war in the United States. On July 30, 1916, German agents, along with Indian and Irish nationalists, destroyed the munitions depot at Black Tom, New Jersey. Seven people, including a baby, were killed in the blast.[161] Even with this, US support for the war was lukewarm at best. President Woodrow Wilson had run for reelection in 1916 on the platform "He kept us out of war." In January 1917, Wilson proposed that the Central Powers and the Allies renounce the imperial alliances that had led to the start of the war. He had hoped to achieve "peace without victory," a concept that was ridiculed by his opponents and dismissed by both Germany and Austria.[162]

However, just over a week later, Germany reinstituted unrestricted submarine warfare. On February 6, U-boats sank fourteen ships. The next day they sank twelve and ten more the day after that.[163] In addition, intercepted cables between Germany and Mexico, later verified by Germany, showed that Germany was secretly trying to get Mexico into the war on the side of the Central Powers.

Finally, the United States had had enough. Wilson asked on April 2, 1917, for Congress to declare war on the Central Powers. Even with all that had transpired over the previous year, the acrimony between the pro- and anti-war factions was palpable. At least one fistfight broke out in the halls of Congress.[164] Patrons at one restaurant heckled and hissed at an antiwar senator trying to eat.[165] Congress did declare war, and the antiwar protestors dejectedly resigned themselves to war.

Now with the threat of unrestricted submarine warfare, which was the overriding reason for the United States' entry into the war, some Americans were still not convinced of the need to get involved in a European conflict. Besides infantry, the United States would need "blacksmiths, farriers, stenographers, wheelwrights, masons—anyone with a skill who could aid the massive mobilization that lay ahead."[166] Wilson therefore created the Committee on Public Information (CPI) to foster popular support for the war. This committee was responsible not only for encouraging support for the war but also for censorship of antiwar sentiment. George Creel, head of the committee, embarked on a "vast enterprise in salesmanship" to emphasize the positive aspects of the war.[167] To do this, he inundated the country with press releases disguised as news stories. Additionally, the CPI utilized speakers called "Four Minute Men" (after the Minute Men of the American Revolution) to give short speeches to the public in churches, men's and women's clubs, and colleges.[168] In total, the CPI "published millions of leaflets, booklets, and posters; organized forty-five war conferences; and employed 75,000 speakers," reaching 315 million people.[169] As Fondren and Hamilton put it, "The CPI wanted the public to cheer, not think."[170]

This propaganda continued during the first two months, as the CPI ran these press releases expounding on the achievements and ideals of America and casting those people from Germany, Austria, and Hungary as just like other Americans. However, this was not getting the desired results, so on Flag Day 1917, Creel embarked on a negative campaign. While still extolling the virtues of the American soldier and the endeavor to make the world safe for democracy, he changed gears and began to paint Germany as a country fixed on world conquest that must be destroyed. One such portrayal is the poster "Destroy This Mad Brute" (figure 11) with the image of an apelike soldier clutching an innocent young woman whom the beast is ready to defile and destroy.

Figure 11. In this 1918 propaganda poster, artist Harry R. Hopps depicts Germany as an enraged gorilla stepping foot on an American shore as Europe lies in ruin in the distance. This enlistment poster issued by the San Francisco Army Recruiting District was intended to rouse its audience to join the cause against the "German militarism" threatening American liberty. In this powerful image, liberty takes the form of woman whom the brute clutches in his one arm, while his other carries a bloody club marked "Kultur." SOURCE: LIBRARY OF CONGRESS, "MAD BRUTE," ACCESSED MAY 20, 2025, HTTPS://WWW.LOC.GOV/EXHIBITIONS/ WORLD-WAR-I-AMERICAN-EXPERIENCES/ ABOUT-THIS-EXHIBITION/OVER-HERE/ RAISING-AN-ARMY/MAD-BRUTE/.

The problem with this approach was that it caused the bigots to believe it was their calling to enforce this appeal to destroy the enemy. Any expression of doubt about the war could result in a beating by a mob. Those who chose not to buy liberty bonds, issued by the Department of Treasury, could end up with streaks of yellow paint on their houses. Pacifistic churches were set on fire. Those suspected of disloyalty were tarred and feathered or, worse, lynched. And while not as physically violent as these deeds, many organizations, both governmental and private, terminated German aliens from their employment, canceled performances of German music, and banned the teaching of German in schools.[171]

In May 1918, the United States Espionage Act of 1917 was expanded into the Sedition Act, which in effect ended the constitutional protection of free speech for the duration of the war.[172] The original Espionage Act had, by a margin of one vote, removed any form of press censorship but did allow for the prevention of distribution of print material through the US Post Office. That act was subsequently amended to prohibit any form of speech that could be construed as anti-American:

> Whoever when the United States is at war, shall willfully cause or attempt to cause, or incite or attempt to incite, insubordination, disloyalty, mutiny, or refusal of duty, in the military or naval forces of the United States, or shall willfully obstruct or attempt to obstruct the recruiting or enlistment services of the United States, and whoever, when the United States is at war, shall willfully utter, print, write or publish any disloyal, profane, scurrilous, or abusive language about the form of government of the United States or the Constitution of the United States, or the military or naval forces of the United States, or the flag of the United States, or the uniform of the Army or Navy of the United States into contempt, scorn, contumely, or disrepute, or shall willfully utter, print, write, or publish any language intended to incite, provoke, or encourage resistance to the United States, or to promote the cause of its enemies, or shall willfully display the flag of any foreign enemy, or shall willfully by utterance, writing, printing, publication, or language spoken, urge, incite, or advocate any curtailment of production in this country of any thing or things, product or products, necessary or essential to the prosecution of the war in which the United States may be engaged, with intent by such curtailment to cripple or hinder the United States in the prosecution of war, and whoever shall willfully advocate, teach, defend, or suggest the doing of any of the acts or things in this section enumerated, and whoever shall by word or act support or favor the cause of any country with which the United States is at war or by word or act oppose the cause of the United States therein, shall be punished by a fine of not more than $10,000

[more than $200,000 in 2024] or the imprisonment for not more
than twenty years, or both.[173]

Thereafter, speeches against conscription or advocating the independence
of British colonies, failure to donate to the Red Cross, or any similar "anti-
American" act could result in one being denounced by friends and neighbors
(including children), arrested, and imprisoned.[174] As Creel himself stated,
"With the existence of democracy itself at stake, there was no time to think
about the details of democracy."[175]

Nor was the speech of English-speaking American citizens the only speech
to be restricted. The Trading with the Enemy Act of 1917 mandated that all
foreign-language publications file daily English translations to ensure that there
were no subversive messages being relayed to anti-American groups.[176] Britain
had its own variation of the Trading Act in the Defence of the Realm Act
(DORA) of 1914.[177] As the United States and its citizens found out, trying to
mobilize the masses by negatively appealing to their patriotism could be a two-
edged sword.

An additional disadvantage to this propaganda was the disenchantment
and cynicism that the public felt when, after the continual escalation of the
amount and content of the propaganda, the government was unable to fulfill
its promises. One such repercussion that haunted the world barely twenty years
later was the feelings the Germans experienced when they realized that the
promises they believed had been given them by President Wilson, if they were
to surrender, failed to materialize. The propaganda presented to the Germans
by the CPI was that President Wilson would ensure Germany would receive
fair treatment upon its surrender. When the extreme opposite occurred, it only
served to fuel the stab-in-the-back feelings of the German army.[178] The Germans weren't the only ones to feel dismay after the war. I delve into this more
in chapter 4.

The national draft proved to be another contentious matter. Proponents
of a draft believed that it "would assimilate immigrants, level class differences,
and instill military discipline in the populace," while detractors viewed it as a
"threat to democracy, a symptom of a coercive federal government that, unable to raise a volunteer force, must instead dragoon an army through force of
law."[179] Wilson didn't want one but realized there was no other way to muster
sufficient men to serve in the war. In addition, they needed to keep men out of
the service whom they needed for positions stateside.

Once the draft was instituted, everything possible was done to make it pal-
atable. Registration Day was set for June 5, 1917, and declared a holiday by
Wilson. And everything was done to make it as much of a holiday as possible.
Churches and fire stations were ordered to ring their bells at 7:00 a.m., which
was when the registration stations opened. There were to be brass bands playing patriotic songs, and families and friends were encouraged to accompany the

soon-to-be new registrants to the registration locations. Additionally, Registration Day was cloaked in overtly religious tones: "Registration Day should be celebrated as a consecration of the American people to service and to sacrifice. It should be a welcome to those registering. It should be a public expression by each community of willingness to surrender its sons to the country," wrote Walter Gifford, the director of the Council of National Defense.[180]

The draft was absolutely necessary if the United States was going to fight in a war. In April 1917, the US Army consisted of 127,588 men—121,797 enlisted men and 5,791 officers—and 181,620 National Guard reserves.[181] As shown in table 3, it was not uncommon to lose that many or more soldiers in a single battle. And prior to the mobilization of the US First Division, there had not been a single division in existence in the army since the Civil War. There was no heavy artillery, few machine guns, and inadequate munitions.[182] In addition, the logistical side of the army was in chaos. The appointment of Major General Peyton C. March as chief of staff finally led to the coordination and streamlining of army resources.[183] General March's top priority was to get as many men to Europe as possible.

All told, the draft turned out to be both highly effective and highly efficient. More than 2.7 million men were drafted into the US Army by the end of the war.[184] Combined with volunteers, which were terminated in August 1918 due to the success of conscription, the size of the army grew from the meager 200,000 prewar soldiers to nearly 3.7 million men by the end of the war.[185] Eventually, all males between the ages of eighteen and forty-five were required to register.

Along with the mobilization of men to fight the war, the American industry had to be mobilized to produce the needed matériel. This was done under the auspices of the Munitions Standards Board, which was established before the war began. The demands placed on this agency as the need for wartime matériel increased led to its metamorphosis into the War Industries Board. With its military and civilian representatives came increased military-civilian coordination, thereby increasing efficiency. This coordination was necessary given the sheer number of arms needed by the US troops. When Lieutenant Colonel Amos Fries took command of the fledgling US Army gas service, one of his first tasks was to order 100,000 gas masks from Britain and another 100,000 from France.[186] Though they had a slow start, American industry was able to quickly recover and reach its potential with impressive results.

THE END
On March 3, 1918, the Germans and Russians signed the Treaty of Brest-Litovsk, thereby ending the war on the Eastern Front. The treaty was a clear victory for Germany, leading to two important consequences: First, German troops fighting in this theater could be transferred to the Western Front. With the troops returning from Italy, forty-four divisions were sent to the west. For

the first time in the war, the Germans had a slight numerical superiority on the Western Front.[187] Second, the Germans could negotiate a peace treaty with France and Britain from a position of strength. France and Britain worried that this would lead to German hegemony in Europe. Rather than risk that outcome, they continued to fight. On March 8, the French prime minister, Georges Clemenceau, told the French National Assembly, "You ask me what my domestic policy is? It is to make war. You ask me what my foreign policy is? It is to make war."[188]

After amassing all their available troops, seventy-six divisions in all, the Germans launched their spring 1918 offensive on March 21. Their goal was to gain as much territory as possible in order to annex it and keep it as part of Germany after their planned victory. However, this didn't happen. By manning a good portion of the front, the American troops freed up a significant number of French troops, who were able to repulse the Germans. This was a huge gamble, as Sun Tzu taught. After the successful defense of the German attack, the Germans were spent militarily. They had rolled the dice and come up short.

On July 18, the Allies launched their counteroffensive, known as the Second Battle of the Marne. Through the use of a creeping artillery barrage laid down by Allied artillery and their new tanks to support the infantry, the Allies were able to break through the German lines. Once through, they didn't stop advancing. As a result of this breakthrough, the British captured 12,000 German prisoners and advanced seven miles on the first day alone. On August 8, 430 British tanks broke through the line near Amiens, France. The Germans were so overwhelmed by the attack that large numbers of German troops either surrendered or reported as ill in order to escape fighting.[189] The German generals knew they had lost the war.

While this loss was the main factor in the German decision to end the war, it was not the only one. The British blockade of Germany was creating great hardships for the German people (see figure 12). On October 29, the head of the Berlin police department gave the kaiser a report stating that the general population now wanted peace at any cost. Profiteering, rationing, murder, prostitution, and theft had all increased significantly in the previous few months. Additionally, failure of the submarine war and the defection of Allies led to even worse morale during this time. These had all greatly contributed to increased social divisions in the city.[190]

Furthermore, the sailors of the German navy grew tired of conditions on ship and the disparity between sailors and officers. When the battleships decided to go into battle against the British and US battleships, something they had done only once previously during the entire war, the sailors recognized the mission for what it was: a suicide mission. On October 29, the crews of three battleships mutinied. By November 6, mutinous sailors controlled Kiel, Lübeck, Hamburg, and Bremen. These mutinies further exacerbated the social unrest of the civilian population, hurling the country toward revolution.[191]

FOOD SHORTAGES IN EUROPE, 1918

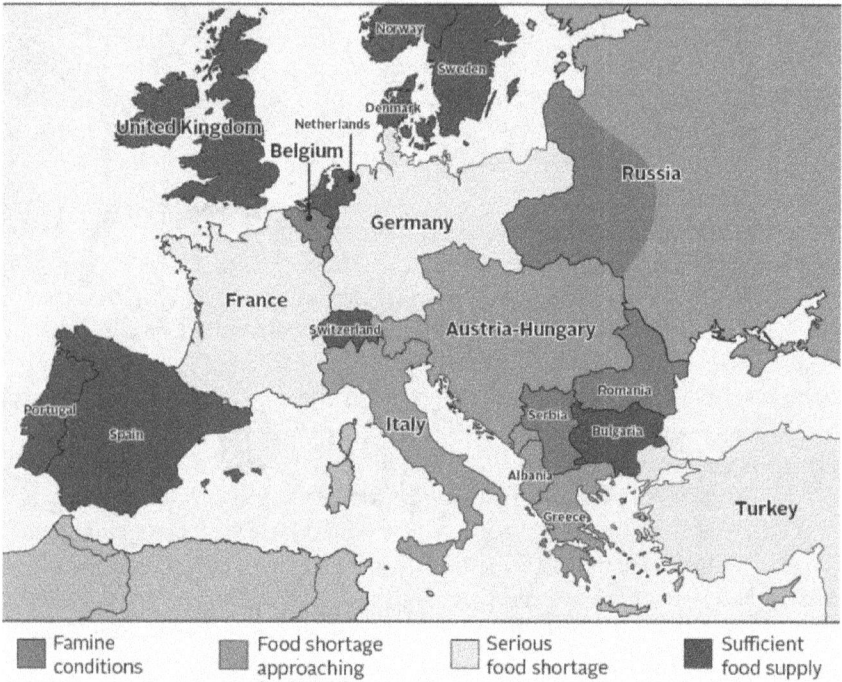

| Famine conditions | Food shortage approaching famine point | Serious food shortage | Sufficient food supply |

Source: *Food Guide for War Service at Home*

Figure 12. Map of Europe, 1918, showing the status of food supply for that period.
SOURCE: MAPS ON THE WEB, ACCESSED MAY 20, 2025, HTTPS://MAPSONTHEWEB.ZOOM-MAPS.COM/POST/90454941981/FOOD-SHORTAGES-IN-EUROPE-AFTER-WORLD-WAR-I-1918.

Additionally, not wanting to admit defeat on the battlefield, the Germans instead blamed left-wing factions for inciting the public to end support of the military. The civilians back home were also supposedly to blame due to their inability to deliver reinforcements and improve their general morale.[192] General Erich Ludendorff, first quartermaster general of the German army, who, along with General Paul von Hindenburg, commanded the army forces and determined overall strategy, initially believed his forces spent. However, three weeks later, he reported that they could continue to hold out against the Entente forces. At least one historian believed this was an attempt to shift the blame for failure to hold back the Entente troops from Ludendorff and the armed forces to the general population.[193] This stab-in-the-back belief was very convenient. Hitler, in his rise to power, would use this premise to gain support from the German people to annul the Treaty of Versailles that ended the war.

To be sure, the German people were, in general, in very low morale. However, this only served as an excuse. But it was a powerful excuse. Regardless, on November 9, 1918, Kaiser Wilhelm II, emperor of the German Empire,

abdicated, and a new government was formed. This new government authorized the signing of the armistice on November 11, thereby ending the war.[194]

The final statistics are staggering: 9.45 million dead; millions more seriously wounded. About 2 million Germans, 1.8 million Russians, 1.4 million
French, 1.3 million Austrians, 1 million from Britain and the British Empire,
116,000 American soldiers, and nearly 2 million from other countries died in
the conflict. More generalized casualty rates show France experiencing a 27
percent casualty rate among men ages eighteen to twenty-seven.[195] In the
United Kingdom, the war left 192,678 widows and 355,211 children entitled
to government pensions.[196]

It is clear that this worldwide conflict was extremely costly, not just in
battle casualties, but in civilian loss of life, as well. Regardless of the cause,
noncombatants suffered just as much as combatants. Unfortunately, in just
twenty-one short years, the world would again begin to experience such casualty rates and associated suffering.

CHAPTER 4

INTERWAR PERIOD, 1919–1937

"A bad peace is even worse than war."

—Tacitus

The end of the Great War brought relief to all involved. The carnage during that cataclysmic event was without precedent in the modern world. The world vowed never again to indulge in such an apocalyptic war and took several important steps toward that end, starting with the Treaty of Versailles in 1919. However, those steps ultimately led to further worldwide conflict. Quintus Ennius wrote, "The victor is not victorious if the vanquished does not consider himself so."[1] This was never more true than at the end of World War I. The interwar years only reinforced what both Tacitus and Quintus Ennius knew.

TREATY OF VERSAILLES

While the German command knew that they had lost the war, the rank-and-file soldier didn't know what the generals knew. The common soldier felt that they, the German soldiers, had been betrayed and that they had actually won the war. All the fighting had taken place on enemy soil, not German. Therefore, it should come as no surprise that any peace treaty forced on Germany likely would be greeted with derision.

The Treaty of Versailles ended World War I. However, it was not the end of the psychological aspects of the war. The treaty has been alternately described as being either too harsh or not harsh enough. As it was, the treaty was harsh enough to warrant widespread German resentment toward it but not harsh enough (in the opinion of the French military leader Ferdinand Foch, among others) to completely ensure that Germany would not be able to circumvent its provisions and regain power.[2] All the treaty did was humiliate Germany, without solving the underlying issues that led to the war. In fact, Foch proved especially prophetic when, at the signing of the treaty, he stated, "This is not a peace treaty, it is an armistice for twenty years."[3] As history showed, it was almost exactly twenty years later, to the month, that the Second World War began.

First, the treaty laid the blame for the war entirely on Germany, even though Austria-Hungary had started the war. Germany came in only after being asked for help by Austria-Hungary, which is not to say that Germany didn't want the war. It did.[4] Regardless, Georges Clemenceau was determined to avenge the French defeat by Germany at the end of the Franco-Prussian War of 1871. This loss humiliated France and resulted in the loss of the provinces of Alsace and Lorraine.[5]

The defeated countries, Germany, Austria-Hungary, Bulgaria, and Turkey, were not represented at the Paris Peace Conference.[6] The first time they were presented with the terms of the treaty was on May 7, 1919, even though negotiations had started on January 18. Germany offered counterproposals on May 29, but all were rejected at the hands of the French delegation, whom the British and Americans viewed as excessively inhumane and looking to grab territory for their country.[7] Germany refused to sign but was told that if they didn't, the war would resume. Germany reluctantly agreed, calling the treaty and forced acceptance a diktat.

Article 231 of the treaty, commonly known as the War Guilt Clause, stated, "The Allied and Associated Governments affirm and Germany accepts the responsibility of Germany and her allies for causing all the loss and damage to which the Allied and Associated Governments and their nationals have been subjected as a consequence of the war imposed upon them by the aggression of Germany and her allies."[8] Even though some German conservatives realized that this accusation was actually very close to the truth, they and many others saw this clause as an insult. Others viewed it as an unjust burden on a new German government.[9] Furthermore, although the Allies explicitly knew and stated that there was no way Germany would be able to pay for all the costs associated with the war (article 232), they still forced the German government to pay those costs:

> The Allied and Associated Governments recognise that the resources of Germany are not adequate, after taking into account permanent diminutions of such resources which will result from other provisions of the present Treaty, to make complete reparation for all such loss and damage.
>
> The Allied and Associated Governments, however, require, and Germany undertakes, that she will make compensation for *all damage done* to the civilian population of the Allied and Associated Powers and to their property during the period of the belligerency of each as an Allied or Associated Power against Germany by such aggression by land, by sea and from the air, and in general all damage as defined in Annex 1 hereto.[10]

These reparations would not be paid off until October 3, 2010.[11]

Other provisions forced Germany to concede Eupen-Malmédy to Belgium; the Hultschin district to Czechoslovakia; and Poznan, West Prussia, and Upper Silesia to Poland, thereby creating the Polish corridor to the sea. Alsace-Lorraine, which the Germans had taken from France in the Franco-Prussian War in 1871, was returned to France. Additionally, the treaty called for the demilitarization and occupation of the Rhineland, special status for the Saarland under French control, and referendums to determine the future of areas in Northern Schleswig on the Danish-German frontier and parts of Upper Silesia on the border with Poland.[12] All German overseas colonies were taken away, becoming mandates of the League of Nations. Danzig (Gdansk), which had a large ethnic German population, became a free city. The army was limited to 100,000 men, with conscription banned. Naval vessels were to be under 10,000 tons, and submarine procurement and upkeep was forbidden. They also were banned from having an air force.[13] In all, Germany lost 68,000 square kilometers (13 percent) of its territory and about seven to eight million (12 percent) of its inhabitants.

Upon examination, most of these land transfers made sense. They were relatively new acquisitions, and most were not ethnic Germans. The one union that did make sense but was prohibited was that of Germany and German Austria. However, this unification would have left Germany with more people than it had in 1914, when the war began. France absolutely forbid this.[14]

While these punishments seem excessive, the following crimes perpetrated by Germany were compiled in the book *Handbook of War Facts and Peace Problems* by Arthur L. Frothingham, under the section "For Those Who Think Germany Has Been Treated Severely." The section contains the provisions of the peace treaty that Germany had planned on presenting to the Allies had Germany been victorious. Consisting of twelve points, the provisions were set forth on June 30, 1918:

1. No armistice until British forces were out of France and Paris occupied.
2. Annexation of Belgium and the channel coast to south of Calais.
3. Annexation of the Briey-Longwy iron region.
4. Annexation of Belfort, Toul, and Verdun, and all French territory east of these forts.
5. Return of German colonies.
6. Surrender by Great Britain of coaling stations, including Gibraltar.
7. Surrender to Germany of the entire British navy.
8. Egypt and the Suez canal to Turkey.
9. Restoration of Constantine as King of Greece.
10. Division of Serbia between Austria-Hungary and Bulgaria.

11. Payment of indemnity of $45,000,000,000 by the United States, Great Britain, and France.
12. Occupation of French territories until agreements were carried out, costs of occupation of being met by the enemy.

In addition, as a matter of course, treaties with Russia and Rumania were to stand, with Poland, Courland, Lithuania, Livonia, and Esthonia practically annexed, and Finland, the Ukraine, and Rumania to be subject kingdoms.[15]

Frothingham continues with a section entitled "Germany's Inhuman Treatment of Her African Colonies," in which he details that German agreements with Native chiefs were viewed as nothing more than "scraps of paper" (identical to the terminology used for the treaty that guaranteed Belgian neutrality when it was invaded by Germany in 1914). He further documents that the rebellions by various tribes against oppressive German rule resulted in the slaughter of the Herero, Hottentot, and Berg-Damara tribes, with Hereros reduced from 80,000 to 15,100 members, Hottentots from 20,000 to 9,800 members, and Berg-Damaras by more than half.[16] In achieving these results, Lothar von Trotha, governor of German South West Africa at the time, issued an extermination order to the Herero:

> I, the great general of the German soldiers, send this letter to the Hereros. Herero are no longer German subjects. They have murdered, stolen, cut off the ears and noses and other body parts from wounded soldiers, and now out of cowardice refuse to fight. I say to the people: anyone delivering a captain to one of my stations as a prisoner will receive one thousand marks; whoever brings in Samuel Maherero [paramount chief of the Herero people] will receive five thousand marks. The Herero people must now leave this land. If they do not, I will force them to do so by using the great gun [artillery]. Within the German border every male Herero, armed or unarmed, with or without cattle, will be shot to death. I will no longer receive women or children but will drive them back to their people or have them shot at. These are my words to the Herero people.[17]

Here, Trotha rolled the clock back on centuries of progress in the rules of war and reverted to the ancient practice of annihilating the losing tribe in a war.

Frothingham also documents crimes for which the kaiser was deemed culpable. The thirty crimes were

1. Massacre of civilians.
2. Killing of hostages.
3. Torture of civilians.
4. Starvation of civilians.

5. Rape.
6. Abduction of girls and women for purposes of enforced prostitution.
7. Deportation of civilians.
8. Internment of civilians under brutal conditions.
9. Forced labor of civilians in connection with military operations of the enemy.
10. Usurpation of sovereignty under military occupation.
11. Compulsory enlistment among inhabitants of occupied territory.
12. Pillage.
13. Confiscation of property.
14. Exaction of illegitimate or of exorbitant contributions and requisitions.
15. Debasement of currency and issue of spurious currency.
16. Imposition of collective penalties.
17. Wanton devastation and destruction of property.
18. Bombardment of undefended places.
19. Wanton destruction of religious, charitable, educational, and historical buildings and monuments.
20. Destruction of merchant ships and passenger vessels without examination and without warning.
21. Destruction of fishing-boats and of a relief-ship.
22. Bombardment of hospitals.
23. Attack on and destruction of hospital-ships.
24. Breach of other rules relating to the Red Cross.
25. Use of deleterious and asphyxiating gases.
26. Use of explosive and expanding bullets.
27. Orders to give no quarter.
28. Ill-treatment of prisoners of war.
29. Misuses of flags of truce.
30. Poisoning of wells.[18]

The punishments laid out in the Treaty of Versailles, viewed as excessively vindictive by the Germans, were ignored almost immediately, with politicians openly campaigning to void the treaty. Indeed, this was a major selling point for Hitler's campaign as he railed against the treaty and its penalties. The "stab-in-the-back" theory, besides blaming the Left (meaning Socialists, Communists, and Jews) for the collapse of German support at home, further blamed these reparations for the hyperinflation that Germany experienced in the early 1920s. While the Weimar Republic tried to run a successful government, they were hamstrung by the limitations imposed on it by the treaty and the resulting inflation and depressions of the early 1920s to early 1930s.

The treaty, with its land transfers, also didn't stop other countries from proceeding with their own agendas. They also wanted to procure land transfers to help their respective countries or to achieve independence based on President Wilson's Fourteen Points. Russia was torn by civil war, as the Communist reds fought the competing Russian factions. Germans were fighting Poles, committing more atrocities as they withdrew from the new Polish territory. The German general von der Goltz was attempting a coup d'état against the Baltic states and claiming the territories for Germany. The Balkan region continued to fight among themselves, as Yugoslavs fought Romanians and Italians, while the kingdom of Montenegro was thrown under the bus and given to the new country of Yugoslavia.

Sudeten Germans were thrown into Czechoslovakia in order to give her more secure borders. As a result, Hungary found itself in a civil war over these attempts by surrounding countries to take portions of her land. Bavaria was briefly Communist before being overthrown by the German Freikorps. India's population rose up against the new British Rowlatt Acts, which allowed for nearly indeterminate detention of alleged suspects. The revolt was quickly and violently put down.

The same thing happened in Korea, as protests and calls for independence were aggressively stopped by Japanese forces. Egypt, upon calling for independence from Britain, was met with martial law. The Greeks invaded Smyrna on the western coast of Asia Minor. And Germany threatened to tear itself apart as radical splinter factions rose up to fight the new government and the imposed armistice and peace treaty. In nearly all these incidents, multitudes of innocents were murdered, either by the factions rising up or by the forces combating the factions.[19]

GENEVA CONVENTIONS
Biological and Chemical Warfare

Before proceeding, a review of the categories of the chemical portion of chemical, biological, and radiological weapons (CBRW) agents is in order. Broadly speaking, chemical agents are classified as choking/pulmonary, blistering, asphyxiants/blood agents, nerve agents, and cytotoxic agents.[20] Choking agents incapacitate or kill their victims by producing intensive irritation or inflammation of the respiratory tract and lungs.[21] These include gaseous chlorine, phosgene, nitrogen oxides, and hydrogen chloride.[22] Blistering agents incapacitate or kill their victims by producing acidic compounds on exposed skin and mucous membranes, resulting in painful, weeping blisters. These include mustard gases, lewisite, and phosgene oxime.[23]

Asphyxiants work by impairing the ability of red blood cells to carry oxygen, causing red blood cells to break down. The best known of these agents is hydrogen cyanide, or Zyklon B, used by the Nazis in their gas chambers. This is also the agent used in "suicide pills." However, carbon monoxide is the more common agent that nearly everyone is aware of.[24] Nerve agents are closely related to many insecticides; both are organophosphates. They work by blocking a crucial enzyme, which causes the nervous system to stop functioning. These agents include tabun, sarin, cyclosarin, and soman.[25] Finally, cytotoxic agents cause damage at the cellular level. Ricin and botulinum toxin are two examples of these agents.[26]

After witnessing the horrors that chemical warfare wrought during the Great War, countries sought to minimize or eliminate the possibilities of such warfare ever happening again and to prevent war in general. And while the threat of large-scale biological weapons manufacturing was still mainly theoretical, the potential threat of this kind of weapon led countries to attempt to prohibit their development. Building on the Hague Conferences of 1899 and 1907, two very important conventions were held in Geneva in the 1920s. The first of these lasted from May 4 to June 17, 1925, and was marked by the creation and signing of the Protocol for the Prohibition of the Use in War of Asphyxiating, Poisonous or Other Gases, and of Bacteriological Methods of Warfare, commonly known as the 1925 Geneva Protocol. This brief, one-page document extended the prohibition of the use of gaseous warfare, grossly ignored in the war just concluded, and extended the prohibition to include bacteriological methods of warfare. It referenced treaties in force at that time that banned their use. In this instance, the Treaty of Versailles from 1919 mentioned specifically the use of poison gas by Germany. Article 171 reads, "The use of asphyxiating, poisonous or other gases and all analogous liquids, materials or devices being prohibited, their manufacture and importation are strictly forbidden in Germany."[27]

While the vast majority of nations signed and ratified the treaty relatively quickly, the United States didn't ratify it until January 22, 1975, when it was signed by President Gerald R. Ford. The United States had introduced a similar proposal earlier, but it included provisions concerning submarines. This proposal was rejected by France, as they objected to the submarine provisions. The US proposal, known as the Treaty Relating to the Use of Submarines and Noxious Gases in Warfare, had it been adopted, read,

> Article 1. The Signatory Powers declare that among the rules adopted by civilized nations for the protection of the lives of neutrals and noncombatants at sea in time of war, the following are to be deemed an established part of international law:
>
> (1) A merchant vessel must be ordered to submit to visit and search to determine its character before it can be seized. A merchant vessel must not be attacked unless it refuses to submit to visit and search after warning, or to proceed as directed after seizure. A merchant vessel must not be destroyed unless the crew and passengers have been first placed in safety.
>
> (2) Belligerent submarines are not under any circumstances exempt from the universal rules above stated; and if a submarine

cannot capture a merchant vessel in conformity with these rules the existing law of nations requires it to desist from attack and from seizure and to permit the merchant vessel to proceed unmolested.[28]

However, even the adopted treaty had major problems. First, it prohibited the use of such weapons in an offensive manner but not in a defensive measure against a country that had used the weapons first. Second, it still allowed for the research and production of such weapons, under the auspices of developing defensive countermeasures. This led to an arms race of sorts, with the ratifying countries, including France, the United Kingdom, Italy, Canada, Belgium, and Poland, beginning research on such weapons soon after ratification.[29] In fact, while allowing for the study of defensive qualities of gaseous agents, the only way to study the defensive potential was to also study the offensive capabilities: "It is impossible to divorce the study of defense against gas from the study of the use of gas as an offensive weapon, as the efficiency of the defense depends entirely on an accurate knowledge as to what progress is being or is likely to be made in the offensive use of this weapon."[30]

Additional limitations to the protocol included lack of prohibitions on the use of these agents against nonratifying parties or their use within a country's own borders in case of a civil conflict, and there were no provisions for the inspection of countries thought to be in breach of the treaty. Thus, while every country claimed after the war that they were not involved in research for new chemical weapons, every one of them was developing the deadliest weapons they could so that they could study the ways to protect against them.[31]

The ability to use these weapons within one's own borders had repercussions during the Iran-Iraq War and the Syrian Civil War. The treaty also did not prohibit the use of herbicides against foreign countries, which arose in the Vietnam Conflict.[32]

Unit 731

Japan took this opportunity to begin its own program of CBRW development. The navy began their chemical weapons research in 1923, while the army followed suit in 1925.[33] In 1928, they began producing mustard gas. By 1934, they were making a ton of lewisite a week, and by 1937, lewisite production hit two tons per day.[34] Additional research led to the invention of several potential weapons, including

> rockets able to deliver ten liters of agent up to two miles; devices for emitting a "gas fog"; flamethrowers modified to hurl jets of hydrogen cyanide; mustard spray bombs that released streams of gas while gently floating to earth attached to parachutes; remote-controlled contamination trailers capable of laying mustard in strips seven meters wide; and the Masuka Dan, a hand-carried antitank weapon loaded with a kilogram of hydrogen cyanide.[35]

In 1932, Major Shiro Ishii toured Europe and returned home convinced that biological warfare was the future. After all, by his reasoning, why would the Geneva Protocols expressly forbid this if not because it was an area worth pursuing? He was able to convince his superiors into letting him begin research in the area and established the Army Epidemic Prevention Research Laboratory, more commonly known as Unit 731. Located near the cluster of villages of Ping Fan, in Manchuria close to the Soviet border, the program had more than 150 buildings on the site as well as 5 satellite camps staffed by more than 3,000 scientists.[36] In building this main complex, they appropriated 840 acres, along with 1,700 structures, and displaced at least 600 families.[37]

The Japanese used captured Chinese troops as well as civilians in their tests to develop the most efficient weapons possible. Research consisted of inoculating these captives with organisms causing cholera, smallpox, botulism, bubonic plague, anthrax, tularemia, typhus, typhoid, salmonella, tetanus, brucellosis, gas gangrene, tick encephalitis, tuberculosis, glanders, and venereal diseases.[38] The test subjects were then left untreated in order to study the effects on humans.[39] It is estimated that upward of 10,000 people were killed as a result.[40] However, the Japanese never were able to develop advanced bioweapons.

The Japanese continued their research in poison gases, realizing that it would make an effective weapon against the rural Chinese. When China presented to the League of Nations proof that mustard was being used against them, nothing happened, and the use of gas continued; one Soviet official estimated that a third of the munitions sent to China were chemical weapons.[41]

Besides individual test subjects, the Japanese researched the development of bioweapon bombs to be used on wide areas of the enemy population through direct infection; contamination of reservoirs and wells; and dropping plaque-infected fleas, food, and clothing into areas not occupied by Japanese troops.[42] More than one thousand such water wells were consequently poisoned. These pathogens included B. anthracis, Vibrio cholerae, Yersina pestis, Shigella species, and Salmonella species. The tests were terminated after 1,700 Japanese died during these trials.[43] However, it is estimated that several thousand Chinese died as a result of these tests. As with chemical weapons, the control of such weapons after deployment is difficult, given the finicky nature of weather and the tendency of wind to change directions without warning.

In an attempt to hide the truth of identities and in order to dehumanize them, subjects were referred to as "logs" (maruta). Close working relationships with other secret facilities, such as Unit 100, led by General Yamada, and the Kwantung army, kept Unit 731 supplied with such logs. The intended targets of these weapons would be primarily Chinese, Soviets, and Mongols. The close proximity of the research facilities in Manchuria to both Russia and Mongolia facilitated the procurement of such test subjects. These locations provided the raw working material for the tests without having to worry about testing on Japanese. The Japanese government had ruled it illegal to conduct human experimentation on Japanese citizen or within the actual country of Japan.[44] This further illustrates the racist beliefs of the Japanese people at that time: Experimentation on Japanese was forbidden, but on any other nationality, it was acceptable.

A significant purpose of Unit 731 was drawn up in the Kantokuen Plan of 1941, the proposed invasion of the Soviet Union. Drawn up by the general staff of the Imperial Japanese Army and approved by the emperor, Japan would have invaded the far eastern portion of the Soviet Union, taking advantage of the German invasion of western Russia. The invasion involved three readiness phases followed by a three-phase offensive to combat and destroy the Soviet army. A significant part of this plan involved the heavy use of chemical and biological weapons.[45]

Prisoners of War

The other convention was the Convention Relative to the Treatment of Prisoners of War. Commonly known as the Geneva Convention of Prisoners of War, this document sets forth in great detail the exact manner in which prisoners of war were to be treated, from the time they become prisoners (article 5) until they were either no longer prisoners (article 75) or they were deceased (article 76). It also provided for the protection of those involved with but not in the military, such as correspondents and newspaper reporters (article 81).

These two treaties formed the basis of the rules of war for the next twenty years. The Geneva Conventions, which are commonly referred to, used these rules as their basis again in 1949. It is from the Geneva Convention of Prisoners of War, article 5, that we get the famous "name, rank, and serial number," which prisoners give to their captors:

> Every prisoner of war is required to declare, if he is interrogated on the subject, his true names and rank, or his regimental number.
>
> If he infringes this rule, he exposes himself to a restriction of the privileges accorded to prisoners of his category.
>
> No pressure shall be exercised on prisoners to obtain information regarding the situation in their armed forces or their country. Prisoners who refuse to reply may not be threatened, insulted, or exposed to unpleasantness of disadvantages of any kind whatsoever.
>
> If, by reason of his physical or mental condition, a prisoner is incapable of stating his identity, he shall be handed over to the Medical Service.[46]

From this it is evident that not only are there requirements mandated of the capturing forces, but there are also requirements expected of the prisoner. However, the implementation of these guidelines was not as universal as would

have been expected. And there were two very important countries not bound
by these rules: Japan and the Soviet Union. While Japan signed the conventions, it did not ratify the treaty, so it was not required to follow it. The Soviet
Union did not sign the treaty, either. This proved instrumental in the war to
come.

WEAPONS DEVELOPMENT

Prior to the signing of the 1925 Geneva Protocol, gas was used as a counterin-
surgency weapon. Britain used it in the Russian Civil War, both on land and
from the air in 1919. Later that year, they used it in Afghanistan to put down
Afghan insurgents, and after that, Spain and France used it in Morocco.[47]
Clearly, gas, at least nonlethal gases, had a legitimate use in daily life. Today
police use smoke grenades and tear gas to quell riots.

Italy, too, engaged in chemical warfare during this period. In an attempt to
expand their "empire," as the other countries were doing, they invaded Ethi-
opia on October 3, 1935. During this time, they dropped mustard gas bombs
on the Ethiopian troops. In spite of signing the Geneva Protocol, the League
of Nations did nothing in response to this use of chemical warfare.[48] The Italians developed aerial spraying as a more efficient way of applying the different
agents to the intended target, and the effects were horrendous. Villagers unable
to get treatment had gangrene set in. Large areas of skin were missing from the
bodies of the afflicted. The carnage was described as a slaughter.[49]

Faced with the evidence of all the research that was happening, the remaining nations, such as Britain and France, had no choice but to continue
their own research, effectively leading to an arms race in chemical warfare.
Britain and France both built new research facilities and factories in which
to develop and produce new chemicals. They also shared their research with
each other to develop adequate defense equipment and protocols.[50] However,
their intelligence on German advances was severely lacking. One summary
conclusion stated, "*It is not thought that any important new war gas has been discovered.*"[51] This statement was grievously wrong.

Not only were these countries conducting research into these chemical
weapons, but they were also actively rearming themselves with them. In 1936,
the French constructed a phosgene factory. In 1937, factories at the Edgewood
Arsenal in the United States were put back into production manufacturing
mustard and phosgene gases. The Soviet Union constructed new chemical
weapons plants, as well. The British, with the help of the Imperial Chemical
Industries, began construction of a new mustard gas factory in 1936, with two
more plants planned. On November 2, 1938, the British government ordered
the production capacity of the nation increased to three hundred tons of mustard gas per week, with a reserve of two thousand tons. The Italians were reportedly capable of manufacturing twenty-five tons of mustard gas and five tons
of lewisite daily and had an unreported capacity for phosgene and chloropicrin.

The Soviet Union "appear[ed] to devote the greatest effort to developing the chemical arm."[52] And the failure of the United States to ratify the treaty for fifty years allowed the country to begin research on all facets of CBRWs. The British and French began to collaborate on a joint chemical weapons policy in May 1939. In it, the British reported that the French believed the German gas-development program was so advanced that its use by the Germans was certain. The thought that either Germany or Italy would refrain from using these weapons was never considered.[53] France was so sure of Germany's plans to use these weapons that they already had a chemical weapon arsenal of 4.5 million grenades filled with mustard gas that would be used as bombs, as well as other weapons filled with phosgene. The British had developed aerial spraying techniques that allowed them to release from an altitude of 15,000 feet a spray containing a mustard variant that was three times as potent as the original and had a very low freezing point, permitting it to be used at higher altitudes, in colder climates, and in cold ocean waters.[54]

As important as the Geneva Protocols were, this treaty ended up doing as little as the League of Nations ultimately did. The dichotomy of such attitudes of starting or continuing research on CBRW while still espousing a desire for world peace is apparent. Nevertheless, this protocol was an important step forward in the protection of noncombatants.

During the 1930s, Germany ramped up its chemical warfare research efforts to prepare for war. In 1935, the government ordered all toxic substances to be reported, effectively making all chemical laboratories part of the Nazi government. Dr. Gerhard Schrader had been researching organophosphates in an attempt to find new insecticides. On December 23, 1936, he found one that was incredibly potent. Upon further research and development, he realized that this was more than just an insecticide. After working with this compound for several weeks, he and his associates were forced to take three weeks away from the material due to the extreme side effects they were experiencing. They were lucky that this was all that had happened to them. They had, in fact, developed the world's first nerve gas, named tabun.[55]

Organophosphates were a little-studied family of compounds, and tabun was an entirely new compound with extremely toxic effects.[56] It worked by blocking the effect of cholinesterase on the neurotransmitter acetylcholine. Without the reuptake of acetylcholine, muscles went into a permanent state of contraction. The muscles controlling breathing, voluntary movement, and defecation contracted violently, leading to frothing at the mouth, diarrhea, jerking, and convulsions and ending in death by asphyxiation. Further research in this group of chemicals led to the discovery in 1938 of a compound related to tabun but ten times more poisonous, sarin, named for the four main individuals involved in its production: Schrader, Ambros, Rüdriger, and van der Linde.[57] Meanwhile, Fritz Haber continued refining the manufacturing process for mustard gas, helping the Soviet Union to develop their first chemical weapons plant and helping Germany with its secret research into and stockpiling of chemical warfare agents, in the direct breach of the Treaty of Versailles.[58]

The United States was rather late to the CBRW table. The British came to the US government to contribute research to this cause. In fact, it was Sir Frederick Banting, who won the Nobel Prize for discovering insulin, who established the first bioweapons research program. This was a nongovernmental endeavor established in 1940 with the help of corporate sponsors.[59] Besides agents that were toxic to man, the United States had been working on developing herbicides and biological weapons, and while there is no proof, circumstantial evidence points to the United States using these agents outside the laboratory. The autumn of 1944 saw a widespread plague of Colorado beetles threatening the potato crop in Germany to the point that Schrader was pulled off the tabun project and tasked with finding an insecticide to combat the beetles. In another case, the 1945 Japanese rice harvests were stricken with blight, forcing them to develop a crop-rotation plan to help ward off the infection.[60]

JAPAN AND CHINA, 1931

Japan's victory over Russia in 1905 had helped minimize the feelings of being slighted by other countries. From the opening of Japan to the outside world until 1918, Japan kept friendly relationships with the Western powers. However, from 1918 on, their views of these powers changed, as they saw them as increasingly racist. Shūmei Ōkawa stated that the purpose of the League of Nations was, "[to] preserve the status quo and further the domination of the world by the Anglo-Saxons."[61] He believed that "Japan would strive to fulfill her predestined role of champion of Asia."[62] The leaders of Japan were denied the opportunity to become equal partners with the West, as they were viewed not as equals but as slightly better Asians, who were not believed to be equal. Forcing the Japanese to subscribe to the 1930 London Naval Treaty merely reinforced their view that the West saw them as inferior. All these slights, whether perceived or real, fueled the entire Japanese populace in their animosity toward the West, especially the United States.

In 1890, the emperor was declared a divine descendant of the sun goddess Amaterasu. As such, while the Japanese viewed their race as superior to all others, they typically viewed the individual as worthless, except as it could serve Japan and the emperor. Japanese children were taught from an early age that the supreme virtue was giving one's life for their country in battle.[63] This indoctrination included unwavering obedience not only to the emperor but also to the members of the government and military.[64] One third-grade language book included a song, "Brave Soldiers," that told of three soldiers who became human bombs by strapping explosives to their bodies and blowing themselves up to clear the way for others, a very ominous portent of the kamikazes used during the latter stages of World War II.[65] A Japanese historian stated, "The goal of our education was only to create men who would fervently throw away their lives for the sake of the Emperor, men who were full of loyalty.[66] The military seized on these beliefs in their efforts to combine them with industry into one synergistic entity engaging in total war. As part of this viewpoint, they taught that any form of surrender was deeply dishonorable. Even enemies were expected to commit suicide, as surrender rendered the enemy as well as themselves a nonentity. Nonentities could be treated by the victorious Japanese as worse than an animal.[67]

Morale was supposedly much more valuable than sheer numbers in considering military might. Because the individual was basically just a tool, this training was enforced with extreme brutality and humiliation toward all members of the military. Soldiers were slapped by superiors until they bled (termed an "act of love") or were forced to wash their superiors' underwear, among other humiliating actions.[68]

Once in the field, orders prohibiting retreat, surrender, and being taken alive by the enemy were standard.[69] The soldier's manual stated, "Bear in mind the fact that to be captured means not only disgracing the Army but your parents and your family will never be able to hold up their heads again."[70] This highly regimented, physically punishing training resulted in two very dramatic outcomes: First, the Japanese soldier fought to the last man. There was no surrender, no retreat. To do otherwise would bring dishonor to not only the soldier but his family as well. Second, the lack of individual worth and brutality of training of soldiers resulted in a complete absence of traditional human inhibitions when these soldiers were finally released from military constraint.

As an additional part of their training, Japanese soldiers were taught that the Chinese were blood enemies, an inferior race, and that Japan would rule the world. Combine these with the view of Japanese superiority, and it becomes understandable why non-Japanese experienced rape, pillage, and murder of men, women, young, old, civilian, prisoner of war—anyone and everyone—at their hands. Commonly referred to as the "transfer of oppression," where individuals treated brutally in turn brutalize those who come under their supervision, this phenomenon was taken to the extreme in this case.[71] As Dan van der Vat writes, "Life was cheap in Japan; those in its services were as unlikely as anyone else to place greater value on an enemy than they did on themselves."[72]

The officers and soldiers submitted to all this because they, too, believed the orders to be divinely issued. The emperor wanted them to act like this. The generals knew they could get away with these kinds of orders for this very reason: The indoctrination of the people led to a very subservient military, in essence another aspect of training, willing to do anything for the emperor, including rape, torture, and murder anyone considered the enemy. The early Japanese victories in 1941 only further convinced them of their moral and racial superiority and encouraged more horrific treatment of their foes. This was revenge against those who had previously slighted them and became the impetus for an all-out race war. Felton writes,

> The Japanese officer corps labeled the Geneva Convention the
> "coward's code" and branded Allied capitulation another Western
> mindset as corrupt and contemptible as democracy. "Remember
> your status as prisoners of war," one Japanese camp commandant
> told his prisoners in 1942. "You have no rights. International law
> and the Geneva Convention are dead."[73]

The Japanese often called themselves the "leading race of the world" and viewed themselves as not only unique among all other races of the world but also superior to these races.[74] The Imperial army divided the nationalities of Asia into master races, friendly races, and guest races and placed the Japanese race as the unquestioned leader of these races.[75] "Guest" races had weakened the native "master" races, necessitating the intervention of Japan to liberate them from their colonial masters.[76] This, along with the need for living space and raw materials, was a primary motive for the Japanese invasion of such surrounding areas as Manchuria and Southeast Asia. One cannot help but notice the similarity between the Japanese "living space" and the Nazi *Lebensraum*, and both were to be accomplished at the expense of other races.[77]

Japan had been experiencing the same economic upheavals as the rest of the world, and by 1931, half of Japan's factories were idle.[78] Domestically, the period from 1931 to 1937 was rife with political discord. Assassinations of political leaders were common. It is against this background that on September 18, 1931, Lieutenant Suemori Kawamoto of the Twenty-Ninth Japanese Infantry Regiment detonated some dynamite next to the railway belonging to Japan's South Manchuria Railway. So weak of an explosion as to not stop a train from passing by moments later, the incident nonetheless served as the pretext for the Japanese invasion of Manchuria. They quickly occupied the area and established the puppet state of Manchukuo.

The "liberation" of the Asiatic nations under the auspices of the Greater East-Asia Co-Prosperity Sphere from the colonial rule of the European nations resulted in the death of 30 million civilians, of whom about 23 million were Chinese.[79] Combining their views of their racial superiority, their belief that the war with China was a holy war, and their idea that Japan was destined to rule the world with a military willing to do anything in the advance of these ideas led to an "ideological witch's brew that had a direct bearing on the treatment of Japan's enemies."[80]

GERMANY

The road to total war was well established in Germany with the ratification of the Treaty of Versailles. Using his gift for public oration and with the Treaty of

Versailles as his text, Hitler set to work convincing the German populace of the wrongs that had been heaped on their countrymen. Much has been written about the Holocaust, so this book does not go into the detail expressed elsewhere. Instead, it focuses on total war's implementation by ordinary people. In other words, how did the Nazi regime convince ordinary people to commit such atrocities, and what are some examples of what these people did in the name of the Reich?

The German tendency toward anti-Semitism since at least the late 1800s is been well documented. During this time, Jews were blamed for problems experienced by rural Germans that were actually a result of a worldwide depression. The burning of synagogues had been occurring in the German provinces since as least that time, if not earlier.[81] Blaming Jews was a common occurrence, as they were seen as responsible for the death of Jesus Christ as told in the New Testament. In fact, this persecution extended back as far as the Middle Ages, when they were blamed for the Black Death of Europe.[82]

While Jews had slowly assimilated into German culture prior to the 1870s, this new upswing in anti-Semitism resulted in in the exclusion of Jews from German society.[83]The prevalence of anti-Semitism in the higher socioeconomic levels was relatively well established, as demonstrated by such people as composer Richard Wagner and novelist Julius Langbehn, along with such military leaders as Generals Werner von Fritsch and Erich Ludendorff.[84] With the unexpected surrender of Germany in 1918, common people as well as disgruntled soldiers needed to affix blame to some group for this loss, and the Jews were the perfect group. In fact, it is hard to imagine the Holocaust occurring without some deep-seated cultural anti-Semitism on the part of Germany.[85]

Part of this racism was based on the German idea of *Volk*. George L. Mosse defines this as the "union of a group of people with a transcendental 'essence.'"[86] In other words, groups of people were linked to the environment from which they originated. Germans, being native to the German countryside, were deep, dark, and mystical. Jews, however, were from the desert wastelands of the Middle East and were therefore considered "shallow, arid, 'dry' people, devoid of profundity and totally lacking in creativity."[87]

Besides believing that Jews were "subhuman," Hitler repeatedly preached that Jews were directly threatening the very existence of Germany. By portraying the Jews as an enemy to Germany and whose sole intent was the elimination of Germany as a sovereign country and the extermination of the German people, Hitler was able to convince the populace that the destruction of the Jews was an act of self-defense for Germany (see figures 13 and 14). He and his top leaders constantly painted the direst of circumstances were the Jews to be allowed to live.

Moreover, such apocalyptic views were backed up by such anti-Germans as Theodore N. Kaufman, an American Jew who published the book *Germany Must Perish!* in which he calls for the sterilization of the entire German race

Figure 13. This vivid poster from the September 1930 Reichstag election summarizes Nazi ideology in a single image. A Nazi sword kills a snake, the blade passing through a red Star of David. The words radiating from the snake's body are *usury; Versailles; unemployment; war guilt lie; Marxism; Bolshevism; lies and betrayal; inflation; Locarno; Dawes Pact; Young Plan; corruption; Barmat, Kutistker,* and *Sklarek* (the last three Jews involved in major financial scandals); *prostitution; terror;* and *civil war.* Original image from Dr. Robert D. Brooks. Image and caption courtesy of Dr. Randall L. Bytwerk, accessed May 20, 2025, https://research.calvin.edu/ german-propaganda-archive/posters/ snake1.jpg.

Figure 14. A poster for the July 1932 Reichstag election. The caption says, "The workers have awakened!" Various parties are trying to persuade the worker to side with them, without success. The small chap in the center with the hat and glasses represents the Marxists (note the Jew whispering in his ear). His piece of paper says, "Nazi barons! Emergency decrees. Lies and slanders. The big-wigs are living high on the hog, the people are wretched."

During the Weimar Republic, a party's position on the ballot depended on its strength. The higher the position on the list, the better the party had done in previous elections. Image and caption courtesy of Dr. Randall L. Bytwerk, accessed May 20, 2025, https://research .calvin.edu/german-propaganda-archive/ posters/liste2a.jpg.

and the partitioning of Germany among its neighboring countries.[88] While having little impact in the United States, the publication served as a major catalyst in Nazi propaganda. Hitler and Joseph Goebbels were quick to take full advantage of this publication as "proof" of the international Jewish conspiracy against Germany, verifying the need for Germany's war against Bolshevism and Jewry. While Goebbels knew the effects of this book in the United States were minimal, he welcomed it as a godsend for Nazi propaganda: "This Jew did a real service for the enemy [Germany] side. Had he written this book for us, he could not have made it any better. I will have this published in an edition of millions for Germany and above all for the front, and will write the forward [sic] and afterward [sic] myself."[89]

By constantly portraying Jews in such negative light, the Nazi Party was able to eventually indoctrinate its citizens on the need for drastic measures against the "evil" Jews. This propaganda came from the top levels of the Nazi Party and was distributed to nearly every German citizen. Daily newspapers carried such "reports," while party speakers were instructed to emphasize the Jewish threat in their speeches at neighborhood meetings.[90] In December 1942, "speakers were told to make anti-Semitism the center of all coming meetings," realizing that as important as the printed word is in advancing propaganda, the spoken word ("word-of-mouth propaganda") is the most important aspect of local meetings.[91] The constant reminder of the severity of the Jews' intentions was imperative. Because the Nazis realized that such written propaganda could become tiring and ineffective after prolonged exposure, special actions were needed to ensure that these threats to Germany were continually passed on by word of mouth.[92]

Given the many quotes by Hitler that the Jewish people would be "finished," "exterminate[d]," and "annihilat[ed]," it was a simple matter to convince the people that this was, in fact, Hitler's goal from the very beginning.[93] Hitler was able to effectively use politics to push his radical racial agenda. And while there remains debate about whether the general population actively embraced anti-Semitism or was merely indifferent to it, the constant exposure to such propaganda was bound to have an effect on the people. At the very least, it appears that the "Germans [knew] enough to know that they did not want to know more."[94]

Prejudice against the Jews was just one aspect of the Nazi desire for racial purity. Laws were passed targeting other undesirables, due to either race or affliction. The 1933 Law for the Prevention of Offspring with Hereditary Diseases specifically targeted hereditary illnesses or disabilities:

> Physicians sterilised those with schizophrenia, Huntington's chorea and epilepsy, as well as so-called "mental defectives," chronic alcoholics and the blind and deaf. The law marked a break with democratic structures of public health provision. The Nazis also targeted

for sterilization groups of "mixed race," Roma and Sinti ("Gypsies"), and those showing "antisocial behaviour."[95]

Hitler further authorized the implementation of a program to identify and eliminate mentally and physically handicapped children to lessen their demands on the state. This program was extended to include handicapped adults before the program began. As with the inclusion of adults, this program was easily expanded to incorporate Jews with mental health problems.[96]

His views on this were classic social Darwinism—removal of the weak to encourage the prosperity and increase of the strong: "If Germany was to get a million children a year and was to remove 700–800,000 of the weakest people then the final result might even be an increase in strength."[97] By the end of the war, the Children's Euthanasia Program had been expanded to include healthy children of unwanted races, including Jews, with more than 5,200 children killed.[98] During the course of Aktion T-4, ending in August 1941, at least 72,000 adult mental patients were eliminated.[99]

The Nazi SS took an active part in this program, helping to transport identified children to the facilities where physicians were trained to perform medical experiments on the children prior to murdering them.[100] Likewise, nearly 400,000 German citizens were forcibly sterilized in order to prevent them from reproducing and passing on their "defective" genes to progeny. This program was apparently quite popular among Germans.[101] Given these actions against his own citizens, it should come as no surprise that Hitler transferred these beliefs and methods to those he considered racially inferior.

Mass media, especially radio, were extremely important to Hitler in spreading his views to the German public. Joseph Goebbels became the head of the Reich Ministry of Public Enlightenment and Propaganda, where he effectively used radio, along with the other media, to spread the Nazi ideology: "The radio will be for the twentieth century what the press was for the nineteenth century."[102] In a speech given at the formation of the Ministry of Propaganda on March 15, 1933, he further stated, "It is not enough for people to be more or less reconciled to our regime, to be persuaded to adopt a neutral attitude towards us; rather we want to work on people until they have capitulated to us, until they grasp ideologically that what is happening in Germany today not only *must* be accepted but also *can* be accepted."[103]

Eugen Hadamovsky, the Nazi national programming director, concurred in a book written in 1934, "Through enormous mass meetings that radio brought to the whole nation, it was able to assure the peaceful development of Germany's domestic life. This was important not only for our domestic policies, but also for our foreign relations. Our leaders always said that radio was the best guarantee of peace."[104] He further acknowledged the difficulty in forming and maintaining human contact between the radio and its listeners, noting,

When a man speaks to a microphone in a small radio studio, he has no sense of the effectiveness of his words. He does not know if any-one is listening at all, or whether the radio audience stops paying attention the moment he begins speaking.

More than that, the peculiarity of radio is that the spoken word is supposed to influence the audience without any visible component.

Everyone knows that in a meeting or theatrical play, one must see the speaker or actor to understand him. No one looks at the back of the speaker. But this is entirely absent in radio.

If one wants the spoken and heard word of the radio to realize a common will, it cannot be done only through transmitters and receivers; instead, a real human connection between sender and re-ceiver must be established.[105]

For this purpose, radio wardens were established to serve as this human link between the radio stations and the people listening. Their role is examined shortly.

In 1924, in the infancy of radio, there were very few Germans listeners. By 1933, approximately five million radio subscriptions existed. These included households where multiple people would utilize one subscription.[106] As with any new technology, the cost of radios was initially high but dropped as the technology matured. Furthermore, Goebbels, realizing the importance of the radio in reaching the public, ordered the mass production of an inexpensive radio in 1933. In 1938, another less-expensive model was introduced.[107] Ad-ditionally, the Nazis continued to expand the categories of those who qualified for exemptions to the monthly radio charge.[108] As a result of these actions, radio use increased dramatically by the middle of the war, with nearly 16 mil-lion listeners.[109] Robert Gellately notes that the motto during this time was "The Führer's voice into every home and factory."[110]

Additionally, whenever a major speech or important announcement was given by a Nazi leader, loudspeakers were erected in such public places as pub-lic squares, factories, offices, schools, and restaurants so that as many people as possible could hear these broadcasts.[111] The radio wardens in charge of this task were also responsible for encouraging the general population to share their radios with others and forwarding programming requests and critiques. How-ever, these radio wardens were also responsible for monitoring listening hab-its.[112] Hadamovsky noted that the radio wardens' task was to make sure that the state retained control over the radio, so that the methods used by the Nazis in controlling the radio stations, thereby controlling the people, could not be used against them in an attempted coup.[113] The radio wardens played an im-portant role in executing these directives.

Maja Adena and colleagues studied the efficacy of radio propaganda and determined that while its effectiveness varied with its content, overall it was very effective. Before the Nazis came to power in 1933, neutral and negative Nazi messaging led to lower vote totals for the Nazis. After Hitler's appointment as chancellor, radio propaganda became exclusively pro-Nazi, which they found reflected in the final relatively "free" elections of March 1933.[114] They found similar correlations with anti-Semitism—increased anti-Semitic radio propaganda led to increased rates of anti-Semitism in the general public, including the number of Jews deported.[115] As noted by David Welch, in 1933, fifty speeches by Hitler were broadcast, and by 1935, his broadcast speeches reached more than 56 million.[116]

Anti-Semitic indoctrination was also prevalent in the German school system during this time. Julius Streicher published a book, *Der Giftpilz*, intended for children and used in some schools. In it, children were taught that just as there are good mushrooms and bad, poisonous mushrooms, so are there good and bad, poisonous people. As the mother in the story tells her son,

> "And do you know, too, who these bad men are, these poisonous mushrooms of mankind?" the mother continued.
>
> Franz slaps his chest in pride:
>
> "Of course I know, mother! They are the Jews! Our teacher has often told us about them."
>
> The mother praises her boy for his intelligence, and goes on to explain the different kinds of "poisonous" Jews: the Jewish pedlar, the Jewish cattle-dealer, the Kosher butcher, the Jewish doctor, the baptised Jew, and so on.
>
> "However they disguise themselves, or however friendly they try to be, affirming a thousand times their good intentions to us, one must not believe them. Jews they are and Jews they remain. For our Volk they are poison."
>
> "Like the poisonous mushroom!" says Franz.
>
> "Yes, my child! Just as a single poisonous mushrooms can kill a whole family, so a solitary Jew can destroy a whole village, a whole city, even an entire Volk."[117]

The children were further instructed that the easiest way to tell if a person was a Jew was by their nose: "One can most easily tell a Jew by his nose. The Jewish nose is bent at its point. It looks like the number six. We call it the 'Jewish six.' Many Gentiles also have bent noses. But their noses bend upwards, not downwards. Such a nose is a hook nose or an eagle nose. It is not at all like a Jewish nose."[118]

Hitler's racism had been thoroughly indoctrinated into the German population, both soldiers and civilians alike. American foreign correspondent William L. Shirer wrote about a conversation with his maid in which she asked him why the French had declared war on Germany. He asked her why Germany made war on Poland, to which she replied, "But the French, they're human beings." Shirer replied, "But the Poles, maybe they're human beings too."[119]

Besides believing that Jews were "subhuman," Hitler repeatedly preached that Jews were directly threatening the very existence of Germany. Jews were behind both capitalism and communism in their goal of world domination. Moreover, such apocalyptic views were backed up by such anti-Germans as Theodore N. Kaufman, who published *Germany Must Perish!*[120]

UNITED STATES

One of the big concerns in the United States during these years was the isolationist movement. Since its birth, the United States tried to keep to its own affairs. Beginning with George Washington's farewell address, in which he so famously advised the nation, "It is our true policy to steer clear of permanent alliances with any portion of the foreign world—so far, I mean, as we are now at liberty to do it," much effort was made to follow that admonition.[121] While of necessity the United States was forced to trade with other countries for the goods they could not produce themselves, it did tend to maintain its neutrality—barring, that is, the conquest of the continent in the United States' quest for Manifest Destiny. The United States had no problem becoming involved with foreign countries when it benefited the country as a whole.

During the 1920s and 1930s, the country veered back to isolationism, having been through the Great War. This was a textbook example of the results of a permanent alliance. Not only had it led the United States into war, but also through having to honor commitments, alliances started the war in the first place. However, that isolationism tended to go hand in hand with a rejection of anything not American. This included those who were "hyphenated" Americans: German-American, Italian-American, Irish-American, and similar.

These attitudes originated in the 1880s, a blowback to the Black presence manifesting during Reconstruction. During that time, the pro-immigration policies of the government were criticized as lowering the wages, thereby threatening the economic security of the White laborer, an issue that arose again in the Trump years of the 2010s and forward.[122] This sentiment persisted throughout those years, especially during the high-immigrant years of 1901 to 1915. Discussions of making English the official language of the United States were also present during those years.

The question for the hyphenated Americans as well as the issue for the country was, Where are their loyalties? Were they first and foremost Americans? Or the hyphenated nationality? In this aspect, the melting pot that was America was expected to melt away all the ethnicities into one homogeneous mass, the entity known as American. Indeed, as Specter and Venkatasubramanian write, "For [Patrick] Buchanan, [Charles] Lindbergh, and other America Firsters throughout history, American exceptionalism depended on homogeneity rather than cultural diversity."[123] The phrase *America first* was common during this time, with such groups as the America First Committee (AFC) coming into existence and the resurgence of the Ku Klux Klan, both of which utilized the phrase as their motto.[124] At this time, the target of racial discrimination broadened itself to now include Jews. And while the Klan was typically viewed as a backward, redneck affiliation, their ideology about immigrants, Jews, and African Americans closely resembled that of the typical White American then.[125]

The isolationist views of Americans during this time were one factor in the failure of the United States to join the League of Nations: The needs of the world should not surpass the needs of the nation. However, even during this time, the isolationists knew that a complete withdrawal from the world stage was not only impractical but also impossible. All of this laid the foundation for the events preceding the Second World War.

The AFC was a populist/fascist American isolationist group organized in 1940 to protest the actions of President Roosevelt as he aided Britain in its fight against Nazi Germany. Some members were avowed pacifists, arguing against any type of conflict. However, most were normal citizens who believed in putting America first and that involvement in foreign wars was in direct conflict with that belief.[126] Charles Lindbergh, famous for having flown across the Atlantic Ocean nonstop, became the face of the AFC. However, Lindbergh held very extreme views. He expressed great admiration for Nazi Germany, calling its air force the strongest in the world and believing that Germany would be victorious in any conflict into which it entered by virtue of said air force. Besides, he believed the real threat lay in the Soviet Union and that a strong Germany would hold the Soviet Union at bay.[127] He wrote, "This present war is a continuation of the old struggle among western nations for the material benefits of the world. It is a struggle by the German people to gain territory and power. It is a struggle by the English and French to prevent another European nation from becoming strong enough to demand a share in influence and empire."[128]

Lindbergh was not the only prominent member of the AFC. Members included Henry Ford, Walt Disney, and future President Gerald Ford. At its zenith, it boasted 800,000 members.[129] As time went on and Britain was subjected to the Blitz of 1940–1941, more and more Americans began to favor intervention to aid the Allies. While Americans became more interventionist, the AFC became more isolationist, and Lindbergh's anti-Semitism increased.

This all came to a head on September 11, 1941, in Des Moines, Iowa, where Lindbergh gave his infamous Des Moines speech, "Who Are the War Agitators?" In it, he accused three groups advocating for the United States to go to war: President Roosevelt, the British, and the Jews. Also contributing to this move but to a lesser degree were the "capitalists, Anglophiles, and intellectuals" as well as the communists. After explaining the reasons for the British intervention, he proceeded to explain the Jewish rationale:

> It is not difficult to understand why Jewish people desire the over-throw of Nazi Germany. The persecution they suffered in Germany would be sufficient to make bitter enemies of any race.
>
> No person with a sense of the dignity of mankind can condone the persecution of the Jewish race in Germany. But no person of honesty and vision can look on their pro-war policy here today without seeing the dangers involved in such a policy both for us and for them. Instead of agitating for war, the Jewish groups in this country should be opposing it in every possible way for they will be among the first to feel its consequences.
>
> Tolerance is a virtue that depends upon peace and strength. History shows that it cannot survive war and devastations. A few far-sighted Jewish people realize this and stand opposed to interven-tion. But the majority still do not.
>
> Their greatest danger to this country lies in their large owner-ship and influence in our motion pictures, our press, our radio and our government.
>
> I am not attacking either the Jewish or the British people. Both races, I admire. But I am saying that the leaders of both the British and the Jewish races, for reasons which are as understandable from their viewpoint as they are inadvisable from ours, for reasons which are not American, wish to involve us in the war.[130]

The speech finished with painting the Roosevelt administration as a power-hungry group foisting the war on the American public to further consolidate their power.

The reaction to his speech was swift and negative. He was accused of anti-Semitism and un-Americanism and of delivering a speech of which Hitler would have been proud. Rather than advance the cause of isolationism, his speech had the opposite effect. It mattered not: In less than three months, the issue was decided for them. However, the issue of anti-Semitism was not resolved.

In the summer of 1942, Herbert Karl Friedrich Bahr, a twenty-eight-year-old German immigrant crossed the Atlantic and sought asylum from the Nazis. However, after being interviewed through a process involving five different US government agencies, Bahr was exposed as a Nazi spy and rushed to trial, where the prosecution called for the death penalty.[131] While cases like this were rare, it only served to prove to the US government just how dangerous it was to let immigrants into this country. In June 1939, trying to flee from Nazi Germany just three months from the start of World War II, the ocean liner MS *St. Louis*, loaded with 937 passengers, most of whom were Jewish, was turned away by the United States and forced to return to Europe. More than a fourth of the passengers, 254 in all, died in the Holocaust. The United States, as a matter of national security, could not risk there being spies aboard.[132] The policy remained in effect, spurred on by people like Attorney General Francis Biddle, who warned President Roosevelt not to give immigrant status to refugees, until nearly the end of the war, in late 1944, when photos and newspaper stories began to document the sad truth of what was happening in Europe.[133]

The United States was not alone in their actions against immigrants. In July 1938, representatives from thirty-two nations met to discuss how to deal with the refugee crisis. The countries all expressed their regrets that they couldn't accept more immigrants, but with the world trying to cope with the Great Depression, there was no way they could take more refugees. In the case of the *St. Louis*, the United States wasn't the only country to turn them away, as Cuba and Canada also refused the vessel. The majority of the passengers had purchased Cuban visas while in Germany, but Cuba decided to revoke all but twenty-eight.[134] In the words of Hannah Arendt, the refugees "were welcomed nowhere and could be assimilated nowhere. Once they had left their homeland they remained homeless; once they had left their state they remained stateless; once they had been deprived of their human rights they were rightless, the scum of the earth."[135]

SPAIN

The Spanish Civil War occurred from 1936 to 1939 between the Nationalists under General Francisco Franco and the Republicans under President Manuel Azaña. The Nationalists, who were supported by the military, landowners, businessmen, and the Catholic Church, were conservatives who favored the traditional social and political order and wanted to overthrow the left-wing government. The Republicans, supported by urban workers, agricultural laborers, and middle-class intellectuals, were classic liberals defending the Spanish Republic government but who also wanted social and political reform.[136]

In 1936, after decades of tension, national elections saw the leftist Popular Front government obtain power. The Right refused to accept this, and on July 17, 1936, a series of military uprisings began, resulting in achieving power and control of several areas of the country. Under the leadership of Franco, the Nationalist government, with Franco as head, was set up in Burgos on October 1, 1936. The Republicans were initially led by Francisco Largo Caballero as premier, followed by Juan Negrin, who remained premier, both in Spain and in exile, until 1945, along with President Manuel Azaña. The Republicans were split between the far-left anarchists and militant socialists and the more moderate socialists and republicans, who wanted the republic to continue to rule, which led to enormous infighting, weakening their ability to formulate a cohesive strategy.[137]

Neither side had the resources to defeat the other, so they turned to for-eign nations for help. The Republicans, being leftists, wanted to avoid Nazi Germany. Therefore, they turned to the Soviet Union, followed by Britain, France, and the United States. The Nationalists, appropriately, turned to Nazi Germany and Fascist Italy for help. Consequentially, the Spanish Civil War turned into a practice run by each side for tactics involving, among others, tanks and airplanes that were used in World War II.

There was intense criticism of the League of Nations for failing to do more to stop the conflict. After the war, the United Kingdom and France recognized the Nationalist regime as the legitimate Spanish government. The Soviet Union did not, leading to further tensions between these countries.[138]

As well as serving as a precursor to World War II, with the resulting mil-lions of casualties, the Spanish Civil War saw an immense number of casual-ties. This number has fluctuated between 500,000 and 1 million, depending on whether victims of bombardment, execution, assassination, malnutrition, starvation, and war-engendered disease are counted.[139] Jeremy Black lists the casualties as at least 300,000.[140]

Both sides were responsible for the torture and execution of civilians as well as prisoners. Antony Beevor points out that the slaughter was different be-tween the sides: The Nationalists focused on the purging of "reds and atheists," which continued for years, while the Republicans reacted mainly in surges of revenge fueled by the release of pent-up fears.[141] The reds did, however, focus their attacks on the Catholic Church. Attacking priests and nuns and disin-terring mummies from the convent vaults were frequent occurrences. This was due, in no small part, to the role the church had in the religious persecutions in the previous centuries, with the Spanish Inquisition just one example of these tortures. In all, of 115,000 clergy, 13 bishops, 4,184 priests, 2,365 members of other orders, and 283 nuns were executed.[142] Both sides were responsible for the murder of clergy, and both sides protested the murder of their sympathetic clergy while saying nothing of the murders of the opposing side's clergy.[143]

The torture of the clergy included castration, disembowelment, and burn-ing alive in their churches. Churches were burned and vandalized, stained-glass windows were broken, and men shaved in the church's fonts.[144] Nonclergy business owners, professionals such as attorneys, factory owners, and similar people were treated as they had treated the public: Those who were fair and did not take advantage of the people with whom they did business were gen-erally left alone. Those who took advantage of their customers or patrons were quickly targeted for punishment meted out by the public.[145] Paranoia led to many ordinary citizens denouncing their friends and neighbors as belonging to the opposition.[146]

Hitler's two primary goals for his air force during the period of German rearmament were "to provide as large an air force as possible . . . and secondly to use this expanded air force for terror purposed in international affairs."[147]

The earliest example of this was the supply of German air forces, under the name of the Condor Legion, to aid Generalissimo Franco. Along with using this conflict as a testing ground for new German weapons, General Hugo Sperrle, commander of the legion, used the new German bombers and fighters in raids against the Spanish towns of Durango and Guernica. Durango, a town of 10,000, was undefended and had no military presence. The raid targeted a local church while mass was in process, killing fourteen nuns, a priest, and most of the congregation. After this, the fleeing civilians were strafed, with a death toll of about 250 civilians. The official explanation for the targeting of Durango was to bomb "military objectives."[148]

Guernica was even more devastated. Bombers dropped 100,000 pounds of incendiary, high-explosive, and shrapnel bombs, and fighters strafed the civilian population, killing 1,654 and wounding 889.[149] An eyewitness account of the bombing made clear that the town was not a military target: The factory that produced war matériel was located outside town and was untouched. So, too, were two barracks left untouched. Moreover, the bombing appeared to be timed for maximum terror: The attack was on a Monday afternoon, the customary market day, with the markets filled with townsfolk. The attack started at 4:30 p.m. and continued until 7:45, when dusk approached. The only countermeasures, the reporter wrote, was that of the clergy as they "blessed and prayed for the kneeling crowds—Socialists, Anarchists, and Communists, as well as the declared faithful."[150] The slaughter was immortalized in the painting *Guernica* by Pablo Picasso (see figure 15).

Germany defended their actions by claiming they were trying to bomb the Rentería Bridge but missed due to high winds. Again, the veracity of this claim was highly questionable: "The bridge was never hit, there was virtually no wind, the Junkers were flying abreast and not in line, and antipersonnel bombs, incendiaries and machine-guns are not effective against

Figure 15. Pablo Picasso, *Guernica*, 1937, oil on canvas, 3.49 meters (11 feet, 5 inches) × 7.76 meters (25 feet, 6 inches), Museo Reina Sofía. Source: Draw Paint Print, accessed May 20, 2025, https://drawpaintprint.tumblr.com/post/39144508954/pablo-picasso-guernica-1937-1374-x-3055.

stone bridges."[151] In fact, targets of decidedly military interest, such as the arms factory of Unceta and Co. and Workshops of Guernica, were left intact, while more than 70 percent of the town's buildings were destroyed or damaged.[152]

Instead, these attacks appeared to be exercises in the terror bombing of civilians, a key tenant of the air power theorists, and seemingly confirmed the effects capable of being inflicted on these populations. Wolfram von Richthofen, chief of staff of the Condor Legion, observed that it was "necessary to achieve 'a triumph at last against enemy personnel and matériel.'"[153] Upon receiving reports of the action, he wrote, "Guernica, city with 5,000 residents, has been literally razed to the ground. Bomb craters can be seen in the streets. Simply wonderful."[154]

As apparent as Guernica may seem as an example of German terror bombing, the case is not so cut and dried. James S. Corum states, "There is no evidence to indicate that the German air attack on Guernica was a 'terror bombing' or that Guernica was carefully targeted to break the morale of the Basque populace."[155] He further explains that Guernica was a military target due to the fact that the two major roads that Basque troops would utilize in an escape from the nearby city of Bilbao intersected in Guernica. Elimination of Guernica would severely hamper the Basque forces in their retreat and prevent them from fortifying that town into a major strong point. Moreover, two Basque army battalions, the Eighteenth Loyala Battalion and the Saseta Battalion, were stationed in the town. Given these parameters, Guernica was a legitimate military target for aerial attack.[156]

Further reasoning for targeting such towns as Guernica is given in the following statement: "We have had notable results in hitting the targets near the front, especially in bombing villages which hold enemy reserves and headquarters. We have had great success because these targets are easy to find and can be thoroughly destroyed by carpet bombing."[157] The reason that the Renteria Bridge was not hit is that the Ju 52s that had been modified to serve as bombers and equipped with rudimentary bombsights were simply not capable of hitting a small target like a bridge. Additionally, von Richthofen never engaged in terror bombing during the Spanish Civil War or later during World War II but was a "ruthless commander who never expressed any sympathy or concern for civilians who might be located in the vicinity of the military target."[158] In fact, von Richthofen's main complaint in the bombing of Guernica was that the Nationalist ground forces did not exploit the effects of the bombing fast enough, taking four additional days to capture the town.

The Great War was supposed to bring about peace. Looking back at the events of the 1920s and 1930s, we see that it didn't. Marshall Foch was correct that Versailles wasn't a peace treaty per se but simply an armistice, and an uneasy one at that.

CHAPTER 5

WORLD WAR II, 1937–1945

"Those who can make you believe absurdities, can make you commit atrocities."

—Voltaire

While Japan had been occupying China since September 18, 1931, hostilities came to a head on July 7, 1937, with the Marco Polo Bridge incident, in which a Japanese soldier was reported missing and the Japanese and Chinese opened fire on each other. While September 2, 1939, is traditionally viewed as the start of World War II, many scholars consider this incident the true start of global hostilities, while others consider the initial Japanese invasion of China on September 18, 1931, as the beginning of the war. For purposes in this book, the Marco Polo Bridge incident is viewed as the start of global hostilities. First, I examine some of the more specific actions taken by Japan, Germany, and the United States. Then I look at some more general topics of the war.[1]

JAPAN

The Marco Polo Bridge incident served as the catalyst for the Japanese invasion of China. While the Japanese had been occupying China since 1931, this incident marked a new level of involvement in the country. However, history was about to repeat itself. As had happened in World War I, Japan was expecting a quick victory over China. China was viewed as an inferior country that would be easily conquerable. However, the Battle for Shanghai set the tone for the Japanese occupation of China.

On August 13, 1937, World War II began with the Japanese invasion of Shanghai. Even though the Chinese had 70,000 troops facing 6,300 Japanese marines, the Japanese expected a quick victory. Instead, Japan eventually sent some 100,000 troops to fight. Ultimately, the Chinese were overpowered by the Japanese, and on November 26, 1937, the battle was over. It was an urban affair eliciting comparisons to Stalingrad some years later.[2] The battle resulted in more than 600,000 refugees fleeing the city.[3] Worse than that, it led to one of the worst atrocities of the war: the Battle and Rape of Nanjing.

Nanjing

The Second World War saw some of the most depraved actions ever imagined. The Holocaust is one such example of the depths to which the human soul can plunge as it attacks its fellow human beings. One of the earliest and most horrific examples of this mentality occurred immediately before the traditional start of World War II with the Japanese occupation of the Chinese city of Nanjing in December 1937. After the fall of Shanghai, the Japanese army continued to Nanjing. The Nationalist Party, or Kuomintang (KMT), under Chiang Kai-shek moved the capital to Chongqing in western China and ordered General Tang Shengzhi to defend Nanjing. This was to no avail, and the general fled Nanjing on December 12, 1937. The Japanese entered the city the next day.[4]

After the Japanese invasion, those who could leave the city did, leaving behind those who couldn't: women, children, elderly, and sick. Unfortunately, the Chinese authorities decided that to keep the Japanese out, they would close all the gates surrounding Nanjing except for one. Instead of keeping the Japanese out, they trapped all the Chinese in the city. This presented the Japanese troops with easy targets, unable to defend themselves.[5] For six weeks, the Japanese troops raped, murdered, tortured, and mutilated the remaining population. Decapitation was a particularly common means of execution (see figure 16). Soldiers would line men up and use them for bayonet practice, then continue practicing on the corpses (see figure 17). Other common methods of torture included live burial, mutilation, spraying gasoline on people and burning them alive, freezing in ponds and using the bodies for target practice, burying victims to their waists and letting dogs eat them alive, skinning alive, and torturing to death with needles.[6] Husbands were forced to rape their wives, while sons were forced to rape their mothers and sisters. Babies were tossed in the air and bayonetted on the way down or thrown into vats of hot oil and water.[7] Competitions were held to see who could kill the most people or kill them the fastest.[8] One such contest was the "Kill One Hundred People with Sword" contest.

Just prior to entering the city of Nanjing, sublieutenants Toshioki Mukai and Tsuyochi Noda agreed to a contest to see who could kill one hundred people with their ceremonial swords. When next they met, on December 10, 1937, Noda claimed 105 kills, while Mukai countered with a "score" of 106. Because they couldn't determine who had reached 100 first, they decided to prolong the contest to see who could get to 150 first. Mukai talked about the contest:

> I'm happy that we both exceeded 100 kills before we found out
> the final score. But I damaged my "Seki no Moroku" on some guy's

helmet when I was cleaving him in two. So, I've made a promise to present this sword to your company when I've finished fighting. At 3 AM, on the morning of the 11th, our comrades used the unusual strategy of setting Purple Mountain of fire, in order to smoke any remaining enemies out of their hiding places. But I got smoked out too! I shot up with my sword over my shoulder, and stood straight as an arrow amidst a rain of bullets, but not a single bullet hit me. That's also thanks to my Seki no Magoroku here.[9]

This contest was picked up by the Japanese newspapers *Osaka Mainichi Shimbun* and *Tokyo Nichi Nichi Shimbun*, which ran four different articles, complete with photos of Mukai and Noda with their swords, from November 30 to December 13, 1937. This helped the two men become heroes of a sort, which Mukai freely admitted to using to find a wife. However, once the war crime trials commenced, both men were identified as having joined in this competition and were executed as war criminals.[10]

In the years after the war, a heated contention arose as to whether the two men were guilty of the crimes for which they died. The left-wing component of this debate pointed to the newspaper articles and the bragging the two men had done after the fact to confirm their guilt. The right-wing portion (those

Figure 16. A photograph of Japanese soldier Yasuno Chikao an instant before he struck off Leonard Siffleet's head, 1943. Source: Rare Historical Photos, "Leonard Siffleet About to Be Beheaded with a Sword by a Japanese Soldier, 1943," accessed May 20, 2025, https://rarehistoricalphotos .com/leonard-siffleet-sword-1943/.

Figure 17. Japanese bayonet practice against a dead Chinese prisoner near Tianjin.
SOURCE: 3CARTIER, "JAPANESE BAYONET PRACTICE AGAINST A DEAD CHINESE PRISONER
NEAR TIANJIN," REDDIT, POSTED 2022, HTTPS://WWW.REDDIT.COM/R/WW2/COMMENTS/WAIEG3/
JAPANESE_BAYONET_PRACTICE_AGAINST_A_DEAD_CHINESE/

who staunchly supported the Japanese government in their action) said that they exaggerated their claims, as men like to do, and that the men only killed a few soldiers during battle. Mukai actually admitted as such to his wife, Chieko, after he had married her, to which she replied, "Oh, so you lied to deceive me, did you?"[11] When it came time to turn himself in, he freely did so, which was used as proof of an action that a guilty man would not have done. They also found eyewitnesses who testified that Mukai had been injured on December 2, 1937, and was physically unable to participate in such a competition, and he did not arrive back at his unit until December 15, 1937. Additionally, they pointed out that Mukai led an artillery platoon at the rear of an infantry unit. Likewise, Noda was an adjutant to a battalion chief, which would keep him from seeing much if any combat action. Finally, they argued that the few men they might have killed were soldiers in battle. If this "contest" actually happened in battle, then neither of them were war criminals, as that is the job of combat units—to kill the enemy.

We may never know exactly what happened during this "contest," but one must consider the mindset of the people involved in this debate. Those on the left tended to want to treat every action taken by every soldier as a crime and convict them of war crimes. They took these beliefs to the extreme position that not only were the soldiers but also the government and even every citizen guilty. On the other extreme, those on the right (revisionists) wanted to believe and prove that every action taken by soldiers that might be construed as a war crime was either grossly overblown or did not occur at all. They believed, in spite of all the proof to the contrary (photos, firsthand accounts, diary entries from Japanese soldiers, and other documents), that none of this happened.[12]

Once the Japanese army entered Nanjing, this debauchery continued, and their actions became even more repulsive. Females of any age were raped, including those who were pregnant or had just given birth. When the soldiers tired of rape, they turned to other activities, such as vaginal impalement by swords, rods, beer bottles, fire crackers, and golf clubs.[13] Total casualties for this orgy of violence are uncertain, but estimates for the number of rapes range from 20,000 to 80,000, while the estimated number of dead range from just a few thousand (by Japanese sources) to more than 400,000, with 260,000–300,000 as a seemingly close estimate.[14] All of this was done on innocent civilians for no military purpose, with the excuse that the Chinese were swine, so the Japanese could do whatever they wanted "to such creatures."[15]

In the case of China, there were more motives for these horrors. Whereas the Japanese were burning buildings for sport, the Chinese were conducting a scorched-earth campaign to deprive the advancing Japanese army of as many supplies as possible. Additionally, the Japanese were aware and scared of the Chinese plainclothes soldiers. Unable to tell if a "civilian" was actually a civilian or a soldier who had shed his uniform but continued to fight, the Japanese treated them all the same and tortured as many as they could.[16] The media did not help in this aspect, as they routinely demonized the Chinese in their attempts to defend themselves. Incidents that happened were almost always blamed on the Chinese and their reckless behavior toward the Japanese.[17]

The Nanjing horrors were merely the opening act in a macabre spectacle that lasted for the next eight years.

Singapore and Alexandra Hospital
With the bombing of Pearl Harbor on December 7, 1941, the Japanese proceeded to invade other locations in their attempt to secure the raw materials needed for their military and their country. One of these locations was the northeastern coast of British Malaya. With the British expecting an invasion from the sea, they were ill equipped to defend against a land-based invasion, and Singapore fell on February 15, 1942. What happened next became known as the *Sook Ching* Massacre.

Lieutenant General Tomoyuki Yamashita, who commanded the invasion, and his chief of planning and operations, Lieutenant Colonel Masanobu Tsuji, wanted to eliminate, or at least greatly reduce, any anti-Japanese resistance that might remain in Singapore. On February 18. 1942, under the command of Yamashita and with planning by Tsuji, the massacre began. The Japanese targeted, among others, the pro-British Straits Settlements Volunteer Force (SSVF), the Chinese-dominated Dalforce, the Malayan Communist Party (MCP), and the China Relief Fund. Additional groups included looters, armed civilians, and in general anyone the Japanese decided posed a threat to them.[18]

The military ordered all Chinese males to report to various screening centers to determine if they were anti-Japanese or not. There was no standardized procedure for this, so local screening centers used such arbitrary measures as Chinese men who wore glasses (possible patriotic Chinese intellectuals), who were from Hainan (MCP sympathizers), or who were English educated (British loyalists).[19] If one failed the screenings, then they were driven off by the military to their execution sites and buried then and there.[20]

Not all of those who failed the screenings died. Guangyu Zheng failed the screening and was taken away to be machine-gunned with others. He was struck in the nose by a bullet but was not dead. After this, the army stabbed all those who were still alive. Zheng survived because the soldiers thought he was dead.[21] A Mr. Chen managed to fall into the sea before he was shot and was rescued by a fisherman.[22]

This massacre continued until March 4, 1942, at which point the military left to invade Dutch-controlled Sumatra. However, the Kempeitai, or Imperial Japanese Army Military Police, stayed behind and continued the operation.[23] Later, the Japanese claimed that only five thousand people were executed during this operation. However, based on the excavated remains found, the Chinese in Singapore put the death toll between 20,000 and 50,000.[24]

Though not involving the death of as many people as the *Sook Ching* Massacre, the Alexandra Hospital Massacre was no less brutal. As the Japanese troops were advancing through the city, they came upon the Alexandra Hospital. The hospital was built to accommodate 550 patients, but because of the fighting in the city, there were 900 in there.[25] The hospital was full of patients, staff, and doctors. A Lieutenant Weston left the hospital with a white flag to signify the surrender of the hospital. He was promptly bayonetted to death by the troops. As the troops entered the hospital, the occupants started shouting, "Hospital!" and pointing to the red crosses displayed throughout the hospital. The Japanese paid no heed. They proceeded to bayonet patients as they lay in their beds as well as a patient who still lay anesthetized on an operating table.[26] At this point, about fifty men had been killed.[27] Now, the troops gathered two hundred men and marched them to a row of buildings, where they were stuffed into three small rooms and left there overnight. In the morning, those who were still alive were allowed to leave in groups of two, believing that they were being allowed to get water. Instead, they were executed.[28] The slaughter would have continued except an artillery shell hit the hospital. All told, some three hundred occupants of the hospital were butchered during the few days the Japanese were there.[29]

Bataan

Although the Japanese claimed to be treating their prisoners of war with the utmost humanity, the opposite was the norm. This was vividly demonstrated at the Bataan Death March. The Japanese invaded the Philippines on December 10, 1941, right after the attack on Pearl Harbor. While General Douglas MacArthur anticipated the American and Philippine troops could easily hold off a Japanese invasion, he was quickly proven wrong. MacArthur did nothing to protect the air force stationed there, and, consequently, the vast majority of the airplanes of the JS Far East Air Force were destroyed while on the ground. The air force ceased to exist, forcing the troops to fight with no air cover.

In March 1942, MacArthur was ordered to Australia. Though he stated that he would return, the remaining troops knew they had been abandoned.[30] After three months of hopeless defense, General Edward King, who had been left behind to command the remaining forces, surrendered, disobeying orders to not do so. The tired, hungry, and diseased remnants of the US Army in the Philippines, some 75,000 in all, were given over to the Japanese.[31]

At Bataan, one Japanese colonel called for the execution of all prisoners in the Philippines on the basis that the Americans were White colonialists, and the Filipinos had betrayed the Japanese, their fellow Asians. Upon surrender, General King offered to move the prisoners on trucks for sixty miles, from their location at Mariveles to San Fernando, where they would take a train the remaining distance to Camp O'Donnell. General Homma Masharu refused, believing King, for having surrendered, to be as worthless as the troops he commanded.[32] The troops were forced to make the trip on foot. The soldiers dared not stop along the way for food or water or falter as they marched, as doing so would result in death from the Japanese soldiers. Locals who tried to give food and water to the prisoners were shot. One prisoner reported seeing two decapitated heads, no other dead bodies, lying on the side of the road every mile marched.[33]

The trip by rail was no better. Rail cars designed to hold forty people were crammed with a hundred. Some people couldn't even touch the floor; they were held suspended by the bodies of the other soldiers crammed in around them. Dysentery and the lack of toilet facilities worsened the problem. More than 20,000 prisoners died due to starvation, illness, and executions on the march from Bataan to their internment at Camp O'Donnell.[34]

At the camp, of those who survived the trek, one in six perished, again due to starvation and disease. The death of all these soldiers was not due to just the sadistic actions of the soldiers. Those actions were partially due to a direct order from the Japanese War Ministry that stated, "Whether they are destroyed individually or in groups, and whether it is accomplished by means of mass bombing, poisonous smoke, poisons, drowning, or decapitation, dispose of them as the situation dictates. It is the aim not to allow the escape of a single one, to annihilate them all, and not to leave any traces."[35]

Manila

The atmosphere of treatment of the enemy had not changed three years later. With the American troops nearing the Philippine capital of Manila, the Japanese proceeded to butcher the locals. The violence unleashed on the civilians of Manila was described as an "orgy of mass murder."[36] During the twenty-nine days of fighting in Manila, tens of thousands of civilians were murdered. Many had their hands bound when they died.

Once again, this was not a spontaneous act on the part of the soldiers. They had been ordered to murder these people: "When Filipinos are to be killed, they must be gathered into one place and disposed of with the consideration that ammunition and manpower must not be used to excess. . . . Because the disposal of dead bodies is a troublesome task, they should be gathered into houses which are scheduled to be burned or demolished. They should also be thrown into the river."[37] Henry Keyes of the *London Daily Express* wrote, "At last the Japanese have matched the Rape of Nanking [sic]."[38]

The city had not gone unmolested during its three years of occupation. The Japanese forces took the city's food supplies and medicine, ransacked department stores, stole farm equipment, and left the crops to rot in the fields.[39] One person wrote in her diary that she was worried about a lump in her stomach, only to realize that it was her backbone.[40]

It did not need to be this way. General Tomoyuki Tamashita was ordered to slow the American's advance. Rather than fight in the city, he had planned on fighting a guerrilla action from the mountains and jungles. However, Rear Admiral Sanji Iwabuchi, whose job was to destroy the port facilities to prevent the Americans from using them, decided otherwise. Again, MacArthur's arrogance led him to believe that the Japanese would evacuate the city, as he believed when the Japanese attacked in December 1941. He ordered his staff to plan a liberation parade.[41]

With the arrival of US forces, Iwabuchi ordered the destruction of Manila. Fires were set, and buildings dynamited. Along with this destruction came the attack on the city's residents. Once Iwabuchi realized he had lost, he ordered, "The Americans who have penetrated into Manila have about 1000 artillery troops, and there are several thousand Filipino guerrillas. Even women and children have become guerrillas. . . . All people on the battlefield with the exception of Japanese military personnel, Japanese civilians, and special construction units will be put to death."[42]

People were herded into buildings that were then dynamited or torched; those who managed to escape were bayoneted or decapitated. They poured gasoline on women's heads and set their hair on fire. In one house, a hole was made in an upstairs floor where the civilians were forced to kneel. A soldier would then cut off their heads with a sword and kick the body into the hole. The bodies accumulated into a pile forming a pyramid (see figure 18). Two hundred people died this way, but there were nine survivors.[43] Women were

Figure 18. A survivor's sketch shows how the Japanese
killed two hundred men over a hole in a floor, toppling
their bodies to the room below. Source: James M. Scott,
"Battlefield as Crime Scene: The Japanese Massacre in Manila,"
HistoryNet, January 12, 2019, https://www.historynet.com/
worldwar2-japanese-massacre-in-manila/.

locked inside a hotel and kept there to serve as sexual targets for the soldiers.
One woman reported being raped twelve to fifteen times—she was unable to
remember exactly how many.[44] Everyone was murdered:

> They slaughtered Russians, Spaniards, Germans, and Indians.
> Troops killed men and women, the old and the young, the strong
> and the infirm. The butchered victims included two Philippine Su-
> preme Court justices, the family of a Philippine senator, and scores
> of priests. "The list of known dead that has come to my attention
> sounds like a Who's Who of the Philippines," Manila attorney Mar-
> cial Lichauco wrote in his diary on February 19. "Judges, lawyers,
> bank directors, doctors, engineers and many other well-known fig-
> ures in public life now lie rotting in the ruins and ashes of what was
> once the exclusive residential districts of Malate and Ermita."[45]

The survivors were so ravaged by the ordeal that they were unrecognizable as male or female.

Japanese Treatment of POWs

Surrender was not an option for the Japanese soldier. One was expected to fight to the last man. Therefore, if an enemy combatant surrendered to the Japanese, they were viewed with absolute contempt. They were not people but things that did not warrant humane treatment as POWs. Only through their deaths could they redeem themselves. The following diary entry, written by an unknown Japanese soldier, typifies this view and exemplifies the typical means of execution:

> The prisoner who is at the side of the guard house is given his last drink of water. The surgeon, Major Komai, and Headquarters Platoon Commander come out of the Officers' Mess, wearing their military swords. The time has come. The prisoner with his arms bound and his long hair now cropped short totters forward. He probably suspects what is afoot but he is more composed than I thought he would be. Without more ado, he is put on the truck and we set out for our destination.
>
> I have a seat next to the surgeon. About ten guards ride with us. To the pleasant rumble of the engine, we run swiftly along the road in the growing twilight. The glowing sun has set behind the western hills. Gigantic clouds rise before us and dusk is falling all around. It will not last long now. As I picture the scene we are about to witness, my heart beats faster.
>
> I glance at the prisoner. He has probably resigned himself to his fate. As though saying farewell to the world, he looks about as he sits in the truck, at the hills the sea, and seems deep in thought. I feel a surge of pity and turn my eyes away. The truck runs along the seashore now. We have left the Navy guard behind us and now come into the Army sector. Here and there we see sentries in the grassy fields and I thank them in my heart for their toil, as we drive on; they must have "got it" in the bombing the night before last; there were great gaping holes by the side of the road, full of water from the rain. In a little over twenty minutes, we arrive at our destination and all get off.
>
> Major Komai stands up and says to the prisoner, "We are going to kill you." When he tells the prisoner that in accordance with Japanese Bushido he would be killed with a Japanese sword and that we would have two or three minutes' grace, he listens with bowed head. He says a few words in a low voice. He is an officer, probably a flight lieutenant. Apparently, he wants to be killed with one stroke

of the sword. I hear him say the word "one"; the Major's face becomes tense as he replies, "Yes."

Now the time has come and the prisoner is made to kneel on the bank of a bomb crater, filled with water. He is apparently resigned. The precaution is taken of surrounding him with guards with fixed bayonets, but he remains calm. He even stretches his neck out. He is a very brave man indeed. When I put myself in the prisoners' place and think that in one minute it will be good-bye to this world, although the daily bombings have filled me with hate, ordinary human feelings make me pity him.

The Major has drawn his favourite sword. It is the famous masamune sword which he has shown us at the observation stations. It glitters in the light and sends a cold shiver down my spine. He taps the prisoner's neck lightly with the back of the blade, then raises it above his head with both arms and brings it down with a powerful sweep. I had been standing with muscles tenses but in that moment I closed my eyes.

A hissing sound—it must be the sounds of spurting blood, spurting from the arteries: the body falls forward. It is amazing—he has killed him with one stroke.

The onlookers crowd forward. The head, detached from the trunk, rolls forward in front of it. The dark blood gushes out. It is all over. The head is dead white, like a doll. The savageness which I felt only a little while ago is gone, and now I feel nothing but the true compassion of Japanese Bushido.

A corporal laughs: "Well, he will be entering Nirvana now." A seaman of the medical unit takes the surgeon's sword and, intent on paying off old scores, turns the headless body over on its back and cuts the abdomen open with one clean stroke. They are thick-skinned, these keto [hairy foreigner—term of opprobrium for a White man]; even the skin of their bellies is thick. Not a drop of blood comes out of the body. It is pushed into the crater at once and buried.

Now the wind blows mournfully and I see the scene again in my mind's eye. We get on the truck again and start back. It is dark now. We get off in front of Headquarters. I say good-bye to the Major and climb up the hill with Technician Kurokawa. This will be something to remember all my life. If I ever get back alive, it was [sic] make a good story to tell; so I have written it down.[46]

After the Doolittle Raid on Tokyo in April 1942, eight members of Doolittle's force were captured, and three of these subsequently were executed. Of the remaining five prisoners, one died in captivity, but the remaining four were finally liberated in August 1945.[47] Additionally, the Japanese murdered 250,000

Chinese civilians in the areas that might have been used as landing areas by the Doolittle planes.[48] After the raid on Makin Island in August 1942, the Japanese beheaded their American prisoners. Some prisoners had vivisections performed on them: "The two prisoners were dissected while still alive by medical officer Yamaji and their livers were taken out, and for the first time I saw the internal organs of a human being. It was very informative."[49]

During the Battle of Midway on June 4, 1942, pilot Ensign Frank W. O'Flaherty and radio-gunner Bruno P. Gaido were attempting to return to the USS *Enterprise* when their fuel tank was damaged in a skirmish with a group of Japanese Zeros. Running out of fuel, O'Flaherty was forced to ditch his plane in the ocean. Unfortunately, they were rescued by the Japanese destroyer *Makigumo*. After these two were interrogated, they realized the Japanese had suffered a catastrophic defeat. The crew bound the two with rope, tied them to weighted fuel cans, and tossed overboard to drown.[50] The same happened to Ensign Wesley Osmus and his radio-gunner Benjamin R. Dodson Jr. as they were trying to return to the USS *Yorktown*. Hit by a Zero, their fuel tank exploded, killing Dodson. Osmus was rescued by the destroyer *Arashi*, where he was interrogated, murdered, and thrown into the ocean. However, in throwing him overboard, Osmus was able to catch the chain railing. The Japanese then used a fire axe to finally kill Osmus, and his body fell into the ocean.[51]

Hospital ships were deliberately targeted. The hospital ship *Manunda* was in Darwin Harbor in Australia when a Japanese dive-bomber targeted the ship, hitting it and killing and injuring the nurses, medial staff, and crew.[52] On March 20, 1943, the commander of the Japanese First Submarine Force stationed at Truk relayed the order, issued directly from the government, to murder all merchant ship crew members after they had been interrogated and their ships finished being sunk.[53] The submarine service was not the only group to receive this or similar orders, as seventy-two crewmen were taken aboard the Japanese cruiser *Tone* and executed on the orders of Vice Admiral Sakonju, who was later executed as a war criminal for these actions. Pleading that the order came from the highest levels of the government had no effect on the outcome of the trial.[54]

These are not isolated incidents, as I show. Because of the horrors involved with Hitler and the Holocaust, Japan's war crimes tend to get glossed over. However, in their own way, they were just as barbaric as the Nazis, if not more so. Where the Nazis targeted primarily Jews, Slavs, and Romani, they treated POWs in a comparatively humane manner. This was not universally true, though. The Nazis conducted experiments on these prisoners using chemical weapons that tended to be extremely invasive and painful, with multiple subjects dying as a result of the tests.[55] The Japanese, though, treated every person, regardless of background, in the same contemptuous, racist manner. The Geneva Convention of Prisoners of War that had just been implemented barely ten years previously was now nonexistent. Prisoners were no longer people to be taken care of and protected. They were now vermin to be eradicated. They were merely another type of lab specimen on which to be experimented.

In the early days of the war, the speed with which the Japanese captured their prisoners was chaotic. For example, the Japanese took an estimated 350,000 prisoners. Of these, 210,000 (60 percent) were taken in the first three months of the war, growing to 290,000 (83 percent) in the first six months of the war.[56] Even if the Japanese had been predisposed to rendering proper care for their prisoners, the sheer magnitude of prisoners would have made proper treatment a near impossibility.

The camps where the prisoners were housed also contributed to survival rates. Prisoners at the Singapore River Valley camp had work schedules considered leisurely, compared to such camps as the Thai-Burma railway. At the Singapore River Valley, they were also able to pilfer food more easily, which kept them better nourished and therefore in better health.[57] Overall survival rates seemed to depend on three issues: transport on a "hell ship," internment at the Thai-Burma railway camp, and enduring a forced march. Avoidance of these three issues greatly increased a prisoner's overall chance of survival.[58]

Even the Thai-Burma railway camp statistics bear analysis. As Sturma points out, death rates of local populations of Burmese, Tamils, Javanese, Malays, and Chinese ran as high as 50 percent, significantly skewing overall rates.[59] Additionally, many prisoners from those groups were in poor physical condition to begin with and were "recruited" with the promises of better food and facilities than what they actually received. Again, deaths among already-ill workers significantly skew the overall statistics.

When all things are factored together, the issue that most adversely affected prisoner survival in Japanese POW camps was the distribution of Red Cross parcels. It was often these parcels that made the difference between life and death. Unfortunately, too many Japanese kept them from themselves.[60] After the Japanese surrendered, the Allies found enough undistributed Red Cross medical supplies to equip a hospital for three full years of the war.[61]

The brutal treatment of others was not limited to active methods. More passive means were common. While the Japanese touted the Greater East Asia Co-Prosperity Sphere as a means of uniting the various Asian peoples against the threat of further White American and European colonization of Asia, the result was typically less than ideal for non-Japanese countries. Promises of independence usually amounted to little more than empty promises. Instead, the countries involved ended up subservient to the Japanese and serving as supply stations for Japan.

One such example, often overlooked in historical writings, is the Vietnamese starvation of 1944–1945. Often attributed to multiple reasons or minimized or completely denied (as with the Rape of Nanjing), figures show that between 1 and 2 million Vietnamese (of a total population of 20 million) starved during this time.[62] The impact of the Japanese occupation in this tragedy is significant.

Not only did Japan demand the bulk of the area's exports in order to feed its own citizens, but also even with flooding that destroyed nearly 50 percent of the rice crops in certain areas in 1944, the Japanese did not lower their demands for exports.[63] Additionally, because they needed more crops besides rice, the Japanese forced the locals to convert their rice paddies into other crops, such as cotton, jute, hemp, and other nonfood plants, further contributing to famine conditions.[64]

Hell Ships

In order to use the prisoners as slave labor, Japan began transporting them by sea as needed. These ships were unmarked and thus unrecognizable as to the nationality or that they were carrying prisoners or sick and injured. Like the rail transport to Bataan, the prisoners were packed so tightly they were not able to move. Left without adequate food, water, ventilation, or toilet facilities, they were forced to stand in their excrement. Many died as a result of this treatment, and others lost their sanity. They would start biting, clawing, and cutting each other, then sucking the blood. Others tried to escape by climbing ladders out of the holds but were immediately shot by Japanese guards.[65]

Because these ships lacked any identifying markings, many were sunk by US submarines or aircraft. When sunk, other Japanese ships would attempt to rescue the Japanese sailors, but the POWs were left to their fate. Japanese records indicate that twenty-five ships carrying 19,000 prisoners were sunk.[66] US records indicate 126,000 Allied prisoners of war were transported on these "hell ships," with more than 21,000 Americans killed from Allied attacks on these unmarked ships. One such ship, the *Ooryoku Maru*, was carrying 1,619 POWs. Of those, 1,290 survived the sinking. After the wreck, they were denied food or medical treatment for two days. After that, they were given five spoonfuls of raw rice daily and a little water. Those too sick to handle transport to their final destination were told they were being taken to a different location but instead were taken to a cemetery, where they were executed. The others would get so thirsty that they resorted to drinking seawater, with many dying as a result.[67]

In another incident, the *Junyo Maru* transported about 2,200 Dutch, British, American, and Australian prisoners, along with 4,500 Javanese slave laborers, on September 15, 1944. Leaving from the port of Batavia on Java, they sailed for Sumatra. Three days later, the British submarine HMS *Tradewind* intercepted the *Junyo* and sank it. Lost were 1,449 of the 2,200 POWs and 4,171 of the 4,500 slave laborers. It was the worst loss of life from the sinking of a single ship to occur during the Pacific War.[68] The survivors continued the voyage to Sumatra, where ultimately only ninety-six POWs and no slave laborers survived the war. Table 4 shows some statistics from these sinkings.

Table 4. Number of Hell Ship Sinking Casualties Reported in Different Sources

Name of Ship and Date Sunk	Intercepted Japanese Radio Messages	Japanese War Prisoners Information Bureau	American POW Information Bureau, November 15, 1947	US War Department Military Intelligence Division (MID), March 19, 1945	Magic Far East Summary, March 29, 1945	US State Department Summary of POW Ship Sinkings, 1944
Shinyo Maru, September 7, 1944	750 aboard	750	654*	667	667	667
Arisan Maru, October 24, 1944	—	1,775	1,640*	1,795	1,778	1,753
Oryoku Maru, December 14–15, 1944	250	942	284 256*	365	363	1,001 (942 of unknown causes, 59 of illness)
Enoura Maru, January 9, 1945	444	—	411 361*	—	—	—
Brazil Maru	925	—	264 366*	—	—	—
POWs who died during transportation from Olongapo to San Fernando, Philippines, December 1944	—	—	16 21*	—	—	—

* From National Archives, Access to Archival Databases, World War II Prisoners of War Data File, December 7, 1941–November 19, 1946, https://aad.archives.gov/aad/.
Source: Lee A. Gladwin, "American POWs on Japanese Ships Take a Voyage into Hell," Prologue Magazine 35, no. 4 (Winter 2003), https://www.archives.gov/publications/prologue/2003/winter/hell-ships.

Comfort Women

Starting in 1932 with the initial invasion of China, Japan began the reprehensible practice of providing "comfort women" to their troops. This was, in fact, forced sexual slavery to provide a sexual outlet for the Japanese troops, and it lasted until the end of World War II in 1945.

Set up in 1931 theoretically to reduce the rampant sexual assaults against Chinese women, to help with Sino-Japanese relations, to prevent the spread of sexually transmitted diseases, and to minimize the chance of spreading military secrets, the number of stations increased dramatically with the Sino-Japanese War of 1937.[69] A *Written Notification of Warning on the Treatment of the Local Population by Military Units and Personnels* dated June 27, 1938, reads,

> According to various reports, the trigger causing such potent anti-Japanese sentiment is the widespread diffusion of news about rapes committed by Japanese military personnel in various areas. In fact, [these rapes] have fomented unexpectedly profound anti-Japanese feelings.
>
> Along with strict controls on soldiers' individual behavior of the aforementioned type, the provision of facilities for sexual comfort as quickly as possible is of great importance, [as it will] eliminate cases in which people violate the prohibition [on rape] for lack of facilities.[70]

Originally trying to recruit three thousand Japanese women to serve in Shanghai, these efforts were denounced as kidnapping unsuspecting women, thereby adversely affecting the honor of the Japanese army. The result of this was another memo requiring that the women recruited had to be prostitutes already, at least twenty-one years of age, and obtain parental consent to travel overseas. At least initially, it appears that the military attempted to employ women who were already used to sex work. However, as the demand for these women increased, it appeared these regulations were ignored. Documentation from some of these stations demonstrate the women to be no more than girls of sixteen or seventeen.[71]

As the war progressed, so did the need for more women. And with increased need for women came the ignoring of such regulations. Women were now routinely lied to about procuring reputable employment; the age of the recruits became increasingly younger; and other nationalities, including Koreans, Taiwanese, Indonesians, Filipino, and Dutch women, were used.

Kim Busŏn was one such girl. As a fifteen-year-old Chinese girl, she was told she was going to work at a rubber factory. Instead, she was taken to Taiwan, then the Philippines. While in the Philippines, she reported servicing thirty to forty soldiers per day. Besides sex, humiliation was also common. Busŏn again reported being forced by some soldiers to recite the Imperial Citizen's

Charter, which states, "I wish to be a citizen of the Empire," thinking it funny to hear a sex slave request citizenship.[72]

Attempts were made to trick Australian nurses into becoming comfort women by lying about the need for their help with prisoners. This coercion typically failed as the nurses became more aware of what was happening. The Japanese usually started with subjecting the Australian nurses to abominable living conditions and deprivation of sleep and food. Then they would entice them with improved housing conditions and food, with the promise of even better rewards should they cooperate. When bribery failed, they were subjected to even worse conditions than before.[73] Thankfully, these Australian nurses were never raped, though why remains a mystery.[74]

Nobody knows for sure how many women were subjected to such monstrous activity, but estimates range in excess of 200,000, with 80 percent of those being from Korea.[75] Of those, fewer than 30 percent were thought to have survived the war. Now, in addition to the injuries they were forced to endure during the war came the insult, as right-wing Japanese tried to paint the comfort women as merely willing prostitutes, and accounts of forced recruitment were circulated as lies. Additionally, any criticisms of the Japanese in all of this was labeled merely anti-Japanese propaganda put forth by Chinese, Koreans, and left-wing revisionists, or the whole matter was a diplomatic dispute between South Korea and Japan.[76] Attempts to build monuments in honor of the comfort women met Japanese efforts to stop the construction, so they were placed on private property owned by Korean American organizations instead of in public locations.[77]

Unfortunately, the Japanese were not the only ones to employ women in this manner. British forces in North Africa had brothels set up by race for officers and enlisted men. Similar to the Japanese, the British used whomever they could find to work in this capacity, among them four Italian women who apparently had been abandoned by the Italians and their brothels. The British also had brothels in Delhi, until public outcry became too much.[78] The US military contemplated using official military brothels but decided against it due to fear of public opinion. However, the public, nonmilitary brothels were well utilized by the GIs.[79]

After the war and during the occupation, the GIs were not as restrained, and mass rape of the Japanese women occurred frequently. A mere three and a half hours after they landed, two marines forced their way into a private house and raped the mother and daughter there. In another incident, a woman was gang-raped by twenty-seven GIs and nearly died. Young women were raped in front of their parents, while pregnant women were raped in maternity wards. In a ten-day period, there were 1,336 reported rapes in just the area of Yokohama. These figures are comparable to other nations, yet only 247 US soldiers were ever prosecuted for rape.[80]

Unit 731

Unit 731 continued its experiments during the war years. They experimented on any person possible. Besides the methods of torture and execution already mentioned, the Japanese also experimented with biological and chemical weapons on civilians and prisoners of war. They sprayed plague bacteria over cities like Shanghai; they released cholera, dysentery, typhoid, and other virulent microbes into rivers and other bodies of water; they introduced these same bacteria into such food as ripe fruit and rice cakes, which they then fed to civilians, enemy military, and prisoners of war. This food was either handed directly to the unsuspecting individuals or covertly placed into a subject's basket of vegetables. The agents doing this work were often Japanese soldiers disguised as civilians.

Other ways of infecting the targets included giving anthrax-laced chocolates to children and administering tainted vaccines to children. They released infected animals, such as rats, dog, horses, and birds, into the villages of their unsuspecting targets. Unit 731 doctors went to the villages to offer their help, abducted those who fell for their subterfuge, and conducted vivisections on them.[81] They found that the most efficient ways of spreading disease was to drop infected fleas from bombs on the population or to introduce the diseases into the water supply of the intended target.[82] Some of the attacks and the areas they affected were the May 1940 bubonic plague attack of Eastern Zhejiang Province; the November 1941 bubonic plague attack of Changde, Chu County, and Jinhwa County; the 1938 and 1939 cholera attacks of Yangjian County in Guangdong; and the July 1939 salmonella and typhoid attacks against the Russians at Nomonhan.[83] A plan was devised wherein a submarine would approach the California coast and launch a plane that would then drop fleas carrying bacteria over San Diego. Planned for September 22, 1945, the dropping of the atomic bombs rendered this plan useless.[84]

The main facility of Unit 731 was established in 1938 and named the Epidemic Prevention and Water Purification Department of the Kwantung Army, in direct contrast to their activities. In the early phases of the war in 1942, plans were drawn up to release one thousand kilograms of plague-infected fleas on the Bataan Peninsula on ten different occasions. However, the Japanese finished operations there before the plan could be implemented.[85] Plans to use biological weapons targeting various cities, including Dutch Harbor, Alaska, were drawn up. However, these plans were only implemented on select Chinese areas.[86]

The township of Rabaul and the island of Truk became major bases for the Japanese during World War II. While the prisoners of war obtained at the beginning of the war were eventually shipped out of those areas, the areas became the major POW encampments for New Britain and eastern New Guinea. It was here that experiments examining the diets of these men began. They were fed 660 grams of peeled cassava root per meal along with a thin, watery soup and some water. This was done for thirty days, during which two prisoners died. Then the experiment was repeated using 660 grams of unpeeled root. The goal of the experiments was to determine if the prisoners would gain weight while being fed cassava root, which they did not.[87] Additional experiments were performed on five POWs on the immunity to malaria. Of those five, two died, and a third nearly did.[88] These were not the only ones experimented on in such manner.

In addition to this type of experimentation, subjects were used in the training of Japanese physicians. Being trained in how to treat wounded frontline soldiers, physicians performed appendectomies, tracheostomies, bullet removal (after shooting the subject first), and wound repair (again after first cutting the patient). After this, the prisoner would be murdered. Further training involved "suturing of blood vessels and nerves, thoracotomy, celiotomy, craniotomy, blood transfusion, various anesthetizations . . . nephrectomy, performed serially on 'six bodies of prepared material.'"[89] While being called physician-training exercises, they were actually experiments in developing hospital surgical techniques and field surgical procedures. Other experiments included exposure to extremely low temperatures and low pressure to mimic high altitude, salt overdose, drinking only distilled water, and IV air injection.[90]

Eight B-29 airmen were captured when their plane went down. Sentenced to execution, they were turned over for experimentation. Two POWs had a lung removed, another had total gastric resection along with heart surgery, while another prisoner had his gall bladder and half of his liver removed. Another prisoner had his facial nerves severed (trigeminal rhizotomy). Of the final three prisoners, two had surgery on their mediastinum, along with the removal of their gall bladders. All the prisoners died during the operations.[91]

Vivisections were common with the subjects of their experiments. As mentioned in chapter 4, subjects were referred to as "logs," with the joke being that the main instillation at Ping Fang was a "lumber mill."[92] Such dehumanization facilitated experimenting on living persons due to the belief that, especially with the Chinese, torturing these people was of no more consequence than "squashing a bug."[93] "Recruitment" vans drove around the streets of Harbin and randomly abducted people when more subjects were needed.[94] While these vivisections were a daily occurrence, those involved required acclimation to the procedures:

> The fellow knew that it was over for him, and so he didn't struggle when they led him into the room and tied him down. But when I picked up the scalpel, that's when he began screaming. I cut him open from the chest to the stomach, and he screamed terribly, and his face was all twisted in agony. He made this unimaginable sound, he was screaming so horribly. But then he stopped. This was all in a day's work for the surgeons, but it really left an impression on me because it was my first time.[95]

The death toll from Unit 731 is mind-numbing. Estimates for death just by vivisection in Unit 731 range from 3,000 to 10,000. Total estimated dead range from 200,000 to 580,000.[96] The victims were primarily Chinese but also included Russian Jews, Mongolians, Koreans, the mentally disabled, criminals, and Allied and Communist spies.[97] The low estimate of three thousand does not include test subjects from Manchuria during the 1930s, subjects from nearby facilities, prisoners, or the time outside 1941–1945.[98] The effects of these biological agents continue to this day. For example, decaying leg disease is characterized by the spontaneous development of raw, festering wounds on the legs. These wounds rarely heal and are prone to secondary infections. The disease exhibits symptoms that are a combination of anthrax and glanders, diseases that did not exist in China previously.[99]

And what of the Japanese perpetrators of the horrors of Unit 731? Nothing was ever done to them. They struck a deal with the US government to exchange their research for their lives. Some of them were captured by the Russians and tried for war crimes in December 1949. All of those charged were convicted and received sentences ranging from two to twenty years. By 1956, all of these had been quietly repatriated.[100] However, the remainder lived their lives as if nothing had ever happened.

The leader of Unit 731, Dr. Ishii, died of throat cancer in 1959. Other physicians went on to careers in universities, hospitals, pharmaceutical companies, and private practice. Some became the governor of Tokyo, a member of the New York Academy of Sciences, the president of the Japanese Medical Association, and the head of the Japanese Olympic Committee.[101]

The information given to the United States proved important in its biological program. However, its human vivisection information proved to be of limited use to the American government. Regardless, this deal is not without precedent. The United States employed a similar approach in pursuing Nazi scientists to further their research into rocket science. The most notable of these scientists was Wernher von Braun.

The entire existence of Unit 731 was covered up by the US government, as they deemed it a matter of national security, but they also feared the backlash from the general public if they knew. Reports came out about the Russian trial, but the United States claimed the whole matter was just another case of Soviet misinformation. Their official statement regarding the testing on US prisoners of war reads, "That former POW's at Mukden received injections, that conditions in the camp were terrible, and that U.S POWs suffered and died, there is no conclusive evidence that biological warfare experiments were performed on Americans there."[102]

Cannibalism

As horrendous as all these crimes were, perhaps the worst was the Japanese military engaging in acts of cannibalism. As the war progressed, more and more

Japanese troops were bypassed in the island-hopping strategy of the US forces. The bypassed troops were not reinforced due to the United States blocking the passage of Japanese forces. The only real chance these forces had of being reinforced was by submarine, and these were too small and too few to do anything other than prolong the troops' starvation. While few in number, there were isolated instances of cannibalism by Japanese forces.

Of all the Allied nations, only Australia recognized cannibalism as a war crime, possibly because they were the only nation to do any real type of investigation into it.[103] Regardless, cannibalism did occur. As to how this could happen, one has only to look back on the training the military received. The indoctrination of the Japanese military had been so thorough as to cause the soldiers to see non-Japanese as nothing more than cattle. They had no qualms about killing and eating other humans because they didn't see them as humans. The Japanese acted as animals, sending a portion of their forces to gather bodies while the remaining soldiers continued fighting.[104]

A war veteran wrote of one incident in an autobiography. He and a companion came across "many" bodies whose thighs had been hacked off. This had occurred between December 1943 and March 1944 as they were retreating through New Guinea. They next encountered a group of four or five soldiers not from their outfit. The soldiers called them over and offered them some meat that they were eating, saying it was from a very large snake they had killed. However, their demeanor was such that he and his companion felt uncomfortable about the situation. First, if it had been from a large snake, they would not be sharing it. They would be hoarding it for later or for men from their outfit. Second, their actions were unnerving in that they kept smiling and grinning at the two, as if wanting them to join them in their transgression, to become "partners in crime," as he describes it. They politely declined the offer and proceeded on their way.[105]

The most detailed narratives of these incidents come from the Australian army and their descriptions of Australian soldiers who were the victims. These victims were typically members of the same squad as the soldiers who found them and whose reports were filed immediately after the fact, while they still remembered the incidents graphically. However, the victims were just as likely to be from the Indigenous population as from Allied prisoners. One villager testified that a group of Japanese soldiers attacked them on April 12, 1945, and took two villagers prisoner. A party of three went after the Japanese and attacked them at the Japanese troops' encampment. The two prisoners escaped, but two of the rescue party were killed. When they went to retrieve the bodies some days later, they found

> flesh had been cut from the chest, thighs, calves, buttocks and back. His shoulders had been cut through and both forearms were missing. The viscera were intact. The top of the head had been cut off, and

the brain removed. The flesh had been cut with some sharp instrument. . . . I found the bone from a man's forearm. It had been in a
fire and shreds of cooked flesh were still adhering to it.[106]

Some situations became so dire that the Japanese killed each other to gain
food:

> We crossed a small creek and entered a clearing in which we
> counted several "lean to" and one shack. There was smoke rising
> from the Jap occupied shack. We knew there were Japs in the shack
> from their voices. We observed a Jap walking from a brush pile just
> outside the shack. We shot him and three others inside the shack.
> The former had blood on his hands which we found had come from
> a corpse lying under the brush pile. We lifted the brush and found a
> corpse of a Jap and it was crudely butchered. The flesh was cut from
> the legs from the knee to the hips. The calves of both legs were cut
> clear to the bone. The Jap had his hands tied and a rope around his
> waist. We checked the inside of the shack and found bloody mess
> gear and a bloody knife, a crude instrument used in the butchering.
> We reported back to camp. A small patrol was sent back under the
> control of Lt. Miller to investigate further.[107]

There are several types of cannibalism, classified by why it occurs. Nutritional is the consumption of human flesh for nutritional value (i.e., the person is starving, as in the cases cited previously). However, there are additional
types.[108] Ritual cannibalism is tied into religious or magical rituals. Pathologi-
cal cannibalism is related to a disease, like in the case of Jeffrey Dahmer.
    Interestingly, cannibalism has a place in Japanese history.[109] Rolando Este-
ban also cites other sources documenting cannibalism during World War II.[110]
Jeanie M. Welch notes, "The war crimes trials involved some of the most heinous charges imaginable, including mass murder of prisoners of war and *cannibalism.*"[111] Welch then cites a case brought before the war crimes tribunal. It
involved the beheading of two American airmen in August 1944 on Chichijima in the Bonin Islands. Lieutenant General Joshio Tachibana of the Japanese Imperial Army ordered the beheadings. In this incident, nine airmen were
shot down over Chichijima. Eight were captured, beaten, and tortured. Finally,
they were all beheaded. After the beheadings, the airmen were dissected, and
parts of their bodies, their livers and thigh muscles, were eaten afterward.[112]
Defending their actions later, Major Sueo Matoba said, "These incidents occurred when Japan was meeting defeat after defeat. The personnel became ex-
cited, agitated, and seething with uncontrollable rage. . . . We were hungry. I
hardly know what happened after that. We really were not cannibals."[113]
    This incident is notable for one other reason besides the documented
cannibalism. As mentioned, nine airmen were shot down, but only eight were

captured. The ninth airman, upon realizing his plane had been shot, dropped his bombs on his target, then headed as far away from Chichijima as he could before crashing. He managed to escape his burning plane, but two others, the radio operator and the gunner, never left the plane. He landed in the water, where he floated for some time until a submarine, the USS *Finback*, surfaced near him and took him aboard. The pilot's name was George H. W. Bush, future president of the United States and, at the time, twenty years old and one of the youngest pilots in the war (see figure 19).[114]

Another incident involved a group of soldiers in the Philippines who refused to surrender. The Japanese were not prepared to hold out for long and lacked provisions. This posed a problem as a group of around 180 headed into the mountain of Basak on June 1, 1945. They did not leave the area until February 14, 1947, when the thirty-three soldiers who were still alive finally surrendered.[115] Their rice ran out on the mountain on June 1, 1946. They had guns, so they could have hunted, but they were ordered not to in order not to give their position away.[116] The lack of proper nutrition led to diseases like beriberi, salt deficiency, malaria, and diarrhea. They also suffered from a lack of protein. This is what drove them to cannibalism. Again, had they been able to hunt or fish, this would not have been a problem.

The first instance of cannibalism by this group occurred on July 17, 1945, when a woman and three children were captured by a group of Japanese men. The woman escaped, but the children were never seen again and were assumed to have been eaten. More abductions occurred in August 1945; September 24, 1945; and October 10, 1945. Other murders occurred on October 5, 1945; December 27, 1945; early 1946; March 14, 1946; and September 16, 1946.[117]

Figure 19. George H. W. Bush, rescued by the USS *Finback*, escaping the horrors of the Chichijima Incident. Source: Kaleena Fraga, "Inside the Chichijima Incident, George H. W. Bush's Harrowing Escape from Cannibal Enemies During World War II," ed. John Kuroski, All That's Interesting, updated November 7, 2023, https://allthatsinteresting.com/george-bush-cannibalized-chichijima-incident.

These are the instances when they actively killed somebody. They also ate from the bodies of people who died by means other than their own hands more than three times.[118]

Balloon Bombs

While the American firebombing of Japanese cities is well known and is discussed later in this chapter, the Japanese also experimented with firebombing the United States. In 1942, with the completion of the Doolittle Raid on Japan, Japanese leaders were desperate to do something that would help them regain the honor that the attack had cost them. The Doolittle Raid was never going to cause significant physical destruction to Japan. That was never the plan. The plan was to give Americans a psychological boost and show that they could take the war to Japan, and hopefully, it would damage the psyche of the Japanese so that they would act irrationally in order to redeem themselves. Both outcomes did, in fact, occur.

As part of this plan for vengeance, the Japanese came up with a plan to launch balloons carrying incendiary material toward the West Coast of the United States and Canada. By launching them high into the sky, the balloons would catch the jet stream and make the trip east, eventually landing in the western forests of said countries and hopefully starting more fires than could be put out. Additionally, explosives would be attached to kill the population and destroy infrastructure. There was also the possibility of arming the balloons with bacterial loads like they had done to China.

Development of a balloon program started in 1933 and progressed sporadically over the next ten years, with little to show for it. However, the feasibility of such a weapon never left the minds of the planners. Research into the jet stream showed that it was strongest beginning in October, with maximum velocities from November through March at altitudes above 30,000 feet (9.1 kilometers).[119] The thought of launching a balloon and having it reach the US West Coast was at least possible.

Further research determined the optimal size of the balloon as well as mechanisms to keep the balloon at the desired altitude. While there were still a multitude of technical issues to be overcome, steady progress was made toward a workable final product. Finally, everything was ready by November 1944. On November 3, at 5:00 a.m., the first set of balloons was launched. Originally planning to launch 15,000 such balloons, Japan was ultimately able to launch 9,3000.[120]

The first discovery of a balloon by the United States occurred on November 4, 1944, about sixty-six miles southwest of San Pedro, California, when a navy patrol craft found some unidentified debris floating on the ocean. Not much thought was given to it until two weeks later, when additional debris was found. During the next four weeks, more balloon fragments were recovered in

Wyoming and Montana.[121] Fear began to creep into the thoughts of the authorities as they tried to imagine possible scenarios and best ways to deal with them.

Given the dry winter conditions of most of the forests of the western United States, such incendiary devices could pose a significant risk of creating an uncontrollable conflagration. The explosive devices, if they were to land in a populated area, could cause as much destruction from the panic inflicted on the populace as that caused by physical damage to homes, businesses, and factories. And finally, the potential for bacteriological and chemical warfare would leave the United States completely unprepared. The United States decided the best way to deal with the situation was to tell the public as little as possible. By doing that, the Japanese would be kept in the dark about the effectiveness of their new weapon.[122]

One report of the discovery of fragments of a balloon bomb in Thermopolis, Wyoming, made it into the papers and back to Japan, proving that the theory was sound and convincing the Japanese High Command to continue with the program. However, that was the only news to make it out of the United States. This was a major public relations coup for the US government, as the balloons were being found from the Arctic Circle at Holy Cross, Alaska, to the border with Mexico at Nogales, Arizona, and as far east as Grand Rapids, Michigan.[123] Radio broadcasters and newspaper editors across the country were asked not to report on any findings. These voluntary censorships were congratulated later by the Office of Censorship:

> Cooperation from the press and radio under this request has been excellent despite the fact that Japanese free balloons are reaching the United States, Canada, and Mexico in increasing numbers. . . . There is no question that your refusal to publish or broadcast information about these balloons has baffled the Japanese, annoyed and hindered them, and has been an important contribution to security.[124]

Of course, such censorship proved a two-edged sword. By not knowing of the existence of these balloons, the general public was at risk of becoming casualties of them. And that is what happened on May 5, 1945, when a woman and five children between the ages of eleven and fifteen near Lakeview, Oregon, were killed while the balloon they found and were dragging from the woods exploded. As a result of this tragedy, Americans were warned against approaching any strange white balloons they might encounter.[125]

Ultimately, they did little damage, possibly because while the forests of California could become dry during the winter months, the wet winter conditions of the Pacific Northwest would make it extremely difficult for any fires to start. However, the potential for their use as vehicles of biological warfare was evident.

The US government was not taking this threat lightly. The threats named here were just some of the possible uses to which these balloons might be employed. In order of importance, the government listed six possibilities: bacteriological or chemical warfare or both; transportation of incendiary and antipersonnel bombs; experiments for unknown purposes; psychological efforts to inspire terror and diversion of forces; transportation of agents; and anti-aircraft devices.[126]

Responses for all these contingencies were planned out. Firefighting crews were stationed in strategic locations, and word was sent to all agricultural organizations, both governmental and private, to be on the lookout for the first sign of any strange disease in crops or livestock. Fighter planes were scrambled as needed to shoot down any balloons spotted, especially over the Aleutian Islands.[127] Other methods of detection, such as radar, were continuously attempted to be developed but with limited success. Analysis of the sand in the ballast bags of the balloons pinpointed one actual launch site.[128] Ultimately, the ability to consistently identify these balloons proved limited, as by mid-April, all balloon sightings had ceased. Their launches from Japan had terminated, and the war was over before the next winter's launch. It would appear that the balloons' biggest success came in tying up valuable resources, which the government hypothesized might be one of Japan's intentions.

Okinawa

All the aforementioned cases demonstrate the inhumanity of the Japanese military. However, the military could be just as cruel toward their own countrymen when justified by the circumstances. Perhaps in no better situation is this exemplified than the Battle for Okinawa in April–June 1945.

Okinawa is a beautiful island, often called the Hawaii of Japan. The largest of the Okinawa Islands, it forms the southernmost portion of Japan as the Ryukyu Islands, with a population of a half-million people. Prior to its annexation by Japan in 1879, during the Meiji period, it was a semi-independent kingdom under control of both Japan and China.[129] It was this mixed heritage that led the Japanese military, and in fact the entire government, to treat its citizens as second-class Japanese, not true Japanese. A mixture of Chinese, Japanese, and Micronesian ancestry, they were referred to as "little brown monkeys."[130] They were seen as pawns—useful when needed but useless when their utility was over.

When the war was raging and the Home Islands needed the help of the Okinawans, Tokyo preached racial unity, but Okinawans continued their traditional peaceful ways.[131] When the war was nearly over and the invasion of the Home Islands seemed imminent, the inhabitants were treated as cannon fodder, a means to slow down the advancing American tsunami. Their mixed ancestry led the military to see them as potentially disloyal, much as Americans

had seen their own citizens of Japanese ancestry during the Japanese internment in the United States.

The casualties after the battle were enormous. Americans suffered 75,362
total casualties, with 12,520 dead or missing and another 36,631 wounded.[132] The remaining casualties were due to combat fatigue, today known as
post-traumatic stress disorder, a direct result of the stress encountered in trying
to capture the island. The Japanese lost 110,000 soldiers in the battle.

Perhaps the most heart-wrenching statistic is the death of between
100,000 and 150,000 Okinawans, a fourth to a third of the island's population.
Other than the Japanese invasion of China in 1937 and the Japanese conquest
of the Pacific in 1941–1942, no other battle of the Pacific theater to this point
involved so many civilians. They had not been evacuated before the battle (see
figures 20 and 21). They were left where they were and told to defend the island or commit suicide. This is the unspoken reality of Okinawa—the death of
so many of its civilians, brought about by direct order of the Japanese military.

"The army had given us two grenades each. They told us to hurl the first
one at the enemy and to use the second one to kill ourselves," recounted Choho
Zuheran. "Lots of my school friends were told to commit suicide by Japanese
soldiers. At school we had been brainwashed [that] to surrender to [US troops]
would be to disgrace the emperor. Many of my friends died because they were
told to. The army never once tried to protect us. They told people to die and
stole food intended for women and children."[133] Having been told by the Japanese military that the US forces would rape and murder every inhabitant of the
island, hundreds, if not thousands, of the island's residents committed suicide.

Figure 20. Okinawa, Japan,
May 9, 1945. Eleven Oki-
nawa civilians huddled
in this hillside cave were
rescued when a passing
marine patrol heard a baby
crying. After being assured
that no harm would come
to them, they emerged
from their hideout. Here a
leatherneck lends a hand
to a mother and baby.
SOURCE TIM QIU, "THE JAPANESE
CIVILIAN TRAGEDY OF THE PACIFIC
WAR," PACIFIC ATROCITIES EDU-
CATION (BLOG), AUGUST 5, 2021,
HTTPS://WWW.PACIFICATROCITIES
.ORG/BLOG/THE-JAPANESE-CIVILIAN
-TRAGEDY-OF-THE-PACIFIC-WAR.

Figure 21. Wounded Japanese soldier emerging from cave to surrender to US Marines. Smoke (likely phosphorus) is visible near the cave. Official Caption: "Rome. 7/25/45— No suicide for him—A U.S. Marine (L) signals his companions to hold their fire as a wounded Japanese soldier emerges from his cave on Okinawa to surrender. Other Japanese in the cave gave up soon after." Source: Tim Qiu, "The Japanese Civilian Tragedy of the Pacific War," Pacific Atrocities Education (blog), August 5, 2021, https://www.pacificatrocities.org/blog/the-japanese-civilian-tragedy-of-the-pacific-war.

They were placed into an impossible position. Entire families would do so, many voluntarily but many also forced by the military. One person, Ikue Miyazato, tried to commit suicide with her friends, but her grenade did not explode.[134]

In November 1944, the Japanese military ordered that "soldiers and civilians must live and die together."[135] In Okinawa, part of the reason was to keep military secrets, such as battle plans, troop formation, locations, and movements, a secret from the Americans. There was, however, one opinion that this was a premeditated "suicide" in an effort to make the war so costly to the Americans that they would be willing to forgo the "unconditional surrender" doctrine under which they were working and at least allow the emperor to retain his position.

The insanity of all of this is summed up by E. B. Sledge:"The firing continued, and the bullets hit the mark. The wounded Japanese subsided into the muddy little ditch. He and his comrades had done their best. 'They died gloriously on the field of honor for the emperor,' is what their families would be told. In reality, their lives were wasted on a muddy, stinking slope for no good reason."[136]

After the war was over, Okinawa remained a US possession until 1972, at the express wish of Emperor Hirohito.[137] Now, the battle wages over the revisionist history (technically, the conservative viewpoint, which maintains the infallible nature of the emperor and all those serving him) that the mass suicides never occurred and that the few that did happen were "spontaneous" acts of servitude toward the emperor. A lawsuit claiming defamation for publishing a book documenting these military atrocities was dismissed in 2008, stating that the information contained in the book was indeed factual.[138]

GERMANY

Now I look at some of Germany's actions during the war. Hitler made it known early that he was never going to allow the Soviets to survive if he successfully invaded Russia. The Slavs were every bit as repugnant to him as the Jews. The push east displaced millions of Russians. The response to this problem was the same as that for the Jews: the elimination of the Russian people.

Previously, Hitler had made various treaties with no intention of honoring them once they served their purpose. Rather than repeating this action and placating the Russians as the Germans invaded (until such time as they had secured all their goals) and then ridding themselves of the Russians, their scorched-earth campaign left behind an embittered enemy set on revenge. Many Russians cheered the Wehrmacht as it rolled through the countryside, seeing them as liberators from the Communist regime of Stalin. They soon came to know, though, the real German presence. The revenge started almost immediately, with partisan bands forming across the country.

In preparation for Operation Barbarossa, the German invasion of Russia on June 22, 1941, multiple orders went out. One, the Directives for the Treatment of Political Commissars, or the Commissar Order, laid forth the manner in which the Soviet commissars were to be treated when encountered:

> When fighting Bolshevism one cannot count on the enemy acting in accordance with the principles of humanity or International Law. In particular, it must be expected that the treatment of our prisoners by the *political commissars of all types*, who are the true pillars of resistance, will be cruel, inhuman, and dictated by hate.
>
> The troops must realize:
>
> 1. That in this fight it is wrong to trust such elements with clemency and consideration in accordance with International Law. They are a menace to our own safety and to the rapid pacification of the conquered territories.
> 2. That the originators of the Asiatic-barbaric methods of fighting are the political commissars. They must be dealt with *promptly* and with the utmost severity.

Therefore, if taken *while fighting or offering resistance* they must, on principle, be shot immediately.[139]

Another was the Decree for the Conduct of Courts-Martial in the District "Barbarossa" and for Special Measures of the Troop, or the Barbarossa Decree. Similar in content to the Commissar Order, it laid out punishable offenses by enemy civilians, as well as punishment for offenses committed by members of the Wehrmacht:

I

Treatment of punishable offenses of enemy civilians . . .

2. Guerrillas are to be killed ruthlessly by the troops in battle or during pursuit.

3. Also all other attacks of enemy civilians against the Wehrmacht, its members and employees are to be fought by the troops at the place of the attack with the most extreme means until annihilation of the attacker.

4. . . . . Against villages from which the Wehrmacht was insidiously and maliciously attacked, collective punitive measures by force will be carried out immediately. . . .

II

Treatment of punishable offenses of members of the Wehrmacht and its employees against the native population

1. For offenses committed by members of the Wehrmacht and its employees against enemy civilians, prosecution is not compulsory, not even if the offense is at the same time a military crime or violation. . . .

3. . . . . The judge orders the prosecution of offenses against civilians through court-martial only if it is considered necessary for the maintenance of discipline or the security of the troops. . . . Extreme care must be exercised when judging the authenticity of the statements of enemy civilians.[140]

Contrary to the myth of the "clean Wehrmacht," the ground forces were given carte blanche to treat the Russians as they saw fit. Further license to kill came in the form of the "Guidelines for the Behavior of Troops in Russia," which told the Wehrmacht soldiers, "Bolshevism is the deadly enemy of the National Socialist German People. Germany struggles against this poisonous ideology and those who perpetuate it. Our struggle demands pitiless and drastic action against bolshevic [sic] agitators, mercenaries, saboteurs, and Jews, and we must eliminate totally all active and passive resistance."[141] As Omer Bartov writes, the orders given to the Wehrmacht

demanded the execution on the spot of all political commissars in
the Red Army; curtailed martial law as regards the rights reserved
for occupied populations; called for the elimination of all partisans,
political activists, and Jews; and ordered the close collaboration of
army units with the *Einsatzgruppen*, the extermination squads of the
SS and SD. Moreover, the army was ordered both to "live off the
land," which meant that it supplied its needs by extensive plunder
of the impoverished Russian population, and to assist in the ruthless
exploitation of the occupied lands in favour of the German population in the rear.[142]

The Germans during their invasion of Russia destroyed everything and everyone they encountered. In response, Stalin, on July 3, 1941, announced via
radio, "The enemy must be hunted down and exterminated," which only added
more fuel to the fire. Hitler replied, "This partisan war again has some advantages for us; it enables us to eradicate everyone who opposes us."[143] This should
come as no surprise because as Manfred Messerschmidt and Anne Halley write,
"Ideological war produced ideological soldiers: men who were ready to kill
women and children, to kill Jews simply because they were Jews."[144]

One such incident of German savagery involving Russian partisans occurred on the night of November 4–5, 1941, when three German soldiers,
including a colonel, were killed in a firefight with local partisans.[145] The notification of the incident stated that the three had been "'murdered in their
sleep' rather than killed in action, and their dispatches 'stolen' instead of
'captured.'"[146]

In retribution, the Germans sent troops to investigate what had happened.
They gathered everybody they could find to question them and find the cul-
prits. When the mother of one of the partisans was identified, she was shot, and
her house burned. While the remaining women and children were released, ten
other men were executed in reprisal, and the entire village of more than 550
buildings was burned to the ground, along with the livestock and food supplies.[147] As severe as this may seem, Anderson states, "By the standards of German practice during the Second World War Baranivka was a very restrained
reprisal."[148] A few days after Baranivka, Germans entered the village of Velyka
Obukhivka, where they were greeted with resistance. In response, the German
commander reported that "since the populace kept secret the presence of the
partisans, the village was burned down and the populace wiped out [shot]."[149]

That the Wehrmacht actively participated in these operations, despite
their protestations after the war, is evident in the letters written home by the
soldiers, in which they expressed their pride in helping defend their homeland
against Jews and beasts. Two such letters, one from Private Fred Fallnbigl, written from the front in mid-July 1941 and the second from a Wehrmacht captain
in mid-February 1943, exemplify the views of these men:

Now I know what war really means. But I also know that we had been forced into the war against the Soviet Union. For God have mercy on us, had we waited, or had these beasts come to us. For them even the most horrible death is still too good. I am glad that I can be here to put an end to this genocidal system.[150]

May God allow the German people to find now the peace of mind and strength which would make it into the instrument needed by the Führer to protect the West from ruin, for what the Asiatic hordes will not destroy, will be annihilated by Jewish hatred and revenge. The belief at the front is unshakable, and we all hope that, as Göring has said, with the rising sun the fortunes of war will again return to our side.[151]

Perhaps the most notable incident was the role the Wehrmacht played in the annihilation of Jews in the ravine at Babi Yar in Kyiv.[152] On September 19, 1941, the German army captured Kyiv, Ukraine. On Friday, September 26, posters went up around the city ordering the Jews to appear near a Jewish cemetery. That Monday morning, September 29, more than 33,000 people had gathered (one source quoted 33,771), consisting of mainly women, children, and the elderly.[153] There, under the direction of Police Battalion 45, the residents were made to wait in a meadow while groups were taken to the ravines of Babi Yar and gunned down with machine-gun fire, within earshot of those waiting their turn.[154] The massacre lasted two days. However, as heinous as this incident was, it was not the largest single massacre of the war instituted by Germans. That occurred in November 1943 with the Harvest Festival (*Erntefest*) and the elimination of 42,000 Jews in the Lublin district of Poland.[155]

Babi Yar was not an isolated event: Reprisals for partisan activities were usually swift and brutal, resulting in similar, if not to the same extent, reprisals against entire villages. Furthermore, in similar situations throughout the Eastern Front, soldiers eagerly volunteered for such duty and had to be ordered not to participate. They often photographed these events, sending the film home to be developed.[156]

On March 22, 1943, another horrendous incident occurred at the village of Khatyn, when it was attacked by the Germans in retaliation for the supposed murder of German Olympic shot putter Hans Woellke. Having been told by General Wilhelm Keitel, head of the Nazi armed forces, "Since we cannot watch everybody, we need to rule by fear," the Wehrmacht was ready to unleash hell upon the village.[157] The Germans descended on the village like they had so many times before. In Belarus, an estimated 629 villages were razed. Another 5,454 villages throughout Russia were burned to the ground.[158] In Khatyn, the villagers, 149 in total, were herded into a large barn, where they were forced to wait for about an hour while gasoline was poured onto stacks of hay around the barn. The barn was finally put to fire, and the villagers burned to death. Some tried to flee the barn but were immediately shot by the German soldiers. Of the 149, only 2 children survived, with 6 people witnessing the incident—1 adult and 5 children.[159] Before they left the village, the soldiers looted it of anything of value.[160]

The SS and Wehrmacht were not the only groups responsible for these deaths. The Uniformed Police, whose primary responsibility was police and security actions in rear areas already secured by the Wehrmacht, frequently helped in the liquidation of Poles and Jews in the course of securing these areas. Such actions were often performed under the guise of antipartisan warfare. Heinrich Himmler, in his notes for a meeting with Hitler, wrote, "Jewish question—to be exterminated as partisans."[161] Police Battalion 41 was frequently called on to help the Wehrmacht combat remnants of Polish forces in the army's rear areas during the invasion of September and October 1939.[162] Likewise, Reserve Police Battalion 101 deported 45,200 Jews to the extermination camp at Treblinka and shot an additional 38,000.[163]

Blacks were another minority to be targeted for extermination. The history of anti-Black racism closely follows the history of anti-Semitism in Germany, with Blacks and Jews being racially linked: "According to a theory popularized in Nazi circles, the Jew and the Negro were in fact related: the Jew was of an impure race, consisting of a hybrid between the Negro and the oriental."[164] Consequently, similar racial laws were instituted against them, and similar punitive actions were taken.[165] Race, to Hitler, seemed to be the most important aspect of civilization. As early as the 1920s, he wrote that the mingling of Aryan blood with that of any other race would lead to the downfall of the Aryan race:

> Historical experience . . . shows with terrifying clarity that in every mingling of Aryan blood with that of lower peoples the result was the end of the cultured people. . . . The result of all racial crossing is therefore in brief always the following: (a) Lowering of the level of the higher race; (b) Physical and intellectual regression and hence the beginning of a slowly but surely progressing sickness. To bring about such a development is, then nothing else but to sin against the will of the eternal creator.[166]

He further wrote that France was an enemy not just because of the presence of the Jews but also because of their acceptance of Blacks:

> For this reason, France is and remains by far the most terrible enemy. *This people, which is basically becoming more and more negrified, constitutes in its tie with the aims of Jewish world domination an enduring danger for the existence of the white race in Europe.* For the

contamination by Negro blood on the Rhine in the heart of Europe is just as much in keeping with the perverted sadistic thirst for vengeance of this hereditary enemy of our people as is the ice-cold calculation of the Jew thus to begin bastardizing the European continent at its core and to deprive the white race of the foundations for a sovereign existence though infection with lower humanity.[167]

Blacks were subject to the same racial-hygiene laws as the Jews, including prohibition from marrying Aryans, and suffered the same penalties for breaking these laws. They were also subjected to similar medical procedures and experimentation, including forced sterilizations.[168] They were carried off to concentrations camps, just like the Jews. However, even in the camps, the Blacks were segregated from the other prisoners. One Black German stated, "They considered us to be subhuman beings like animals, chimpanzees."[169] Additionally, Black POWs were often executed, with one commander stating, "Kill all colored prisoners on sight, because they stink."[170] Robert W. Kesting lists numerous such incidents involving the murder of Black POWs.[171] Perhaps one reason this minority has been overlooked is that the total number of Black casualties is estimated at 55,000, less than 1 percent of the total number of Jews killed during the Holocaust.[172]

German civilians also weren't safe from the Nazi regime. Robert Gellately points out that Gestapo records from Würzburg in Lower Franconia, an area not known for its support of Nazism or anti-Semitism, show that of 210 cases of "race defilement," "being friendly to the Jews," or comments in opposition to the official policies about the Jews, 70 percent of those cases were reported by "ordinary Germans" (i.e., Germans who did not work for the government or were not registered Nazis). As Gellately correctly points out, the Gestapo could not have enforced the Nazi rules on racial hygiene without the express help of the ordinary citizen.[173]

In all, the total number of civilians dead is estimated at 20,946,000. Poland was the hardest hit of any of the nations involved in the war, suffering a casualty rate of 15 percent, including 5.8 million dead, predominantly Jews. Europe's 9.2 million Jews were reduced by nearly two-thirds.[174] This is proportionately greater than those civilians killed by the regimes in either Communist China or the Soviet Union.[175]

While the majority of Nazi atrocities were directed toward Jews and other undesirables, there were other targets along the way. However, oftentimes the group responsible was the SS, the Nazi stormtroopers. The SS was known for its fanaticism in enforcing Nazi policies. Formed in 1925 to serve as Hitler's personal bodyguards, its numbers were greatly increased in 1929 when Heinrich Himmler took over as the head of the organization. Himself a fanatical follower of Hitler, he expected nothing less from his members. Such fanaticism was evident in the oath of loyalty they were required to take: "I vow to you,

Adolf Hitler, as führer and chancellor of the German Reich, loyalty and brav-ery. I vow to you and the leaders you set before me, absolute allegiance unto death. So God help me."[176] The Waffen-SS served as the combat branch of the SS, with their zealotry leading them not only to fight without the thought of surrender but also to commit barbarities in doing so.

On June 10, 1944, immediately following the Allied landings at Nor-mandy, at about 2:00 p.m., 120 members of the Der Führer regiment, Waffen-SS tank division Das Reich, came into the town of Oradour-sur-Glane and proceeded to encircle the town, blocking all points of egress. All the res-idents, as well as those men working in the fields, were called into the town center. A few hid, but most complied with the order. An hour later, the women and children were forced into the church and shut inside. The SS searched the town for hidden weapons, while a soldier set up a machine gun. They then opened fire on the residents. After covering the bodies with straw, they were set on fire. The village was then systematically searched. Buildings were set on fire, and villagers attempting to flee were shot dead.

By morning on June 11, Oradour had ceased to exist, reduced to a smol-dering pile of rubble. Of the 642 people dead, 393 were residents of Oradour proper, 167 were from surrounding hamlets, 58 were from other villages, and 24 were from unknown locations. The few who survived were hidden by the other dead bodies lying on top of them. Those who escaped the inferno climbed out the top windows of the building and were only wounded when shot or some-how hid when the villagers were initially rounded up.[177]

Another incident occurred on December 17, 1944, in the heat of the Battle of the Bulge, when elements of the First SS Panzer Regiment, First SS Panzer Division Leibstandarte SS Adolf Hitler (LSSAH), captured 113 US sol-diers. After assembling them in a field near the town of Malmedy, they were machine-gunned down. Eighty-four were murdered; survivors played dead or were buried under the corpses, as in Oradour.[178]

Given the reputation that the SS had acquired as a result of massacres like these two, it should come as no surprise to know that GIs tended not to take prisoners of SS members. Two different brigades, the Dirlewanger (later known as the Thirty-Sixth Waffen-Grenadier-Divison der SS) and Kaminski [the Twenty-Ninth Waffen-Grenadier-Division der SS (Russische 1)], were es-pecially notorious for being comprised of ex-Einsatzgruppen, Russian POWs, and released criminals, whose records led several of their members to be exe-cuted.[179] In this aspect, the Wehrmacht was not "clean" from atrocity but was more so than the SS, and the American troops tended to be more civil toward Wehrmacht prisoners than they were toward the SS.

Though the Nazis are known primarily for the death of six million Jews, Nazi physicians and scientists performed experimentation, dubbed "research," on prisoners also. These experiments ran the gamut of physical and chemi-cal experimentation. For example, tests on the exposure to chemical weapons, including mustard gas and phosgene, were followed by the application of various agents in an attempt to identify a treatment for gas exposure.[180] Dr. Josef Mengele, chief physician at the Birkenau prison camp, experimented extensively on twins. A more exhaustive list of experiments was given at the Nuremberg Trials of the Nazi doctors:

> "The inmates [of concentration camps] were subjected to cruel experiments; victims were immersed in cold water until their body temperature was reduced to 28° C., when they died immediately. Other experiments included high-altitude experiments in pressure chambers, experiments to determine how long human beings could survive in freezing water, experiments with poison bullets, experiments with contagious diseases, and experiments dealing with sterilization of men and women by x rays and other methods." There are many references to these and other experiments in the judgment, and to the responsibility of the various Nazi leaders and organizations.
>
> The experiments, which will be dealt with in more detail in a later article, concern toleration of high altitudes, resistance to freezing, malaria infection, mustard gas, treatment with sulphonamides and other substances of artificially inflicted wounds, the regeneration of tissues in artificially inflicted wounds, the potability of sea water, epidemic, jaundice, methods of sterilization, epidemic typhus, effects of poison, the treatment of burns by incendiary bombs. Further crimes consist of the murder of 112 Jews to produce a skeleton collection, and the various "euthanasia" programmes for removing scores of thousands of Poles with tuberculosis and millions of "useless eaters."[181]

Weindling and colleagues total the victims of Nazi experimentation at 27,759.[182] This is, however, from one study only, and their results should be considered a bare minimum. These experiments were the focus of the Nuremberg Medical Trials, which opened on December 9, 1946. Twenty-three German physicians and administrators were tried for war crimes and crimes against humanity. Sixteen doctors were found guilty, with seven sentenced to death. The executions occurred on June 2, 1948.[183] Some of the medical staff were also tried after the war. Of 265 accused, 9 perished during the war, and 72 faced trial; of those who went to trial, 25 were executed, 4 were imprisoned for life, 31 were convicted and received sentences of varying lengths, and 12 were acquitted. Of the remaining, 125 were never captured or tried, and 20 committed suicide, leaving 39 whose fates are not known.[184]

Anti-Semitism

A common theme in Germany at this time was the need for "living space"—*Lebensraum*. While this notion was popularized by Adolf Hitler, it had been talked of since at least 1900, when it was mentioned by Ludwig Woltmann.[185] As cities became more and more crowded, the rural areas were viewed as appropriate living areas to help restore the traditional living situations and values.[186] Of course, the acquisition of these lands and elimination of inferior races and peoples would come at the expense of those living in the lands east of Germany—Slavs, Jews, and Romani.

Similar to the Jews, Slavs and Romani were both viewed as inferior, so Hitler held no qualms about eliminating them.[187] As such, while war on the Western Front was a more typical conflict, tending to observe established norms, the Eastern Front was not just a war of territorial conquest but also a war of annihilation. Few prisoners, military and civilian, were taken. Consequently, the Nazi goals in the West against France and England were radically different from the goals in the East against Russia: France would be removed as a world power, and Britain would be left a naval power but with no further intervention in Continental European politics. Both countries would be allowed to exist but in different hierarchal positions. However, as Russia was a country of inferior Slavs, it would no longer exist. Thus, Hitler's war against Russia was a more gruesome "crusade" to exterminate the Slavs and colonize their land.[188]

The results of these viewpoints were that Germany's war aims and Nazi racial goals were inseparable, making World War II fundamentally different from World War I. It also explains the disparity in civilian deaths between the two wars: Civilians accounted for nearly 60 percent of the total casualties in World War II, whereas in World War I, the total was 5 percent.[189] Indeed, Hitler bragged about the upcoming brutality in a speech on August 22, 1939, to the commanders of the armed services, indirectly comparing himself to Genghis Khan: "Our strength is in our quickness and our brutality. Genghis Khan had millions of women and children killed by his own will and with a gay heart. History sees only in him a great state builder."[190] And while decrying the descendants of that famous Mongol, Hitler was quick to implement variations of the strategies and tactics he made famous against the modern descendants of Khan.[191]

Of course, the elimination of Jews and other undesirables did not occur overnight, nor was the decision to do so made quickly either. Originally, Jews and other undesirables were to be moved east, out of Germany. As Germany obtained more eastern territories, these populations continued to be moved further east or to other countries, such as Madagascar, much like the Native Americans during the US westward expansion through the late 1800s.[192] As the war progressed with the conquest of Poland and occupation of Russia, it became increasingly obvious to the Nazis that there would be far too many Jews to simply move and that more drastic measures would be required to remove the Jews. The solution had already been instituted; it need only be applied to a different target.

In the summer of 1943, Erna Petri, the wife of a high-ranking Nazi officer, saw six frightened children by the side of the road. She realized that they must be Jewish children who had escaped from the nearby death camp. She took them to her house, fed them, and helped them feel comfortable; then she took them outside and shot each one in the back of the head. How could a person who had just comforted these children then turn around and execute them? In her own words, "I had earlier been so conditioned to fascism and the racial laws, which established a view towards the Jewish people. As was told to me, I had to destroy the Jews. It was from this mindset that I came to commit such a brutal act."[193]

While the European Jews suffered the majority of persecution, Jews in other areas were by no means safe. The Nazis teamed up with the Arab nations to target the Jews in those areas. The alliance between Muslims and Nazis seemed only natural, as both viewed Jews as their avowed enemy. Dr. Zeki Kiram, a Muslim publicist from Syria, stated, "Is this man [Hitler] not called by God? To save the German people from the snare set in the name of humanity by the Jews and their various organizations. These Jewish organizations, which externally seem to bring blessings, are in reality pursuing destructive goals."[194] As Klaus-Michael Mallmann and Martin Cüppers write, "It was not despite but because of their virulent anti-Semitism that Hitler and the Germans gained the sympathies of the Muslims of the Middle East."[195]

With the rapid advance of General Erwin Rommel in North Africa in 1942, security police and SS Einsatzkommando groups were sent to prepare for the task of dealing with the Jews.[196] The capture of Tobruk seemed to signal the implementation of plans to exterminate the Jews in the Middle East.[197] In preparation for this, on the evening of July 7, 1942, Voice of Free Arabism, a Nazi propaganda radio station in the Middle East, broadcast the program *Kill the Jews Before They Kill You*:

> A large number of Jews residing in Egypt and a number of Poles, Greeks, Armenians, and Free French have been issued with revolvers and ammunition [to fight] against the Egyptians at the last moment, when Britain is forced to evacuate Egypt. . . . In the face of this barbaric procedure by the British we think it best, if the life of the Egyptian nation is to be saved, that the Egyptians rise as one man to kill the Jews before they have a chance of betraying the Egyptian people. It is the duty of the Egyptians to annihilate the Jews and to destroy their property. You must kill the Jews, before they open fire on you. Kill the Jews, who have appropriated your wealth and who are plotting against your security. Arabs of Syria, Iraq, and Palestine, what are you waiting for? The Jews are planning

to violate your women, to kill your children and to destroy you. According to the Muslim religion, the defense of your life is a duty which can only be fulfilled by annihilating the Jews. This is your best opportunity to get rid of this dirty race, which has usurped your rights and brought misfortune and destruction on your countries. Kill the Jews, burn their property, destroy their stores, annihilate these base supporters of British imperialism. Your sole hope of salvation lies in annihilating the Jews before they annihilate you.[198]

While the Jews were the predominant minority to be targeted, they were by no means the only one. The Romani and Sinti were other heterogeneous groups that presented problems similar to Jews and the "purity" of the race (i.e., "mixed race" and "pure breeds").[199] The mixed-race Romani were viewed as an outside race prone to commit crimes and were subject to the same treatment as the Jews, including registrations, deportations to concentration camps, slave labor, and random murders.[200] In short, the mixed-race Romani were deemed antisocials, or *Asoziale*.[201]

However, an exception to this was the fact that at least some Nazis, including Heinrich Himmler, viewed the "purebred" Romani as worthy of collecting as an "exotic and romantic collector's" item.[202] Dr. Robert Ritter, the head of the Racial Hygiene and Biology of the Population Research Unit, concluded that only about 10 percent of all Romani were of pure blood. These "racially pure" Romani were not deemed a threat to the Aryan (German) people and were generally not affected by the laws for the remaining Romani.[203] However, approximately 15,000 Romani in Germany were exterminated during Nazi rule, and between 250,000 and 500,000 total Romani were killed.[204]

UNITED STATES
US Japanese Internment
Racism has traditionally been a major cause of war. However, in World War II, this correlation reached new levels, due both to the amount of racism involved and to the technology with which to prosecute these beliefs. White prejudice against non-Whites in the United States has been previously explored with the American Indian Wars of the 1800s and slavery until the 1860s, but US racial views toward the Japanese are perhaps not so well-known.

The arrival of Japanese immigrants began in 1884 and continued through the late 1880s. By the early 1900s, the Japanese population in and around Los Angeles County totaled 72,157. Quotas were enacted, and finally, immigration from Japan was halted completely in 1924.[205] Following the mass immigration of the Chinese, the anti-Asian sentiment was well established. Caleb Foote wrote that by 1900, mass meetings demanding their exclusion from activities and segregating their schoolchildren were common. At this time, the California legislature had seventeen anti-Japanese bills before it, which were only defeated after intervention by Theodore Roosevelt.[206]

After Pearl Harbor, anti-Japanese sentiment reached a frenzy. Upstanding citizens were suddenly seen as fifth-column spies beholden only to the emperor. For this very reason, California congressman Leland Ford called for the forced relocation of all Japanese Americans to inland concentration camps.[207] The US Department of the Treasury froze the assets of all Japanese, whether resident aliens born in Japan or US citizens, while the Department of Justice arrested 1,500 resident-alien civic and religious leaders suspected of posing an immediate threat to the United States.[208] Though some of the initial sentiments toward the Japanese were such that people were calling for tolerance, these sentiments quickly turned against the Japanese.[209] A *Los Angeles Times* editorial noted,

> A viper is nonetheless a viper wherever the egg is hatched. . . . So, a Japanese-American born of Japanese parents, nurtured upon Japanese Traditions, living in a transplanted Japanese atmosphere . . . notwithstanding his nominal brand of accidental citizenship almost inevitable and with the rarest exceptions grows up to be a Japanese, and not an American. . . .Thus, while it might cause injustice to a few to treat them all as potential enemies, I cannot escape the conclusion . . . that such treatment . . . should be accorded to each and all of them while we are at war with their race.[210]

Americans at that time felt they were at war with a *race*, not a *country*. The paranoia surrounding the Japanese was so great that Lieutenant General John L. DeWitt, in charge of the US Army's Western Defense Command, making him responsible for the defense of the coastline from California to Washington and all of Alaska, ordered the cancellation of the Tournament of Roses Parade and the Rose Bowl football game scheduled for January 1, 1942.[211] DeWitt, in 1943, stated, "I don't want any of them [persons of Japanese ancestry] here. They are a dangerous element. There is no way to determine their loyalty. . . . It makes no difference whether he is an American citizen, he is still a Japanese. American citizenship does not necessarily determine loyalty. . . . But we must worry about the Japanese all the time until he is wiped off the map."[212]

At one point, singer Woody Guthrie broke up a mob in Los Angeles who had smashed the front window of a bar owned by Japanese American citizens. He took his guitar and began singing, calming the mob such that by the time the police arrived, the mob had dispersed. Along the Tidal Basin in Washington, DC, four of the three thousand cherry trees planted nearly thirty years earlier were chopped down because they had been presented by Japanese citizens.[213] And while five thousand Germans and Italians had been gathered under circumstances similar to the Japanese, the Japanese had one distinct disadvantage over the Germans and Italians: their facial features. This fact alone served as the equivalent of painting bull's-eyes on them as being "others."[214]

Chemical warfare utilized the metaphor of insects and vermin needing to be exterminated. This was especially true of the American attitude toward the Japanese after Pearl Harbor. At once backward and intellectually stunted and yet intelligent and cunning, they were nonetheless vermin infesting the Pacific and threatening to take over the world with their infestation. Figure 22 is a very graphic example of this sentiment. This was much more the case with the Japanese than the Germans. This also accounts for the attitude toward trophy hunting in the Pacific theater and Germany. German trophies tended to be guns, rifles, flags, medals, and similar items. Japanese trophies often consisted of teeth or skulls. As much as the Americans disliked the Germans, the GIs tended to not desecrate the bodies of their soldiers.

At first, the Japanese of the West Coast were encouraged to voluntarily re-locate to camps further inland. Then on February 19, 1942, President Franklin D. Roosevelt issued Executive Order 9066, which authorized the forced relocation of all persons viewed as threats to national security. While not naming any specific group, the order was clearly aimed at the Japanese Americans on the West Coast. Upheld by the US Supreme Court, eventually some 122,000 men, women, and children were forcibly removed from their homes and sent to

Louseous Japanicas

The first serious outbreak of this lice epidemic was officially noted on December 7, 1941, at Honolulu, T. H. To the Marine Corps, especially trained in combating this type of pestilence, was assigned the gigantic task of extermination. Extensive experiments on Guadalcanal, Tarawa, and Saipan have shown that this louse inhabits coral atolls in the South Pacific, particularly pill boxes, palm trees, caves, swamps and jungles.

Flame throwers, mortars, grenades and bayonets have proven to be an effective remedy. But before a complete cure may be effected the origin of the plague, the breeding grounds around the Tokyo area, must be completely annihilated.

Figure 22. In 1945, US Marine Sergeant Fred Lasswell praised efforts to annihilate "Louseous Japanicas." Note that the proposed solution for the eradication of this "insect" consists of annihilating them in their home of Tokyo. Original by Fred Lasswell, *Bugs Every Marine Should Know*, Leatherneck 28 (March 1945): 37. SOURCE: EDMUND P. RUSSELL III, "'SPEAKING OF ANNIHILATION': MOBILIZING FOR WAR AGAINST HUMAN AND INSECT ENEMIES, 1914–1945." JOURNAL OF AMERICAN HISTORY 82, NO. 4 (MARCH 1996): 1506, HTTPS://DOI.ORG/10.2307/2945309.

concentration camps in remote, desolate areas in Wyoming, California, Utah, Arizona, Colorado, Idaho, and Arkansas.[215]

Karina V. Korostelina gives four possible reasons for the forced evacuation: military necessity, the protection of those evacuated, political and economic pressures, and racial prejudice.[216] The military gave two reasons for the evacuations: military necessity and the protection of the Japanese population.[217] While the removal of a select few might have been warranted (there had been spies at Pearl Harbor giving details of the military schedule and other pertinent information), political pressure and blatant racism were the overwhelming reasons for this situation. The attempt to portray this exodus as protective custody from actions that might be taken by anti-Japanese mobs was easily seen for what it really was: a flimsy excuse. Author Floyd Schmoe states,

> The reason for evacuation considered most valid for many personas is that of "protective custody"—the Japanese must be taken into camps and guarded for their own protection. But what a breakdown of the Anglo-Saxon conception of justice in a democracy such thinking betokens. . . . The very words "protective custody" (*Schutzhaft*) were "made in Germany," not here. How could it accord with American justice that if a man were dangerous to his neighbors they should be put into custody rather than he?[218]

To further illustrate the hypocrisy shown in this line of thinking, California congressman Leland Ford stated,

> To prevent any fifth column activity . . . all Japanese, whether citizens or not, be placed in inland concentration camps. As justification for this, I submit that if an American born Japanese, who is a citizen, is really patriotic and wishes to make his contribution to the safety and welfare of this country, right here is the opportunity to do so, namely, that by permitting himself to be placed in a concentration camp, he would be making his sacrifice, and he should be willing to do it if he is patriotic and working for us.[219]

Besides Schmoe, there were a few others who saw that the relocation had less to do with security and more to do with politics. FBI director J. Edgar Hoover said the Western Defense Command's intelligence was flawed by "hysteria and lack of judgment" and that the push for the relocation of the Japanese population "is based primarily upon public and political pressure rather than on factual data."[220] Likewise, author and cofounder of the American Civil Liberties Union Roger Baldwin believed that "military necessity had less to do with their unprecedented treatment than race prejudice."[221]

Life in the camps was anything but pleasant. One resident remembered, "Every place we go we cannot escape the dust. Inside of our houses, in the laundry, in the latrines, in the mess halls, dust and more dust, dust everywhere."[222] The buildings, made of wood and tar paper, were meant to be temporary at best; some thought they would be lucky to last the duration of the internment.[223] George Takei of *Star Trek* fame was four years old when his family was relocated to Camp Rohwer in Arkansas. Eighty-three years later, he still bore the scars of the ordeal: "She gathered rags and tore them up into strips and braided them into rugs so that we would be stepping on something warm. . . . The horse stalls were pungent, overwhelming with the stench of horse manure. The air was full of flies, buzzing. My mother, I remember, kept mumbling 'So humiliating. So humiliating.'"[224] As a child, he was able to find pleasures that only a child could find ("Camp Rohwer was a strange and magical place. We'd never seen trees rising out of murky waters or such colorful butterflies. Our block was surrounded by a drainage ditch, home to tiny, wiggly black fishies. I scooped them up into a jar. One morning they had funny bumps. Then they lost their tails and their legs popped out. They turned into frogs!"), but the trauma remains, as it does for most, if not all, survivors.[225]

Alaska Native Relocation Camps

The horror of forced relocation did not stop with the Japanese Americans. Alaska Natives were also forced to abandon their homes for their own "safety." In 1935, in an address before a committee of the US House of Representatives, General Billy Mitchell stated, "I believe that in the future, whoever holds Alaska will hold the world. I think it is the most important strategic place in the world."[226] While the US military may not have been as convinced about its importance as General Mitchell was, Alaska was still seen as at least marginally strategically important in its ability to control Pacific transportation routes. Looking at a map of the world from a top-down perspective (see figure 23), one can see how much closer the Northern Hemisphere countries are than from a traditional head-on view (see figure 24).

When the Japanese invaded Alaska in 1942 as a diversionary part of the Midway attack, the military quickly realized its importance. In an effort to avoid civilian losses during the battle that would ensue, the military forcibly evacuated the Alaska Natives who populated the Aleutian Islands. Like the Native Americans in the Lower 48 (continental United States), the Alaska Natives had been treated horribly by White colonists. The first Russian explorers arrived in 1741 and, finding the area rich in sea otters, promptly colonized the area. In doing so, they decimated the Aleut population by 80–90 percent, from 10,000 to about 2,000.[227] The rest adopted Russian surnames and converted to Russian Orthodoxy, which became the center of their lives. They were, however, allowed to keep their Native language. When the United States purchased Alaska from Russia in 1867, the government promptly banned the

Figure 23. View of Alaska from the North Pole showing proximity to other countries. Source: Arctic Ice, "Brief Introduction," April 12, 2013, https://arcticicesea.blogspot.com/2013/04/ale-o-co-tu-chodzi.html.

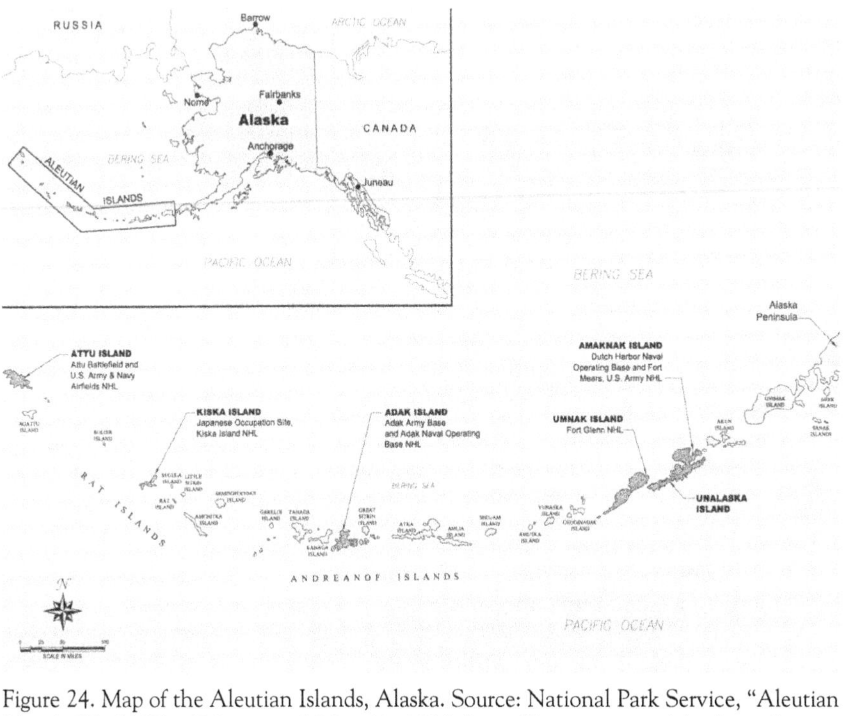

Figure 24. Map of the Aleutian Islands, Alaska. Source: National Park Service, "Aleutian Islands World War II," accessed May 21, 2025, https://home.nps.gov/aleu/learn/history culture/places.htm.

speaking of both Russian and Aleut as well as their art and music, and Methodist missionaries again attempted to convert the "heathens" to their brand of Christianity.[228] The banning of the Aleut language in these early days combined with the forced relocation resulted in the death of the Aleut language.

By the time World War II broke out, the Unangan (Aleut for "we the people") were being supervised by multiple government agencies: the Department of the Interior (DOI), the Office of Indian Affairs (OIA), the Fish and Wildlife Service (FWS), and the military. Therefore, when the Japanese attacked the Aleutian Islands starting on June 3, 1942, no relocation plan had been developed. The resulting evacuation was chaos. The military believed that at least the women and children should be evacuated. John Collier, commissioner of the OIA, was torn, as he did not want to forcibly remove the Alaska Natives, but he did not want to leave them in an active war zone either. Alaska Territorial Governor Ernest Henry Gruening was of like mind that they should not be relocated, stating that the "dislocation resulting from forced evacuation would be a greater damage and involve greater risks to the ultimate welfare of the people than the probably [sic] risks if they remain where they are."[229] Brigadier General Simon Bolivar Buckner, who led the Alaska Defense Command, agreed with Gruening, stating that it "would be 'a great mistake' to evacuate all Unangan beyond Unalaska as 'evacuating them would be pretty close to destroying them, that they now live under conditions suitable to them' and they would 'deteriorate' if they encountered 'the white man . . . [and] fall prey to drink and diseases.'"[230] He was proven absolutely correct.

The OIA decided that the evacuation plans should be carried out by the US Navy. However, Rear Admiral Charles S. Freeman had no desire to do so and refused to accept any responsibilities for the decision-making process. The entire question was moot as the Japanese began their invasion on June 3, 1942. The residents of the island of Atka had been prepared to evacuate, but with nobody to coordinate the process, they watched as the military burned their village to the ground to prevent its use by the Japanese. They were then forced onto a military ship and removed from their home, bringing with them only one suitcase of personal belongings, along with some blankets.[231] They didn't even have time to get any of the religious icons from their church. Those items they could not bring with them were lost forever.[232]

The evacuation ships were floating death traps of overcrowded and unsanitary conditions. A baby girl died of pneumonia and had to be buried at sea.[233] They were loaded into the cargo hatch and subjected to the cold. "No privacy, nothing. Just like animals," recalled Ella Kashevarof. Her husband replied, "She's right. My wife is right. We were treated like animals."[234] In total, 881 Unangan were forcibly relocated.[235]

Once onboard the transports, there were still no concrete plans for where to take the evacuees. Finally, the OIA decided that the Unangan should remain in southeast Alaska in sites where they could be self-sufficient through their own efforts and that villagers should be kept together. They were taken to facilities near Wrangell and just north of Ketchikan. However, these facilities were barely habitable. They were abandoned canneries, a herring saltery, and a gold mine camp. They were rotting, with no plumbing, electricity, or toilets (the "toilets" available were honey buckets). There was little potable water (drinking water was the color of tea), no winter clothing, and barely edible food.[236] Tuberculosis, influenza, measles, and pneumonia were rampant. Ten percent of the evacuees, eighty-five total, died in the camps.[237] The Unangan women wrote to the OIA telling of their living conditions: "We the people of this place wants[sic] a better place to live. We drink impure water and then get sick. . . . We got no place to take a bath and no place to wash our clothes or dry them when it rains. . . . We live in a room with our children just enout [sic] to turn around in. We use blankets for walls just to live in private."[238] The OIA didn't care: "Under war conditions, they could not expect to enjoy the comforts and conditions as they existed on the Pribilof Islands."[239]

Besides the turmoil of the forced evacuation, the Aleuts had to deal with the culture shock of the dramatic change in environment. The Aleutians are barren, treeless, windswept islands. Southeast Alaska is a temperate rainforest. The trees were a constant reminder of the dread and abnormality of the forced relocation, leading many of the evacuees to feel claustrophobic and depressed.[240]

Life in the camps brought another trial: exposure to the White people and all their customs, habits, and prejudices. Many of those who died were the village elders, whose deaths left the remaining villagers without the memory of their traditions. And as was typical of anywhere the Whites went, alcohol followed:

> "Before evacuation we didn't have any liquor store or bars," [Andronik] Kashevaroff [one of the Aleut boys] told me [Eva Holland]. "But when we got to Juneau you start seeing signs—open liquor, open liquor. And bottles in the liquor store, all in the windows. There's where people started getting liquor."
>
> "More and more people started drinking?" his daughter Bonnie asked.
>
> "Yeah," said Ella Kashevarof [Andronik's wife]. "They learned."[241]

While the official reason for the relocation, the protection of the Aleuts from the Japanese invasion, was valid, it was not the only one. Racial prejudice, as with the Japanese, played a significant role. One camp had a swimming pool nearby. The Alaska Natives were prohibited from using it. The men, some as young as fourteen, were forced to return to their homes in the middle of the war to help with the annual seal hunt. The Unangan questioned how it was necessary to evacuate their homes in the Aleutians and have it still safe enough to engage in seal hunting. The US government deemed the benefits of the seal hunt, meaning the profits to be had, outweighed the risk to the Unangan men.[242] FWS officials decreed that those who did not participate would not be allowed to return to the islands after the war: "'Disobedient Aleuts' would permanently lose 'all privileges as an island resident.'"[243] When they finally returned home, there were no homes to return to—their houses had been commandeered and ransacked by the soldiers and civilian contractors who were no longer there.[244] OIA reports stated,

> All buildings were damaged due to lack of normal care and upkeep. . . . Inspections revealed extensive evidence of widespread wanton destruction of property and vandalism. . . . Contents of closed packing boxes, trunks, and cupboards had been ransacked, clothing had been scattered over floors, trampled and fouled, dishes, furniture, stoves, radios, phonographs, books, and other items had been broken or damaged.[245]

Many did not have villages to return to or were not permitted to return to their villages; the cost of restoring the destroyed communities was too great, as claimed by the government.[246] As with the Japanese who were forcibly relocated from the West Coast, the scars and trauma among the remaining survivors and their families continue to this day.

SOVIET UNION
Germany
When the tables had turned and the Russians were advancing against the Germans, reclaiming their lost territories, the vengeance exploded. Stephen G. Fritz describes it quite graphically:

> In an explosion of violence, Soviet troops had enacted a first, bloody revenge on German civilians, with scores of women raped and murdered, often in the most gruesome fashion, stores plundered, and houses burned. Having suffered a whole range of German atrocities for three dreadful years, and having seen firsthand the awesome destruction of the scorched-earth retreat, Soviet soldiers engaged in an orgy of revenge that, although perhaps understandable, was, nonetheless, deplorable.[247]

As the Soviets moved further west, the rumors surrounding their behavior and what to expect when finally meeting them grew. As is true of any type of foreign advance, the vanquished exhibited various emotions. Signs were placed alongside the road that read, "Tremble with fear, fascist Germany, the day of reckoning has come!"[248] Many tried to flee, with long lines of civilians hauling whatever they could while escaping the coming wave of violence, including looting, pillaging, and mass rape of all the women and murder of any man who tried to protect them. Mothers carried their dead babies, refusing to believe they were dead and losing their minds because of it. Hospitals were turned into death houses and houses of forced sexual violation.[249] Unfortunately, most of these stories turned out to be true. Estimates of casualties from this violence are 100,000 dead in the easternmost provinces of Germany in the first few months of 1945, with 1.4 million women raped. Even Russian soldiers who wanted to believe otherwise, who tried to believe that it wasn't the rank-and-file frontline soldier committing these horrors but the rearguard, the NKVD who followed who were responsible, had to admit otherwise as the evidence piled up. There were reports of girls as young as twelve being raped, of men watching the Soviet soldiers rape their wives, of nursing women having to beg for a break from the rape in order to nurse their children.[250] Incidents like these would continue to occur for months after Germany's surrender.

Russia's entrance into Germany proper allowed them to conscript Germans into work crews. On February 6, 1945, an order was issued to "mobilize all Germans fit for work from seventeen to fifty years of age and to form labour battalions of 1,000 to 1,200 men each and send them to Belorussia and the Ukraine to repair war damage."[251] The attitude of the Soviet soldiers toward these conscripts was described as schadenfreude, taking joy or satisfaction at another's misfortune, as one corporal lining up the German recruits yelled at them in broken German, "To Siberia, fuck you!" [252]

The Soviet attitude toward German prisoners of war was no better than their treatment of the general German populace. This bears closer examination, as it sheds light on the Japanese treatment of prisoners. Like Japan, the Soviets refused to ratify the 1929 Geneva Conventions. And similar to Japan, the Soviet view of surrender was that the act was treasonous. Stalin's Order No. 270, issued on August 16, 1941, stated that any Red Army troop who allowed himself or herself to be captured was a traitor and was to be executed. The Red Army knew this, just as they knew of Stalin's Order No. 227, issued July 28, 1942, the famous "Not a single step back" order, forbidding upon pain of death the retreat of any Soviet military member for the same reason: To do so is treason.

These two orders exemplify the same mentality demonstrated by the Japanese—the individual is nothing, just a tool of the government. As Michael Sturma writes, this introduces the issue of reciprocity, or, in lay terms, "Do unto others as you would have them do unto you." Prisoners of war are treated in a humane manner because we expect, or at least hope, that our actions will be reciprocated. The signatories of the Geneva Conventions did so because not only did it seem the humane course of action to take but also because they hoped that their prisoners would be treated in the same, humane fashion. The Japanese and Russians knew that to surrender would mean to be labeled a traitor. Knowing that death awaited them upon release from a prison camp or that they shamed their families, there was no benefit in becoming a prisoner. And without this morality, there was no reason to take prisoners or to treat them humanely, as they expected no reciprocity.[253] Because they expected to die, so, too, should the opposing side. With this attitude in mind, it becomes easy to see why between February and mid-April 1943, approximately 100,000 German prisoners died.[254]

This saturnalia of violence did not wait to start until the Germans invaded Russia. Even before this, Stalin was implementing his own purges. After the Soviet invasion of Poland in September 1939, they began a program of eliminating the Polish leaders and intelligentsia. In April 1940, they executed about 15,000 Polish officers and other intellectuals, then proceeded to bury the bodies in the Katyn Forest near Smolensk. The Germans discovered the graves in April 1943 and made it well known what they had found via Radio Berlin. Stalin denied everything, and when the Russians reoccupied the area, they convened a study that concluded that the Germans were responsible. When the Poles demanded an independent investigation, the Kremlin accused them of siding with the "fascist aggressors" and broke off diplomatic relations. The United States did nothing through all of this. It wasn't until April 13, 1990, that Tass, the Soviet news agency, announced that a joint venture with the Soviets and Poles had proven that it was, in fact, the Soviets who had committed the crime. Benjamin B. Fischer gave the following list of casualties:

> Those who died at Katyn included an admiral, two generals, 24 colonels, 79 lieutenant colonels, 258 majors, 654 captains, 17 naval captains, 3,420 NCOs, seven chaplains, three landowners, a prince, 43 officials, 85 privates, and 131 refugees. Also among the dead were 20 university professors; 300 physicians; several hundred lawyers, engineers, and teachers; and more than 100 writers and journalists as well as about 200 pilots. It was their social status that landed them in front of NKVD execution squads. . . . In all, the NKVD eliminated almost half the Polish officer corps—part of Stalin's long-range effort to prevent the resurgence of an independent Poland.[255]

Both the British and Americans tried to suppress the story for fear of alienating the Soviets.

Berlin became little more than a brothel for the red soldiers. Soviet Marshall Zhukov had stated, "Woe to the land of the murderers. We will get our terrible revenge for everything."[256] Indeed they did.

Crimean Tatar Relocation
While the United States was interring Japanese Americans and Alaska Natives, the Soviet Union was dealing with the Crimean Tatars. The Tatars and Russians have a long history, sometimes good but often rocky.

The Tatars first came to Crimea in the 1200s as Batu Khan, a grandson of Genghis Khan, conquered the area. In the 1300s, they converted to Sunnite Islam. They began to trade with the Russians, and by the 1400s, they dominated the area. They remained dominant until the 1700s. In 1774, Catherine the Great invaded Crimea to prevent the Ottoman Empire from controlling the area. In 1783, Catherine annexed the peninsula and encouraged Russians and Ukrainians to come. During this same period, tens of thousands of Crimean Tatars were forcibly removed from Crimea and taken into the Ottoman Empire to serve as slaves. However, the Russians treated the Tatars as allies, with their citizens serving as commercial and political agents, teachers, and government administrators of the area.[257] Nonetheless, there were several mass migrations of Tatars from Crimea to Turkey from 1783 to 1856 and the end of the Crimean War. Most of these migrations were the direct result of persecution of Tatars by the Russians who were cohabiting the area.

After the Russian Revolution, the new Soviet government supported the various ethnic minorities by allowing semiautonomous rule to ethnic districts and the establishment of ethnic schools as well as newspapers and magazines in the Crimean Tatar language.[258] However, Stalin quickly put an end to these freedoms. When the German Wehrmacht invaded in 1941, they recruited Tatars for local police forces and low-level government positions to make up for their shortage of manpower. They stoked ethnic tensions by favoring the Tatars over the other minorities of the area through the release of Tatar POWs, excusing them from labor duty, relieving them of tax duties, allowing the open practice of their religion, and allowing education in the Tatar language.[259]

While these policies convinced some Tatars to help the Germans, only a few did so. The Germans continued to practice their scorched-earth policy against Russia, which did little to ingratiate themselves with the Tatars. Moreover, the Germans conscripted approximately 10 percent of the Tatar population to fight for them.

The Soviets did their part to keep the Tatars loyal to their government by extolling the patriotism of the Tatars in the newspapers as well as vilifying the Germans and their actions. One article in the Crimean newspaper *Krasnyi Krym* read, "Brother-Tatars! You are in the occupied territory among the enemy. You see and feel the horrors of the Fascist occupation. The Germans send your sons to the frontline. They rape your daughters; they turn you into powerless slaves. They condemned you to starvation and death."[260]

The Tatars peacefully coexisted with the Russians for centuries.[261] The few Tatars who collaborated with the Nazis were portrayed as having severe character flaws. However, as the war progressed and the Soviets began pushing the Nazis back and recapturing their previously lost territories, these glowing reports quickly diminished. Instead, the few Tatars who did collaborate went from being portrayed as errant children to being characterized as enemy accomplices in need of severe punishment. Additionally, it was thought that future generations of Tatars should be punished for the acts committed by these traitors.[262]

With this turn of opinion came the vilification of the Tatars and the return to pounding them into subservience to the Soviets. Whereas before, there had been only a few Tatars who had collaborated with the Nazis, now there were numerous traitors: "Many Crimean Tatars betrayed their Motherland, deserted from the army and joined the army of the enemy, participated in the voluntary Nazi divisions, [and] participated in the barbaric and cruel killings of the Soviet people."[263] Levrentii Beria, head of the Soviet Secret Police NKVD, continued to portray the Tatars as terrorists aiding and abetting the Nazis and no longer simply needing punishment but requiring removal from Crimea: "Considering treacherous action of the Crimean Tatars against the Soviet people and considering unfeasibility of the further residency of Crimean Tatars on the border of the Soviet Union, NKVD asks for your consideration of deportation of all Crimean Tatars from the territory of Crimea."[264]

On May 18, 1944, the deportations began, removing the Tatars to Uzbek Soviet Republic in Central Asia (see figure 25). Gone was any mention of the previous heroics provided by Tatar soldiers. Instead, they were blamed for any and all hardships encountered by loyal Soviets, including horrible conditions in the Red Army and lack of basics needed for daily survival by the whole populace. About 188,000 Tatars, 9,620 Armenians, 12,420 Bulgarians, and 15,040 Greeks were deported, plus a few thousand from other groups.[265] Of the total 228,392 persons deported, about 20 percent (44,878) died from starvation, disease, and exposure between 1944 and 1948.[266]

CHEMICAL AND BIOLOGICAL WARFARE

The Germans had much less use of gas during the war than Japan. There has been much debate over the reasons Hitler did not resort to chemical warfare during the conflict. The consensus seems to be that he had been exposed to mustard gas in October 1918. Knowing firsthand how it felt to be the victim of a gas attack, he possibly thought the use of gas was too inhumane, even for his enemies. There is, though, another possibility.

In May 1943, in his headquarters in East Prussia, Hitler broached the idea with Albert Speer, minister of armaments, and Otto Ambros, Hitler's chemical warfare expert, of using gas against the Russians on the Eastern Front. Upon hearing the idea, Ambros tried to quash it by saying that the Allies could outproduce the Nazis in chemical weapons. Hitler retorted that the Allies didn't have any idea about the Nazi nerve gas tabun, to which Ambros responded that he had intelligence showing that they did. However, his intelligence was faulty, and it was indicative of the Allied development of the pesticide DDT, not nerve agents. Regardless, Hitler dropped the subject.[267]

Of course, as mentioned in chapter 4, the Allies had no clue that the Nazis had such a devastating new weapon. There were rumors of such a material, and

423 100
кримських татар
було депортовано

46, 2 % загинуло в
перші роки вигнання

*Дані самоперепису,
проведеного Національним
рухом кримських татар

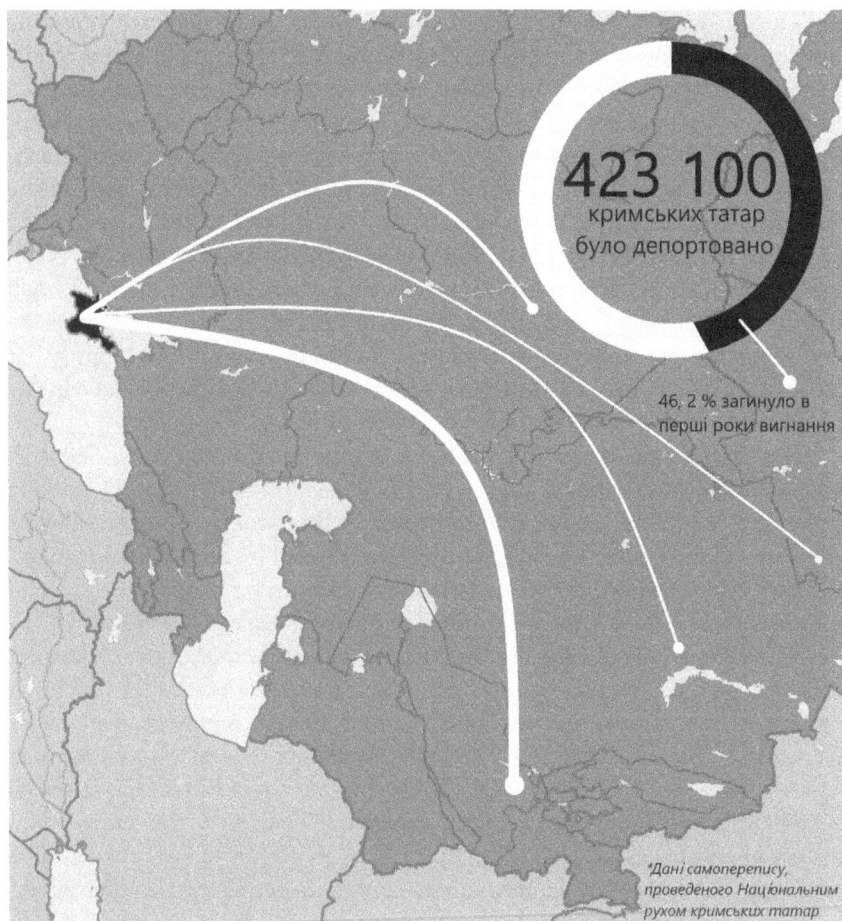

Figure 25. Destinations of the deported Tatars. Image by Devlet Geray—Own work, CC BY-SA 4.0. Used with permission. Source: Wikimedia Commons, "Deporation of the Crimean Tatars," accesed May 21, 2025, https://commons.wikimedia.org/w/index.php?curid=90835122.

the Allies had even developed something similar. But when presented with concrete proof of the existence of tabun through the capture on May 11, 1943, of one of the chemists working on the project, they refused to act on the information given them.[268]

US general Omar Bradley commented on how relieved he was that he never could smell any gas during the D-Day invasion of Normandy. Any indication of the use of gas would have led to a major decision in how to proceed. British general Bernard Montgomery was so sure that the Nazis had nothing new that he left all the landing forces' anti-gas equipment back in England. This included individual soldiers' gas masks. Had Germany decided to use tabun against the Allied landings, there is no doubt that they would have been stymied, which would have dramatically altered the outcome of the war.[269]

Of course, there was one enemy upon which he had no qualms about using gas: the Jews. One of the gases used in World War I was hydrogen cyanide, a well-known pesticide.[270] With a few modifications, the new compound was named Zyklon, later named Zyklon-A to distinguish it from other compounds in the same family. The substance was a potent insecticide, as its base chemical, hydrocyanic acid, had been developed to kill lice and was used in the fumigation of submarines, barracks, and prison camps.[271] Further modifications, including an eye irritant used as a warning as to the lethality of the compound, led to the creation of Zyklon-B. Remembering its original use, it was a small step to try it against humans.[272] It wasn't until the execution of Jews using traditional means, such as shooting, was found to be too time consuming that other, more efficient means of murder were looked for. The answer came in the form of Zyklon-B, first tested on Soviet prisoners of war.[273]

Research into new tools for warfare was not limited to the Axis. The British had established their own bureau for biological warfare research. Initially conceived of in 1934, again as a direct result of the Geneva Protocols drawing attention to the matter, the organization was headed by Sir Maurice Hankey. Research was originally limited to defensive work only, but as time went on and the country went to war, research turned to offensive biological weapons.[274]

The distinction of creating the first biological weapon in the West lies with Britain. This weapon was anthrax, which was intended to infect such animals as cattle but only as a retaliatory weapon in case of biological attack by Germany.[275] By the summer of 1942, planning and production of an anthrax bomb had reached the testing stage. To do this, a team of British scientists occupied Gruinard Island off the coast of Scotland. Their work there focused on the use of anthrax as a biological weapon. Though nothing came of the project other than many dead sheep, they did succeed in rendering the island uninhabitable for fifty years due to its soil being contaminated with anthrax spores.[276]

Despite the initial poor showing, research in the use of anthrax as a weapon continued. A year and a half after Gruinard, the Allies succeeded in constructing a bomb capable of using anthrax as its payload. Code-named N, the anthrax, its bomb, and the entire project was given the same level of security as the Manhattan Project's construction of the atomic bomb.[277] The theoretical bombing run determined that six British Lancaster bombers could carry enough anthrax to kill everyone in a square mile area.[278] The official report stated, "We cannot afford to not have N bombs in our armory."[279] The Chiefs of Staff Committee agreed. Unfortunately for the British, they lacked the manpower and industrial capability to manufacture the number of bombs needed for the endeavor. They had to turn the entire project over to the Americans and beg for whatever scraps the United States would give them. Once production began, the entire project was handed back to the British. The new plant was capable of producing more than 50,000 bombs per month.[280]

This entire project was designed for one reason only: to kill as many people as possible. And while the world was lucky that the bombs were never
used, the plan was drawn up for their utilization. It speculated that with the
right payload tonnage and the right number of bombers carrying them, the
planned trial cities of Berlin, Hamburg, Stuttgart, Frankfurt, Wilhelmshafen,
and Aachen would experience a 50 percent mortality from anthrax inhalation,
with many more dying from contact with the pathogen.[281] One estimate stated
that if used, Berlin would be uninhabitable for forty years.[282]

Anthrax was not the only microbe studied. Botulism was also studied but
not to the same extent as anthrax. Sir Paul Fildes, a British pathologist and
microbiologist, was chosen to lead the Biology Department at Porton Down,
the British government's research and development organization for chemical
and biological weapons. One of the areas in which he was working was the use
of BTX, better known as botulism, as a weapon. He and his team had devised a
weapon by modifying antitank grenades so that they could carry a load of BTX.
They hoped that the shrapnel from the explosion would carry BTX with it and
impregnate the victims with the bacteria.

BTX is an acetylcholine-release inhibitor, which prevents the muscles
from contracting.[283] As mentioned in chapter 4, nerve agents work the same
way. Those infected with BTX usually die of respiratory failure or cardiac arrest
from paralysis of the associated muscles. Death usually occurs a week after exposure to the bacteria. One example stands out.

Reinhard Heydrich was head of the Nazi security service, the Sicherheitsdienst (SD), and the Reich protector (*Reichsprotektor*) of Bohemia and Mora-
via, the part of Czechoslovakia annexed into the Third Reich on March 15,
1939. Ruthless and cruel, he governed the protectorate with an iron fist as he
shut down nearly all resistance cells.

On May 23, 1942, Heydrich, who usually traveled without an armed escort, was in Prague when one small Czech resistance cell prepared to assassinate him. As he drove to their position, one of the resistance tried to open
fire on Heydrich with a submachine gun, but the gun jammed. At this point,
a grenade was used, but it only managed to wound Heydrich with some shrapnel. He was taken to a hospital, where it was deemed necessary to operate on
his wounds. The surgery was a success, and Heydrich was expected to make a
complete recovery.

However, there was one wound about three inches deep that contained
a lot of dirt and little splinters. One week after the attempted assassination,
Heydrich died, to the surprise of everyone. The cause of death was ruled sep-
ticemia. The death report contained one important line: "Death occurred as
a consequence of lesions in the vital parenchymatous organs cause *by bacteria
and possibly by poisons carried into them by the bomb splinters*."[284] In dying, he
exhibited all the signs of BTX poisoning: "extreme weakness, malaise, dry skin,
dilated and unresponsive pupils, blurred vision, dry coated tongue and mouth,

and dizziness when upright . . . progressive muscular weakness with facial paralysis, and weakness of arms, legs, and respiratory muscles."[285]

The retribution for Heydrich's death was swift and merciless. The entire
town was razed, the men shot, the women and children trucked off, and at least
10,000 Czech citizens arrested. And while rumors to the contrary abounded,
the German biological research program was nowhere near as advanced as that
of Britain.[286]

Gas was never used in Europe, but the question arises of whether it could
have been, given the proper circumstances. In the summer of 1940, France had
fallen, and the British Expeditionary Force had been defeated and miraculously
saved at Dunkirk. Nothing stood in the way of a German invasion of England.
On June 15, 1940, only two days after the evacuations at Dunkirk, Sir John
Dill, chief of the Imperial General Staff, presented a memo entitled "The Use
of Gas in Home Defense." In it, he championed the use of gas against Germany
should it invade England. Most of the generals to whom he presented dismissed
the memo primarily because they did not want to be seen as a country willing
to renege on their promise to not be the first to use gas: "We should be throwing away the incalculable advantage of keeping our pledges and for a minor
tactical surprise."[287]

However, on June 30, Prime Minister Winston Churchill made it clear
that if Germany invaded England, England would use every gas available to
thwart the invasion. England immediately gathered its meager stocks of gas
and was ready to use them by the end of the first week in July.[288] The gas was
not needed, but production was increased immediately. Bombers were taken to
Scotland to equip them to spray mustard gas on the invaders, and more gas was
ordered from the United States.[289] Given that England had lost so much heavy
equipment, such as artillery and tanks, its military was in no position to repel
an invasion. They would need every advantage available.

However, this did not mean that people, both military and civilian, were
not exposed to it. In what can only be described as hell on earth, a catastrophe unfolded in a port in southeast Italy. In November–December 1943, First
Lieutenant Howard D. Beckstrom, a chemical warfare expert, was ordered to
escort a secret cargo of one hundred tons of mustard-gas bombs to its potential
final distribution point. The shipment arrived in the port of Bari. The port was
very important and busy, and upon arriving, Beckstrom found it heavily con-
gested with Allied shipping. Rather than unload the cargo immediately so that
it could be transported to a secure storage area, the ship was told to moor and
wait its turn.

Four days later, Beckstrom was still awaiting his chance to unload his
cargo. At 7:30 p.m. on December 2, one hundred German Ju 88 bombers arrived and began dropping their bombs. The result was the sinking of seventeen ships (see figure 26). Backstrom's ship, the *John Harvey*, was one of them.
Explosions damaged the tanks carrying the mustard gas, and the gas began to

Figure 26. The harbor of Bari, December 2, 1943. Source: HistoryPorn, "Allied ships burn during the air raid on Bari, 2 December 1943. One of these, the Liberty freighter John Harvey, was carrying thousands of mustard gas shells . . . ," Reddit, posted 2021, https://www.reddit.com/r/HistoryPorn/comments/nwasp3/allied_ships_burn_during_the_air_raid_on_bari_2/.

leak. Escaping the ship, the gas began to fill the harbor. In doing so, it mixed with the burning oil from the other damaged ships. This led to the gas spreading throughout the harbor and into the city itself. People were breathing in the deadly gas, and those sailors in the water were floating in a mixture of oil and mustard gas.

Because the cargo was secret, nobody knew what they were dealing with, other than Beckstrom and the men accompanying him, and they were all killed in the attack. This left the hospital to deal with wounded men without knowing what they were treating. Upon rescue from the frigid waters, they were given blankets to keep warm and then told to wait their turn. Harris and Paxman explain, "The opportunity for burn and absorption must have been tremendous. The individuals, to all intents and purposes, were dipped into a solution of mustard-in-oil, and then wrapped in blankets, given warm tea, and allowed a prolonged period for absorption."[290] The next morning, the results of this exposure began to manifest. Out past the harbor in the open ocean, the US destroyer *Bistera* had been rescuing some of the sailors. By morning, not only the rescued sailors but also their Samaritan benefactors found themselves nearly totally blind. It was still eighteen hours before they could enter the harbor.

By the time the catastrophe was over, of 628 hospitalized, 83 sailors were dead, as were many more civilians in town. Thousands more had fled the town to escape the hell that was unfurling. The aftermath endured for weeks as townspeople who were afflicted with the exposure dealt with the repercussions.[291] Because it was a secret shipment, it was left to Lieutenant Colonel Stewart Francis Alexander, a chemical warfare specialist, to independently diagnose the situation. He traced the point of origin back to the *John Harvey*, and divers recovered bomb fragments that were definitively identified as one-hundred-pound American mustard bombs. On December 11, 1943, Alexander notified headquarters with his findings. Upon learning of the incident, Churchill immediately censored any and all mention of mustard gas, including in patients' charts.[292]

The Joint Planning Staff (JPS) was tasked with determining the feasibility of using chemical or biological weapons in the war effort. After presenting the advantages and disadvantages of chemical warfare, the JPS advised against using them. They did not, however, rule out the use of biological weapons. In fact, the only downside to using this weapon was the fact that US production of anthrax was behind schedule.[293] They stated that while the German government could probably keep the populace in line after a gas attack, there was no guarantee that the British would act similarly. Churchill finally acquiesced but not without leaving the door open to revisit the subject should circumstances change significantly: "The matter should be kept under review and brought up again when things get worse."[294] Churchill was firmly convinced that every weapon available should be used to win the war. He had no qualms about breaking the previously signed treaty. But he was in a distinct minority, and he recognized it.

Biological warfare was another story. The arguments against using gas did not apply to biological weapons. It was the service chiefs who had advocated for their inclusion in the report. Initially, the main roadblock to using N was that US production was so far behind. But that was 1943. What about 1944, when US production would have been where it was needed? And ethically speaking, what was the difference between using the atomic bomb and using anthrax bombs? It was only because the war progressed so rapidly that anthrax was not used in Germany.[295]

In the United States, things progressed similarly but in reverse fashion. It was the military that advocated for the use of these weapons against the Japanese. They determined that a complete utilization of gas would have a decisive result in the war and endorsed its use in the battle of Iwo Jima. They wanted to saturate the island in poison gas. The only thing that stopped its use was President Roosevelt and his morals. He repeatedly refused to use gas in any way. By the time of Roosevelt's death in April 1945, the atomic bombs were nearly ready to use, thereby negating the need for chemical weapons.

What stopped the use of chemical and biological weapons in World War II? Fundamentally, it was the fact that at every instance in which its use was contemplated, the war turned so as not to need its use.[296] The British and

Russians both feared its use by the Nazis, and the Nazis feared its use by the Allies in retaliation for the Nazis' use.[297]

THE WAR AT SEA
The Atlantic
German U-boat doctrine and tactics were the same as in World War I: wolf-packs, or attacking vulnerable targets with multiple submarines so that when one boat could no longer attack, others could take its place and press home the offensive, followed by elimination of the survivors. Keeley writes, "What is submarine warfare at sea or strategic bombing in the air but guerrilla [read "primitive"] warfare by new technological means in new mediums?"[298] The targeting of survivors did not begin immediately with hostilities. That tactic grew out of circumstances beyond control of the Germans.

The Treaty of Versailles had left Germany without any kind of fleet. Among other naval vessel types, the construction and possession of submarines were prohibited. In protest, as the Allies attempted to take possession of the fleet, Germany scuttled the entire fleet. They succeeded in sinking fifty-two ships in this manner, including ten battleships, five battle cruisers, five light cruisers, and thirty-two destroyers.[299] While they had lost their fleet, they would not endure the limitations imposed on them for long.

Indeed, Germany continued research on subs through a shell company for the German company Krupp. The company, IVS, worked out of Holland and constructed various boats in Finland and Spain. A submarine school was established in Turkey where Turkish submariners were trained on three subs sold to their country.[300] This secret rearmament was no secret, but nothing of import was done about it. On November 15, 1932, a plan to expand the navy, in defiance of the treaty, was passed. Included in the plan was an expansion of the submarine forces to consist of three half-flotillas of sixteen submarines.

Additionally, a submarine school disguised as an antisubmarine school was established.[301] Led by the former submariner Karl Dönitz, the fledgling submarine force blossomed under his leadership. He firmly believed that in the upcoming war with England, submarines would make the difference between victory and defeat.[302] With a fleet of three hundred medium (five hundred tons) U-boats, he believed that he could achieve victory over Britain in a year and a half by means of a naval blockade of the country.[303] His plans were adamantly vetoed. England was an "ally," and no war games with it as the enemy were permitted. The Anglo-German Naval Treaty imposed limits such that even if his plans were approved, Germany would only be able to build forty-eight submarines, due to already approaching the limits of tonnage allowed. Germany had agreed to the Submarine Protocol, which forbid surprise attacks on most merchant ships. And finally, senior officers didn't believe his plan would work, thinking that larger boats in distant waters were needed.[304]

With the approach of the war, Hitler realized that England might become an enemy, and plans were altered accordingly, including the build-up of the submarine forces. However, the rules imposed on the subs made their use almost impossible.[305] Such was the situation when, on September 3, 1939, U-30 mistook the passenger liner SS *Athenia*, a British ocean liner bound for Canada with 1,103 passengers, including 311 Americans, for a legal target. After realizing his mistake, he failed to break radio silence to inform Dönitz of the error. He also didn't attempt to aid the ship and its passengers. The response to this incident was as expected: The Germans were barbarians engaging in unrestricted warfare.

The Germans, under Hitler's orders, categorically denied their actions. The incident resulted in the loss of 118, including 28Americans, some of whom died when a rescue boat's propellers churned up a lifeboat.[306] This incident resulted in even more restrictions on submarines. However, with the passage of time, Hitler would remove, one by one, these restrictions.

Events took a turn for the worse in August 1942, when a group of four U-boats were heading toward South Africa. Coming upon a former Cunard liner displacing 19,700 tons, one of the boats proceeded to torpedo it. However, when the survivors began to abandon ship, the boat commander realized that the ship was carrying some 1,800 Italians along with Allied troops, RAF airmen, and officers, including their wives and families—in total, about 3,250 people. The boat's commander decided that he had to initiate rescue operations. He sent out an uncoded message in English telling of the situation and promising to not attack any vessels that came to help. Nobody came.[307]

The operations proceeded for more than two days, when an American aircraft approached. Unsure how to proceed, the pilot radioed for instructions. Unable to get any response from superiors, the pilot's commander issued the order to sink the sub. This, even though the rescue operations were clearly displaying a Red Cross flag. The repercussions were to be expected. Dönitz issued the following order:

1. Efforts to save survivors of sunken ships, such as the fishing [of] swimming men out of the water and putting them on board lifeboats, the righting of overturned lifeboats, or the handing over of food and water, must stop. Rescue contradicts the most basic demands of the war: the destruction of hostile ships and their crews.
2. The orders concerning the bringing-in of skippers and chief engineers stay in effect.
3. Survivors are to be saved only if their statements are important for the boat.
4. Stay firm. Remember that the enemy has no regard for women and children when bombing German cities![308]

Hitler, in conversation with the Japanese ambassador, expressed his belief that survivors should be eliminated to wear down the Allied war effort, to which the ambassador agreed and stated that they would be doing the same. The reports of British destroyers machine-gunning the survivors of a German minelayer only strengthened his resolve. Citing the need to reduce the number of enemy sailors to man the vessels, the desire to possibly decrease the number of enemy personnel serving, and the fact that Churchill had ordered the bombing of German cities with their women and children inhabitants, the official policy was that no survivors would be cared for.[309] Those inclined to help enemy survivors would only do so after ensuring all of their own crews had been rescued. The unofficial policy continued on: While the enemy was on his ship, he was the enemy, and laws and policies regarding the enemy were followed. Once he entered the water, he became a fellow seaman in distress, and those laws and policies took precedent.[310]

The only actions for which a U-boat commander would be convicted of war crimes occurred in March 1944. After attacking the Greek *Peleus*, under charter to the British Ministry of War Transport, *U-852* decided to dispose of the survivors. Machine guns, machine pistols, and hand grenades were passed up to the bridge, and the boat proceeded to cruise through the wreckage, firing their weapons and lobbing grenades. All the blood in the water attracted sharks, who proceeded to feast on the dead and wounded. The crew did not handle the situation well but were reminded, "If we are influenced by too much sympathy we must also think of our wives and children who die as the victims of air attack at home."[311] For five hours, the boat continued its killing spree. Only three crewmen survived the slaughter and remained adrift for forty-nine days.

Later during that same voyage, *U-852* became so damaged that the crew was forced to ground it and destroy it with demolition charges. The crew was captured and taken back to London for trial. Nobody denied the fact that the slaughter had happened. The reason given for the carnage was that the crew were just following orders and the commander ordered the destruction of all remaining evidence of the wreckage; the murder of the survivors was either accidental or the result of overzealous crewmen attempting to destroy all traces of the wreckage. The captain, Kapitänleutnant Heinz Eck; the second watch officer, Lieutenant August Hoffmann; and the doctor, Walter Weisspfennig, were all found guilty, sentenced to death, and executed by firing squad on November 30, 1945.[312] This punishment was seen as appropriate by the vast majority of current and former German submariners.

This was the only known massacre of unarmed ship survivors committed by the German U-boat service during the Second World War. Hitler had wanted Dönitz to engage in such actions, but Dönitz rebuffed the idea on the grounds that doing so was beneath the dignity of the navy. Indeed, the massacre was widely condemned throughout the German military after the war.[313] However, as Felton points out, condemning the action does not mean that the action could not happen, only that it did not happen as often as with the Japanese.[314] But the propaganda put forth by the Allies of U-boats machine-gunning defenseless survivors remains to this day. Felton points out that the U-boat service was probably the least brutal of all the German armed forces.

After the war, Admiral Dönitz, Germany's leader at the end of the European portion of World War II, was tried for war crimes along with other Nazi leaders. However, unlike those defendants, many of whom were hanged for their crimes, Dönitz was sentenced to a mere ten years in prison. The Allies tried to convict him for the memo referenced above, but Eck at his trial refused to incriminate Dönitz, insisting he had acted on his own accord, even though blaming Dönitz in the "I was just following orders" defense would have saved his life. Try as they might, they could not get any viable proof to convict him. His only "punishment" was to be censured by the tribunal for issuing such an ambiguous order.[315] And as Bennett points out, this ambiguity may have been Dönitz's way of keeping Hitler from interfering in the submarine arm's affair while still allowing the captains to follow the law of the sea.[316]

Unique cases arose in the Allied submarine service, just as they had in the U-boat service. I cover the case of the USS *Wahoo* shortly. The British had their own incident involving the HMS *Torbay*. On the night of July 9–10, 1941, the *Torbay*, skippered by Lieutenant Commander A. C. C. Miers of the Royal Navy, came across four German vessels disguised as caiques. The first vessel was destroyed in a blazing inferno, and the crew either abandoned ship or were killed. The second vessel was approached shortly after and boarded. The crew had attempted to portray themselves as Greek but were in fact German. One tried to throw a grenade but was shot, as was another who tried to shoot the boarding party with a rifle at point-blank range.

The caique was filled with demolition charges, while the crew was forced into a rubber raft, then promptly shot to prevent them from returning to their vessel before the demolition charges had blown. Miers made no attempt to conceal the orders he had given: "Submarine cast off and with the Lewis gun accounted for the soldiers in the rubber raft to prevent them from regaining their ship."[317] Meaning: They were gunned down by the Lewis gun.[318]

The flag officer in charge of submarines, Admiral Max Horton, was horrified at the incident and wrote in a letter to his superiors, "As far as I am aware, the enemy has not made a habit of firing on personnel in the water or on rafts, even when such personnel were members of the fighting services. Since the incident referred to in *Torbay*'s report, he may feel justified in doing so."[319] Miers received a strongly worded letter telling not to repeat his actions, but he was never court-martialed for his actions.

Nonetheless, the British were initially slow in combating the U-boat menace. The fall and early winter of 1940 were known by the German submariners as the "happy time" due to the great success they had against British shipping. However, the British did adapt as they implemented convoys with destroyer escorts. Further developments, like improved radar, the use of sonar, and direc-tion-finding equipment for the shore units to locate U-boat transmissions, all helped to effectively reduce the threat of the German U-boats to practically zero. At the end of 1941, a large convoy was en route from Gibraltar to the British Isles. Having cracked the U-boat codes, the convoy traveled well out-side their normal route, causing the subs and corresponding aircraft to waste two days finding it. When the subs finally did find the convoy, the British easily polished off four of the subs.[320]

But the end of 1941 also saw the entry of the United States into the war. This was a feast for the U-boats because the United States was slow to adopt the convoy, and the navy did not use either British or American intelligence. In just two months, the Germans nearly doubled the tonnage sunk.[321] On Jan-uary 2, 1942, Dönitz ordered five German large, Type IX boats to begin Oper-ation Paukenschlag ("Roll on the Kettledrums"). On January 13, the attacks began. On January 19, U-123, commanded by Korvettenkapitän Reinhard Hardegen, reported sinking eight ships in a twelve-hour period. In all of Janu-ary, twenty-nine ships were attacked, with sixteen sunk.[322] February was even better for the Germans, who attacked thirty-seven ships and sank nineteen.[323]

It wasn't that the United States disagreed with the convoy principle. It was like every other aspect of the US war machine in that it was totally unpre-pared to institute antisubmarine warfare (ASW). It wasn't until May 14 that the first convoy was launched.[324] And the give-and-take of the battle contin-ued until May of the following year, when a battle between Convoy ONS 5 and the U-boats of Groups Star, Fink, and Amsel battled from April 29 to May 6, resulting in thirteen merchant ships and eight submarines lost.[325] Loss rates such as these favored the Allies. Germany did not have the ability to replace that many subs quickly enough to continue the war. After additional failures, U-boat command relocated the groups.

Victory did not come easily. Totals vary from source to source, but in the period from America's entry into World War II through December 1942, losses of convoys in American waters amounted to 609 vessels, for 3,122,456 gross register tons (GRT), at a cost of 22 U-boats sunk.[326] Total number of vessels lost and total tonnage for the comparable period of September 1939 through August 1942 were 1,904 vessels lost and 9,235,113 GRT.[327] In all of World War II, Germany lost nearly 800 U-boats and 19,000 crewmen.[328] The issue Germany had to face was that they were never going to be able to match the US industrial output. That, along with her military presence, was simply too much for the Germans to compensate. The Americans were able to adapt and, in doing so, nullified the U-boat menace once and for all.

The Pacific
US submarine warfare in World War II began at 5:52 p.m. Eastern Stan-dard Time on December 7, 1941, when Admiral Harold Stark, chief of naval

operations, issued the order, "Execute against Japan unrestricted air and sub-marine warfare."[329] This order was significant for one very important reason. Besides the obvious fact that it had been issued before a formal declaration of war against Japan had been issued, it authorized the use of "unrestricted" war-fare. That specific word meant that the American fleets could target civilian as well as military ships. Civilian crews onboard such ships were no longer non-combatants. They were as much of a target as were Japanese military personnel. All Japanese shipping, from the smallest fishing boat to the largest battleship, were viable targets. This order was in direct violation of international law and apparently issued without consultation with President Roosevelt or any other civilian leader.[330]

It was not an extemporaneous decision on the part of the navy. They had, in the course of their planning during the preceding twelve months, come to the decision that unrestricted warfare would be a necessity.[331] Moreover, this was targeting of civilians as combatants years before the decision was made to take similar action with the strategic bombing of Japan and its civilian in-habitants. The fact that the United States had long been an advocate for the prohibition of unrestricted warfare was not lost on those in command.

The training of the Japanese navy was no different from the training of any other member of the Japanese military. It was brutal to the point of being cruel. The pent-up anger at having been so savagely trained led to the release of these feelings against the helpless survivors of ship sinkings. There was no mercy shown to any survivors. However, the American submariners proved a little more flexible. While civilians were viewed as legitimate targets, in that their shipping of materials directly benefited the Japanese war machine, the at-titudes regarding the survivors as well as the size of the targeted ship changed.

In the early years of the war, the Americans tried to avoid smaller Japanese vessels, such as trawlers and sampans. They were simply too small to bother with, especially from the moral aspect that their contributions to the war machine were likely to be too insignificant. The larger vessels were easier to see as simply targets, just as with the strategic bombing campaign and the targeting of cities. The smaller the vessel, the more intimate it was. This is not to say that smaller targets weren't attacked, but those happened later in the war. Smaller vessels were targeted using the submarine's surface guns as opposed to their torpedoes, which were used on larger vessels (e.g., large merchant vessels and warships). In the early stages of the war, it was simply too dangerous to attack a vessel while on the surface, where it could be detected by other vessels or aircraft.

As the war progressed, the larger targets became fewer and fewer. As this happened, the submarine crews were more likely to turn to the smaller targets simply out of boredom.[332] Additionally, some crews that specifically avoided targeting smaller vessels were criticized for their military worth when they returned to base. Upon returning to sea, they began targeting everything.[333]

Oftentimes, when skippers or crew questioned the morality of targeting smaller craft, they were told simply, "Let your conscious be your guide."[334]

Rescuing the survivors proved problematic. Submarines were not made to transport extra personnel. Most of the time, the survivors were simply left in the water. To do otherwise would be to invite attacks from the enemy. However, more than once, the sub's skipper and crew took it upon themselves to eliminate any survivors by opening the deck guns, along with any weapons carried onboard, on the defenseless survivors. One such incident involved the submarine USS *Wahoo*.

On January 26, 1943, the *Wahoo* sank the Japanese troop ship *Buyo Maru* off the coast of New Guinea. The captain ordered the deck guns to fire on the survivors. The effect of the 4-inch and 20 mm rounds was "rounds [ripping] through timbers and flesh and bone, and the sea stained red, attracting sharks."[335] They were given orders to destroy any life rafts they spotted. One survivor was seen playing possum. The captain ordered, "Shoot the Sonza bitch," so the Japanese sailor was put out of action with a Tommy gun.[336] Upon returning to base, the captain, Dudley "Mush" Morton, was given a hero's welcome and eventually awarded the Navy Cross, while the *Wahoo* was awarded a presidential unit citation. The after-action report made it clear what had happened to the survivors, and it was met with approval.[337] This appears to be the only such incident to have occurred in the Pacific US submarine service.

The US submariners fought the early battles under one enormous disadvantage to the Japanese: defective torpedoes. Add to that the fact that article 22 of the London Naval Treaty of 1930 forbade submarines from sinking merchant vessels "without having first placed passengers, crew and ship's papers in a place of safety," which made submarine warfare impossible. The early submariners were working with not one but both hands tied behind their backs.[338] Because of the treaty limitation, submarine skippers were not taught to attack merchant ships. Instead, they were to attack enemy heavy ships (cruiser, battleship, or carrier). Additionally, in order to conserve torpedoes, they were not to fire in spreads but instead single torpedoes.[339] The problem with the skippers' method of attack was only solved with time, in order to weed out the ineffectual commanders. The torpedo problem was not solved until the correct people finally believed the submarine crews that it was the torpedoes' fault, *not* the crews'.

Japanese submariners fought under the same concept as Americans: They were to focus on sinking the enemy's heavy ships. In the days following Pearl Harbor, there were no heavy ships to attack. However, their strategy remained the same. Herein lies the major difference in results between American and Japanese submarine warfare: The Japanese never altered their submarine doctrine and focused on sinking US heavy warships throughout the war, or they engaged in such wasteful operations as refueling seaplanes, ineffectual coastal shellings, and transporting supplies to the Japanese army. In other words, everything other than that at which they excelled.[340] In comparison, the Americans

soon overcame their cautious training and became more aggressive in search-
ing out targets rather than waiting for them to come to the submarine. And
more importantly, they began targeting merchant ships carrying supplies bound
for both Japan and her troops abroad. Even so, capital ships remained priority
number 1, while the targeting of merchant ships happened as circumstances
presented.

In total, the US submarine service destroyed 1,314 enemy warships in the
Pacific, which represented more than half of all Axis power warships, for a total
of 5.3 million tons of shipping.[341] Total Japanese losses from all US sources, including subs, aircraft, and mines, totaled 2,728 vessels and 9,736,068 GRT.[342]
This, at a loss of 375 officers, 3,131 men, and 52 submarines, represented 16
percent of the officers and 13 percent of the men serving in the submarine
service.[343] Along with the submarines, in 1945, B-29s flying from the Mari-
anas laid 13,102 mines around the Japanese home islands. Ships leaving Japan
were sunk outright; repair yards were blocked, preventing the repair of damaged ships; and major ports were blocked, forcing the use of inferior ports.[344] By
the end of the war, Japan was starved for resources. Between December 1941
and August 1945, while the Japanese lost the aforementioned 9.7 million GRT
shipping, they could only replace 2.5 million tons. By March 1945, oil imports
to Japan had been reduced to essentially nothing. Had Japan not surrendered
when they did, there likely would have been mass starvation of the Japanese
populace during the winter of 1945–1946.[345]

Still, it could have been worse. As mentioned, capital ships continued to
be the focus of their searches. The United States had broken the codes de-
tailing the shipping routes and schedules. Additionally, the US radar had advanced to the point that enemy ships could be identified before they had even
broken the horizon, allowing the subs to stay on the surface while they tracked
the enemy vessels. These two advances would have allowed the United States
to eliminate all Japanese shipping with little risk to the subs. Instead, they were
ordered to look for capital ships. This was an exercise in futility, as during the
entire war in the Pacific, the only such vessel to be sunk was one cruiser, the
Kako, sunk by a US sub, the *S44*, in August 1942.[346] And even when Admiral
King warmed to the idea, Admirals Nimitz and Spruance were doubtful of the
feasibility of the plan, and MacArthur wouldn't even consider it as it would
deny him the glory of returning to the Philippines.[347]

Ultimately, the same reasoning that led the United States to utilize strategic bombing, firebombing, and atomic bombs against Japan also led its submarine crews to target any enemy vessel as a legitimate target of war, whether
that be in the sinking of the vessel or the elimination of the survivors of said
attacks.[348] The difference was that the aviators were acting under direct orders
against specific targets. The submarine skippers were given wide latitude for
the vessels they chose to attack and the state in which the survivors were left.
Captain Morton said, "Anyone who has not witnessed a submarine conduct a

battle surface with three 20 mm and a four inch gun in the morning twilight with a calm sea and in crisp and clean weather just 'ain't lived.' It was truly spectacular."[349] On the other extreme were those who did whatever possible to minimize the plight of the survivors. After sinking the five-thousand-ton cargo ship *Teisen Maru*, shots were fired at the USS *Flasher* from a life raft. When asked if they should return fire, Captain Reuben Whitaker ordered the *Flasher* clear of the area stating, "The survivors are in enough trouble already in the middle of the South China Sea—besides, they hadn't hit us."[350]

On the lighter side of warfare is the event of the emergency appendectomy performed while submerged in enemy waters off Rabaul. Seaman First Class Darrel D. Rector had just been diagnosed with acute appendicitis by the boat's pharmacist's mate first class Wheeler B. Lipes. The submarines were too small to warrant a full-time physician onboard, so pharmacist mates assumed that role, tending to small, non–life-threatening incidents. Knowing Rector would die without treatment, skipper Lieutenant Commander Burlingame, after assurances that Lipes could indeed perform the procedure, descended to twenty fathoms, and Lipes gathered everyone possible to clean the wardroom to serve as the operating room, sterilizing everything possible. With everything ready, the operation began. Five hours later, the procedure was finished and the boat, USS *Silversides*, took off in pursuit of quarry. Finding a destroyer, they began to be depth charged, which shook the boat so severely that the recovering Rector was tossed from his bunk. He was still able to make a full recovery.[351] This was not the only emergency appendectomy performed by a pharmacist's mate. At least three such procedures were required during the war.[352]

This does, however, introduce another important point. While technology had made it easier to kill and to kill in greater numbers, it was also helping prolong life and ameliorate the effects of war. The Second World War was the first war in which battle fatalities exceeded disease fatalities during wartime.[353] Improved sanitation as well as personal hygiene reduced camp pollution and its effects on soldiers. Military personnel were immunized against smallpox, typhoid fever, cholera, plague, tetanus, yellow fever, and typhus.[354] Tetanus, which had been a major cause of postsurgical mortality, with rates ranging from 20 to 58 percent, dropped to eleven total cases during the war and only four deaths.[355] Immunization made it so there was not a single case of yellow fever among US servicemen during the war. And penicillin, the miracle drug, was produced in such quantities that by D-Day (June 6, 1944), 100 *billion* units of the drug were being produced each *month*, leading to death from infection to be nearly zero from that point on.[356] By 2007, there were more fatalities from suicide (113) than there were from disease (63).[357]

THE AIR WAR AND STRATEGIC BOMBING

Hugh Trenchard, originally a commander of Britain's Royal Flying Corps during World War I, bluntly stated, "An aeroplane is an offensive and not a defensive weapon," and sought to wear down the enemy through constant ae-rial attacks.[358] In this early stage, though, he did not envision the role of the bomber as did Douhet. When General Jan Christian Smuts submitted a re-port in 1918 calling for an independent air force to bomb enemy industrial and population centers, Trenchard felt that such a creation would divert resources from other essential services and believed the idea that victory could be ac-complished solely through airpower was merely a "bare assertion."[359]

These beliefs continued to develop over the next decade. In 1928, he wrote that rather than destroy the opposing ground forces, air forces should "paralyse from the very outset the enemy's production centres and munitions of war of every sort and to stop all communications and transportations." Further expanding on this theme, he continued, "The great centres of manufacture, transport and communications cannot be wholly protected. The personnel . . . who man them are not armed and cannot shoot back. They are not disciplined and it cannot be expected of them that they will stick stolidly to their lathes and benches under the recurring threat of air bombardment."[360] Given his po-sition during that time as chief of air staff, these beliefs formed a major basis for RAF doctrine before and during World War II. He helped establish the RAF Staff College in 1921, and most of his doctrine was incorporated into the curriculum there. Some individuals realized that these effects had never been quantified and therefore questioned the actual potential for aerial bombard-ment, along with questioning the ethics of targeting civilians, but they were unable to implement their opinions to any appreciable extent.

The US opinion tended to follow a similarly ambivalent approach. The thought of affecting the civilian population through bombing was paired with the realization that a strong government could potentially counteract this ef-fect.[361] However, one person who vehemently echoed Trenchard was General William "Billy" Mitchell. He argued for the air attack of vital centers supply-ing food, transportation, and other supplies, and he suggested that the entire population of an enemy, including women and children, should be targeted be-cause they contribute to the war effort. He also emphasized the indirect effect of bombs on these populations by creating such nervousness as to stop work during the day and prevent sleep at night. Like Douhet, he believed that aggressive bombing would actually make wars shorter, thereby saving lives and serving as a deterrent to future wars.[362] Such beliefs slowly were codified in the US Army Air Corps (USAAC) doctrine: "War is essentially a conflict of moral forces. A decision is reached not by the actual physical destruction of an armed force, but by the destruction of its belief in ultimate victory and its will to win."[363]

From these beginnings arose the belief in the invincibility of the bomber, until it achieved a near-religious status. This cult of the bomber was summa-rized by former British Prime Minister Stanley Baldwin in 1932: "The bomber would always get through."[364] While the Americans at least acknowledged that enemy fighters and antiaircraft artillery could pose a problem to the bombers, they still believed, even though the Spanish Civil War indicated otherwise, that massed bombers armed with multiple machine guns would still emerge successful and unscathed from any mission they undertook.[365] This blind faith caused the United States to basically ignore fighter development, which they later regretted. The advent of such fighter planes as the Hawker Hurricane and Supermarine Spitfire for Britain and the Messerschmitt Bf 109 for Germany, which were mounted with eight machine guns and could fly faster than three hundred miles per hour, soon proved this error in thinking.

As evidenced in these writings, one of the major early theses was that civilians would not be able to psychologically endure the effects of an aerial bombardment. Only certain mindsets would tolerate such treatment: those who were of sufficient breeding or sufficient training. The general population, it was felt, had neither of these qualities. A popular opinion was that the moral character of humanity was already in decay from the effects of the city environment on the urban population. Very little would be required to push them over the edge into a mental psychosis. Given the novelty of aircraft, air warfare, and the lack of reference points on which to base a thesis, these extreme theories seemed entirely viable. The fear of ever-increasing technology, manifested in the use of powered aircraft and poison gas during World War I, combined with vivid imaginations naturally led to a worst-case scenario of the combination of the two and the easy destruction of civilians.

Trenchard wrote in 1916, "The moral effect produced by a hostile aeroplane . . . is all out of proportion to the damage which it can inflict. The sound policy which should guide all warfare in the air would seem to be this: to exploit the moral effect of the aeroplane on the enemy, but not to let him exploit it on ourselves."[366] Echoing these sentiments, Captain B. H. Liddell Hart argued that civilians could not tolerate a deviation from normal routines and advocated the use of aerial bombardment to "[dislocate] their normal life" and force their surrender.[367] The airplane had eliminated the boundary between military professionals and civilians and led to the realization that war had progressed from local fronts into ever-expanding areas of accessibility. The fear of complete cataclysm fueled both pacifists, who demanded complete disarmament, and militarists, who called for complete armament to secure the only protection available.

Along with these views of humanity was the opinion that not only were people vulnerable to such attacks but also that civilization was becoming increasingly vulnerable to disruption due to the increasing specialization of society and industry. Frequently using body metaphors, civilization was seen as a body, and the elimination of any important part, such as the heart or brain, would lead to the death of the entire body. Likewise, disruption of any key industry would lead to a collapse of the affected society.[368]

These assumptions about civilians demonstrate a certain amount of class, sexual, and racial bigotry. The fact that the majority of civilians were young, old, female, and non-White correlates with the theorists' ideas that they would not be able to withstand such exposure, while the White male elite could. Also, targeting civilian populations, while resulting in fewer casualties overall, would result in a proportionately greater casualty rate among those same classes of people, thereby saving the White male elite from such carnage and reinforcing their "rightful" place in leading the countries. Such views were exemplified by the edict of Kaiser Wilhelm II of Germany in 1915 that no attacks were to be made against the British royal palaces, and they were repeated by Lord Cherwell, advisor to Prime Minister Churchill, when he rejected the bombing of the mansions of Nazi leaders.[369]

Although the typical reasons for waging war were present, World War II was fundamentally different from previous wars in that this war was often fought for racial and other reasons of prejudice and not just military and political reasons. The use of race can make a normally reprehensible act seem completely logical, especially when combined with derogatory racial epitaphs: Japanese were monkeys, Jews were vermin, Germans were rats, and Russians were swine. By using such animalistic terms and other impersonal phrases, like attacking metaphorical body parts instead of living people, and combined with technologies allowing combatants to strike people who are not visible from 20,000 feet or higher, the ability to wage war on such a scale of totality reached unprecedented levels during World War II.

However, not everyone accepted the theorists' belief in the vulnerability of civilians. Officially, before World War II, Luftwaffe doctrine prohibited the bombing of civilian populations for the sole purpose of attempting to induce terror. Luftwaffe regulation 16: *Luftkriegführung* [*The Conduct of the Aerial War*], published in 1935, states, "Attacks against cities made for the purpose of inducing terror in the civilian populace are to be avoided on principle," except for retaliation against similar enemy attacks.[370] The express reason given is that the "retaliation attack requires exact knowledge and understanding of the thought patterns and moral attitudes of the enemy population. Selection of the wrong time, combined with a poor estimate of the desired effect upon the enemy can in some circumstances result in an increase in the enemy's will to resist."[371] Furthermore, in 1937, the German army, navy, and air force manuals included a section specifically prohibiting terror raids.[372]

Even Winston Churchill, who openly endorsed the bombing campaign against Germany, seemed ambivalent about the effects of terror bombing. During the latter years of World War I, he commented that it was doubtful that air raids against Germany would cause the government to surrender: "Nothing that we have learned of the capacity of the German population to endure suffering justifies us in assuming that they could be cowed into submission by such methods."[373] Toward the end of World War II, however, when he realized that such total bombing of Germany would result in an "utterly ruined land," especially after the bombing of Dresden, he started to distance himself from his earlier support of such tactics due to the possibility of postwar political back-lash and instructed Bomber Command to avoid terror bombing for the sake of terror bombing.[374] This ambivalence is shown by his statement to Sir Archibald Sinclair, secretary of state for air: "It is not decisive but better that [sic] doing nothing, and indeed is a formidable method of injuring the enemy."[375] Only experience would prove whether civilians would be resilient or descend into anarchy when faced with aerial bombardment.[376] The views of civilian vulnerability played a major role in the prosecution of World War II. The theory of civilian vulnerability and weak morale was proven wrong through the air war of Germany against England in 1940, the bombing of Germany from 1943 to 1945, and the firebombing of Japan in 1945.

The theories of Douhet were also the basis of the Allied doctrine of strategic bombing in World War II. The basic premise in preparing for US entry into World War II was that a war of annihilation through the destruction of enemy forces, compared to one of attrition, was the fastest way to victory.[377] Even in their preparations for a possible total war, the US Army estimated that only 10 percent of a country's population could be mobilized without disrupting its economy.[378] In preparing for such possibilities, the USAAC Air War Plans Division (AWPD-1) of 1941 followed Douhet's theories and stated,

> The basic conception on which this plan is based lies in the application of air power for the breakdown of the individual and economic structure of Germany. This conception involves the selection of a system of objectives vital to continued German war effort, and to the means of livelihood of the German people, and tenaciously *concentrating all bombing* toward the destruction of those objectives.[379]

Furthermore, the US Air Corps Tactical School (ACTS) stated strategic bombing to be the

> direct attack against the most important elements of an enemy's warmaking capacity, for example, his industries, communications, and the morale of his civilian populations, as opposed to the units and equipment of his armed forces. The object of such bombing, which is the product of an age in which the distinction between soldier and civilian had disappeared, is to undermine the enemy's war effort.[380]

There were a few vocal opponents to the cult of the bomber. Captain Claire Chennault wrote that the P-26 then in use by the USAAC was obsolete compared to fighters from other countries. Likewise, he stated that the bomber should not be the "first exception to the ancient principle that for every weapon

there is a new and effective counter weapon."[381] His retirement due to a physical disability left the bomber worshipers firmly in command of the USAAC.

Defeating the enemy by defeating only its military forces was still possible
but no longer a credible option. Defeating the enemy required eliminating its
potential to make war. The bombing of Germany began as soon as Britain and
the United States could start and continued throughout the war. In the case of
Britain, while the case for bombing enemy targets had been advocated for years,
aerial attack was basically the only means to pursue an offense against Germany
in the early years of the war. With the elimination of Allied ground forces from
Continental Europe and the loss of needed weapons during the evacuation of
the Continent, the only weapon left was an air assault against Germany. Churchill wrote on July 8, 1940, "We have no Continental army which can defeat
the German military power. The blockade is broken and Hitler has Asia and
probably Africa to draw from. . . . But there is one thing that will bring him
back and bring him down, and that is an absolutely devastating, exterminating
attack by very heavy bombers from this country upon the Nazi homeland."[382]

Given the theories of Trenchard and the appointment of Sir Arthur
"Bomber" Harris as the head of the RAF Bomber Command, it seemed almost inevitable that strategic bombing would be the main British offensive
weapon during the early years of the war. The first such raid occurred on the
night of May 11–12, 1940, when RAF bombers attacked Mönchen-Gladbach,
Germany. This was followed by another raid on May 14–15, with the official
doctrinal consent for strategic bombing given on May 15.[383] However, these
bombing raids did not start until after Sir Samuel Hoare, First Viscount Tem-
plewood, replaced Sir Howard Kingsley Wood as secretary of state for air (air
minister) on April 3, 1940. While Wood was responsible for increased airplane
production, which helped Britain to be able to produce as many military aircraft as Germany by the start of World War II, he was also an obstacle in the
implementation of strategic bombing.[384] Wood refused to let the RAF bomb
German arms factories, as he considered the factories private property. In all
likelihood, this was a direct result of Wood having worked as an industrial in-
surance specialist, which led to his knighthood in 1918, before he was elected
to Parliament. His failure to allow the early strategic bombing led one historian
to refer to him as a "ninny."[385]

Britain quickly realized that they did not have the ability to target individual sites, such as factories, nor did they have the ability to adequately protect their bombers on daylight raids, so they switched to the nighttime area
bombing of Germany.[386] In 1942, the British issued Directive 22 to Bomber
Command, calling for the targeting of residential neighborhoods. The directive
listed as targets the "morale of enemy civil population, in particular industrial
workers," with targeted areas to be "built-up areas, not for instance, the dockyards or aircraft factories."[387] This was followed by the incendiary bombing of
residential sections of the cities of Lübeck, Rostock, and Cologne.[388] In January

1943, at the Casablanca Conference, the United States and Britain issued the following statement: "Your primary objective will be the progressive destruction and dislocation of the German military, industrial and economic system and the undermining of the morale of the German people to a point where their armed resistance is fatally weakened."[389]

The United States, in contrast, continued to express its confidence in the ability to hit such small targets and defend its bombers while doing so. However, they were not completely averse to area bombing when necessity dictated. The British nocturnal aerial bombings most closely resembled total war through the indiscriminate bombing of German cities, and the combination of US daylight precision and British nighttime area bombing resulted in the near devastation of Germany. The United States built replicas of German (and Japanese) housing in the desert of Utah in 1943 in order to bomb them and study the results as well as test the new M-69 napalm bombs to determine their effectiveness. The Allies used nearly any and every means at their disposal to defeat the Axis powers.

Even though Germany had little strategic bombing capability (Hitler once stated, "The munitions industry cannot be interfered with effectively by air raids. . . . Usually the prescribed targets are not hit."), Hitler remained fixated on aerial bombardment and remarked, "The decisive thing is that the English will stop if their cities are knocked out, and for no other reason. . . . I can only win this war if I destroy more of the enemy's [cities] than he destroys of ours, by teaching him the terrors of war."[390] This dichotomy seemed especially interesting given the German development of long-range bombers (the Gotha and Giant) during World War I and the resulting raids on London. However, the severe losses they incurred during these raids helped convince Germany that air power was better served in a tactical rather than strategic role.[391] Thus, while Hitler's vision of air power in general was severely limited to its use as a tactical weapon in conjunction with land forces, he appreciated its strategic potential as a weapon of terror and used it as a threat against smaller countries so they accepted German rule before the actual start of World War II.[392]

Hitler continued with this practice during the invasion of Poland in 1939 and Holland in 1940. Such cities as Warsaw, Rotterdam, and Belgrade were bombed, not just in support of the invading Wehrmacht, but also to instill terror and panic in the civilian populations. Warsaw was targeted to quickly finish the Polish campaign. James S. Corum stated that the attack on the city had many characteristics of an indiscriminate terror raid, with the Luftwaffe haphazardly throwing thousands of small incendiary bombs out airplane cargo doors, resulting in published reports of between 20,000 and 40,000 dead.[393] In the case of Rotterdam, the heart of the old city, which held little military value, was bombed mercilessly, even as the Dutch were surrendering to German authorities.[394] Approximately eight hundred civilians were killed in the attack.[395] During the attack on Yugoslavia in 1941, Belgrade was subjected to a bombardment of nearly one and a half hours, resulting in the death of more than 17,000 inhabitants.[396] The very name Operation Punishment, the air portion of Operation 25, the attack against Yugoslavia, indicates just how ruthless and total the war against that country was.

While the Italians generally assumed a subservient role to Germany during World War II, they, too, had their experiments in terror bombing. During Mussolini's invasion of Abyssinia in 1935, their air force routinely used high-explosive and gas bombs (specifically mustard gas) against villages, horsemen, Red Cross hospitals and ambulances, and retreating troops. Vittorio Mussolini, son of the Italian dictator, who flew in these attacks, described the effects of bombing the horsemen as "blooming like a rose when my fragmentation bombs fell in their midst. It was great fun."[397]

With the honing of such techniques and weapons, many German cities experienced the destruction of more than half of their respective housing, along with thousands of civilian deaths.[398] Among notable targets were Hamburg in July–August1943 and Dresden in February 1945. As with the case of Germany's Operation Punishment against Yugoslavia, the code name for the attack on Hamburg gave an indication of the goal of the raid: Operation Gomorrah. The bombing resulted in multiple fires coalescing into a huge firestorm that incinerated the city. The firestorm became self-nourishing, sucking in people and objects and creating temperatures high enough to burn the asphalt in the streets.[399] Any attempt at controlling the fire was useless due to its size and resulting hurricane-force drafts. Six square miles of the city's core were burned, destroying 49.2 percent and damaging an additional 30 percent of the city's housing, and 40,000 to 100,000 people died.[400]

Likewise, the firebombing of Dresden resulted in a firestorm that gutted the city. In this case, unprecedented accuracy of bombing over the city with the high explosives and incendiary bombs, combined with a major influx of Eastern Front refugees, led to a death toll of 25,000 to 35,000 people.[401] The justification for choosing Dresden as a target was primarily to speed up the end of the war. Dresden was home to several important industries and was a major transportation hub for highway and railway traffic.[402] Combined with the fact that the Allies had just survived the Battle of the Bulge, many military leaders were convinced that the Nazi war machine still had plenty of life left in it, and a quick resolution to the war, commonly predicted in the latter months of 1944, was nowhere in sight. While the tactical results met Allied expectations, the mission did little to actually weaken the German civilian resolve.[403] In fact, Albert Speer, Nazi minister of armaments and war production, stated that it was the American attacks on industrial targets that caused the breakdown of the German armaments industry, and Luftwaffe field marshal Erhard Milch said that had the Allies started their attacks against petroleum refineries earlier, the war would have been over earlier by the same amount of time.[404] These opinions were echoed by the United States at the conclusion of the European war:

"The most serious attacks were those which destroyed the industry or service which most indispensably served other industries."[405]

The final statistics are staggering. As stated by the *United States Strategic Bombing Survey* [USSBS] (*European War*),

> 485,000 residential buildings were totally destroyed by air attack and 415,000 were heavily damaged, making a total of 20 percent of all dwelling units in Germany. In some 50 cities that were primary targets of the air attack, the proportion of destroyed or heavily damaged dwelling units is about 40 percent. The result of all these attacks was to render homeless some 7,500,000 German civilians.[406]

While both world wars resulted in great loss of life and World War I was viewed as the "war to end all wars" due to its horrific casualties in the trench warfare of the Western Front, World War II was fundamentally different in the disparity in civilian deaths: Civilians accounted for nearly 60 percent of the total casualties in World War II, whereas in World War I the total was 5 percent.[407] The total number of World War II civilians dead due to various reasons is estimated at 20,946,000. This is proportionately greater than those civilians killed by the regimes in either Communist China or the Soviet Union and demonstrates the totality of the conflict.[408]

The psychological results of civilian bombings were far less than had been predicted by prewar bombing advocates. Civilian morale was more resilient than previously assumed. This was due to many factors, but two major ones, first, "in a police state, it [is] hard to translate popular dissent into political pressure"; and second, civilians, unlike combat troops, when subjected to bombardment, are able to seek shelter in other areas.[409] Troops are often forced to endure such experiences with little shelter and no means of seeking adequate shelter. Therefore, because the circumstances between military and civilians subjected to aerial bombardment are different, "their psychological state does not parallel the psychological state of the soldier."[410] The USSBS (*European War*) specifically refutes Douhet's theory of civilian morale collapse:

> The mental reaction of the German people to air attack is significant. Under ruthless Nazi control they showed surprising resistance to the terror and hardships of repeated air attack to the destruction of their homes and belongings, and to conditions under which they were reduced to live. Their morale, their belief in ultimate victory or satisfactory compromise, and their confidence in their leaders declined, but they continued to work efficiently as long as the physical means of production remained. The power of a police state over its people cannot be underestimated.[411]

Similar results occurred in Britain during the Blitz of 1940–1941. While there were incidents of individual hysteria, panic, and death from the shock of the situation, and while many more either fled London or turned the Underground into makeshift bomb shelters, the vast majority of the population continued about their daily life. As Ian Patterson describes it,

> As long as the bombing of cities was an impending threat, happening only somewhere else, to other people, in Spain or in China, it was capable of filling the imagination and creating almost limitless apprehension. When actual German planes started dropping high-explosive and incendiary bombs on London, the threat materialized. After a while it almost became dull. What had been exceptional became tolerable.[412]

This also refuted the supposed inferiority of individuals of "lesser" stature compared to the entitled White males. Women, non-Whites, and the poor were just as able to endure the hardships of war as was the socially elite military class of White males.

The Japanese refusal to surrender was a major reason behind the US firebombing missions and dropping the atomic bombs. The United States had every reason to believe that an invasion of the Japanese home islands would be necessary. The invasion of Iwo Jima resulted in 6,000 dead and 25,000 wounded Americans, while only 200 of the 20,000 Japanese stationed there were taken alive, and most of these were wounded.[413] Likewise, the Battle for Okinawa resulted in 49,000 dead or wounded Americans.[414] Casualty estimates for the invasion of the home islands, Operation Downfall (consisting of Operation Olympic and Operation Coronet), were 1.7 to 4 million US soldiers, including 400,000 to 800,000 dead, and 5 to 10 million Japanese deaths.[415] These figures do not include the potential number of Chinese, British, and Russian casualties, nor do they include the probable killing of prisoners of war the Japanese held.

The Japanese, however, fully planned on casualties of 20 million in such an invasion.[416] The introduction of the kamikaze during the Battle of Leyte Gulf in October 1944 further strengthened the opinion that the Japanese would stop at nothing in attempting to defend themselves. The kamikaze targeted military ships and not civilians, which the United States viewed as complete fanaticism that had to be defeated with any method available. However, the hypocrisy of denouncing such determination on the part of the Japanese while elevating it to mythological status during such American events as the Battle of the Alamo or Custer's Last Stand is apparent.[417]

On Saipan and Okinawa, the Japanese military was not the only group to demonstrate such fanaticism. Civilians, either voluntarily or through force, offered their lives for the emperor, with between 8,000 and 22,000 committing suicide on Saipan.[418] One woman did so while giving birth to her child, with only the baby's head having passed when she died.[419] US planners understood that an invasion of the Japanese home islands would be a bloody undertaking. The United States tried to implement the same procedures against Japan that they had against Germany. Early plans called for the identification of "key" and "priority" targets, in keeping with the concept of strategic bombing in Germany.[420] However, the conditions over Japan were different enough (including bad weather and the jet stream at 25,000 feet) that precision bombing had very poor results. In particular, the jet stream, traveling at more than two hundred miles per hour, made the B-29s useless: Bombing runs downwind resulted in the planes traveling so fast that the crews didn't have time to sight the targets, while runs upwind resulted in the planes flying backward. Additionally, the Norden bombsights were unable to compensate for the wind speed in crosswind runs.[421] The solution was straightforward, but revolutionary: decrease the bombing altitudes from 25,000–30,000 feet to 5,000–10,000 feet; fly at night and single instead of in formation; and carry heavy loads of incendiary bombs. In other words, the United States switched from their concept of precision bombing to the British concept of area bombings.

The idea of firebombing Japanese cities had been brought up before the war. In 1932, Billy Mitchell wrote that in the case of war with Japan, "Incendiary projectiles would burn the cities to the ground in short order."[422] Immediately prior to the attack on Pearl Harbor, General George C. Marshall, US Army chief of staff, said, "If war with the Japanese does come, we'll fight mercilessly. Flying fortresses will be dispatched immediately to set the paper cities of Japan on fire. There won't be any hesitation about bombing civilians—it will be all-out."[423] In fact, the revolutionary aspect of this plan was not the use of incendiaries; their use against Japanese industry and civilian workers had been planned since October 1943.[424] Instead, it was the low-level bombing runs and the removal of defensive weaponry and crew, which resulted in a doubling of the B-29 payload, that was revolutionary.[425]

As destructive as the Allied bombing of Germany had been, the bombing of Japan was even more so. The results of this change were devastating: 1,667 tons of bombs were dropped on Tokyo in this first attack, resulting in the destruction of 15 square miles of Tokyo's most densely populated area, along with an estimated 185,000 casualties, although this figure was later lowered to approximately 80,000–100,000.[426] In total, 66 cities were attacked by incendiaries, resulting in 178 square miles being razed, 22 million people (30 percent of the entire population of Japan) being left homeless, 900,000 people killed, and 2.2 million total civilian casualties. Compare this to Japanese combat casualties of 780,000 for the war.[427] All of this was accomplished with minimal US losses: The March incendiary missions lost only 0.9 percent of the American participants.[428] This was a near-textbook implementation of Douhet's theory of bombers causing maximum damage on civilian populations.[429]

The effects on morale were examined from the European theater. Examination of the Pacific theater and the USSBS *(Pacific War)* further supports the fact that low morale does not translate into action against the government:

> Progressively lowered morale was characterized by loss of faith in both military and civilian leaders, loss of confidence in Japan's military might and increasing distrust of government news releases and propaganda. . . . Until the end, however, national traditions of obedience and conformity, reinforced by the police organization, remained effective in controlling the behavior of the populations. . . . It is probable that most Japanese would have passively faced death in a continuation of the hopeless struggle, had the Emperor so ordered.[430]

In fact, morale was not listed as an objective in the planning of the firebombing raids, indicating that this was not their main purpose and the resulting civilian casualties. When lowered morale was mentioned, it was in the context of lowering industrial production, not inciting civilian rebellion or demanding surrender by their government.[431] Additionally, the USSBS *(Pacific War)* stated that the effects of bombing on Japanese society were "uncertain."[432]

As much as World War II involved death on a scale previously unheard of, there were limits to the means the belligerents would use. However, the issue of why these self-imposed limits exist is usually more difficult to explain. For example, Kyoto, Japan's fourth-largest city, was never intentionally targeted for any type of bombing due to cultural and religious reasons, even though Army Air Force planners recognized its importance.[433] Additionally, poison gas was never intentionally used during World War II except in very limited cases, as in the cases of Italy and Japan previously mentioned. Reports that it had been used in the bombing of Guernica were erroneous.[434] Douhet originally proposed the use of explosive, incendiary, and poison-gas bombs. Because poison gas was never used in combat to any appreciable degree, one is left to wonder how history would have differed if it had been. While the use of poison gas by ground troops could potentially prove as lethal to the army using it as to the army against which it was used, the possibility existed to do as Douhet had theorized and drop such bombs from airplanes, keeping the air crews isolated from its effects. Indeed, the question can be asked, What if Hitler had ordered its use during the Blitz of London? Or, What might have happened had gas been used at Iwo Jima or, even more disturbing, had been used with the incendiaries on Tokyo, as Douhet had envisioned? While these questions can never be answered, one can get an idea of the implications by remembering that while Germany did not use poison gas as a military weapon, high-ranking Nazis had no qualms about using it as an effective means of mass extermination of the Jews in concentration camps.

President Roosevelt was presented with the recommendation that poison gas be used in the invasion of Iwo Jima, based on the facts that there were no civilians present on the island and that the United States was not bound by the treaty banning its use, but he refused on moral grounds.[435] Such constraints had been imposed earlier but often for more bigoted reasons—to protect the ruling class at the expense of the working masses, as discussed previously.

Some argue that war can never be justified, that there is no such thing as a just war.[436] However, in such morally based discussions, the rationale for total war is ultimately reduced to the intentions of the belligerents. Whereas the Germans and Japanese waged most of their wars because of perceived racial superiority, the United States and Britain used similar tactics in a defensive response to the actions of Germany and Japan. The Soviet Union falls between these two extremes: They were belligerent toward Poland and Finland early in the war but switched to a defensive mode with the German invasion of their country.

The morality of the Allies using means they had previously decried is explained by one fact: They did not instigate the war but did everything possible to end it. Had Hitler not wantonly killed other racial groups, both as a result of combat and racial pogroms, the Allies, including Russia, would not have been forced to retaliate. The same held true for Japan. The only reason the United States used incendiary and nuclear weapons was that the Axis aggressively attacked neutral countries (including the United States) in its quest for territory, resources, and racial supremacy and refused to surrender in the face of overwhelming odds.

Common opinion among the US troops at the time was that they wished the "bomb" (the atomic bomb) had been available earlier to use in Germany in order to have ended the conflict there sooner.[437] In all probability, had atomic weapons been available to use against Germany, they would have been used (because that was the original intended target) and for the same reason as in Japan: to end the war a quickly as possible with the minim loss of American lives.[438] Given the examples presented, if one is to consider the intentional targeting of civilian populations for aerial bombardment as "terror" bombing and not just "strategic" bombing, then it appears that the Allies were guiltier of such actions than Nazi Germany. Again, the main difference is that Germany instigated the hostilities, whereas the Allies responded to those hostilities in an effort to end them.

Not surprisingly, Germany denounced the Allied bombing of Dresden in February 1945, while the United States and Britain neither condoned nor apologized for it but accepted it as another tool to win the war.[439] Sir Arthur Harris, chief of RAF Bomber Command, had no remorse over this attack because, he felt, the alternative was increased Allied casualties similar to those of the First World War.[440] In September 1939, at the commencement of European hostilities, President Roosevelt appealed to the governments engaged to publicly announce that they would not bomb civilians, following the Hague Rules of Air Warfare of 1907 (which were never ratified and therefore not legally binding) that prohibited the bombing of civilian populations and properties.[441] However, as the conflict continued, he firmly supported the concept of strategic bombing of military targets and fully condoned its preparation: "Against naked force the only possible defense is naked force. The aggressor makes the rules for such a war; the defenders have no alternative but to match destruction with destruction, slaughter with greater slaughter."[442] The continued use of terror tactics by the Axis powers along with the Japanese attack on Pearl Harbor furthered his conviction of the need for bombing and, ultimately, the use of nearly any means possible to quickly end the conflict.[443] The following statement to the German population, signed by Arthur Harris, reflects this reasoning:

> We are bombing you and your cities, one by one, and ever more terri-
> bly, in order to make it impossible for you to go on with the war. . . .
> Obviously we prefer to hit your factories, shipyards, and railways. It
> damages your Government's war machine most. But nearly all these
> targets are in the midst of the houses of those of you who work in
> them. . . . Therefore we hit your houses—and you—when we bomb
> them. We regret the necessity for this. But this regret will never stop
> us; you showed the world how to do it.[444]

One major ethical problem, as demonstrated in this quote, is that military targets and civilian targets are often not clearly delineated. Industries of military value are often intermingled among civilian areas, as in Guernica, Warsaw, Rotterdam, and Belgrade. The Geneva Convention, while forbidding terror bombing of civilians, allowed for the targeting of vital war industries with the probability that civilian casualties would result, along with the breakdown of civilian morale.[445] The bombing of such cities as Warsaw and Rotterdam served as justification for the commencement of British bombing raids on Germany, but the morality of these acts by the Nazis is not as well defined as it might seem.

The bombings of Germany forced the dispersal of industry to smaller sites more difficult to target, and this was even more so in Japan. There, certain industrial chores were outsourced to individual families to be done in their homes. This "home manufacturing" played a major role in Japanese industry and, as mentioned previously, served as a major factor in initiating the firebombing raids against Japanese cities. Of all the cities in Japan, only six were initially designated as industrial and therefore worth attacking.[446] Evidence of the work being done in homes came after various firebombing missions, when drill presses were seen protruding from the remains of the burned houses.[447]

In Warsaw, four surrender demands were made, and leaflets were dropped over the city calling for Polish surrender or suffer the consequences of an air attack. Only after the Poles ignored these warnings did the Germans bomb the city.[448] Furthermore, the French air attaché in Warsaw verified that the German air attack was in accordance with established laws governing military conduct in war and that all the civilian casualties had been located near valid military targets. Finally, the initial casualty reports appear to have been grossly exaggerated; if those numbers were accurate, then the death rate there would have surpassed the death rates of Hamburg, Germany, in August 1943 and Dresden, Germany, in February 1945, commonly acknowledged to be the most lethal bombing raids of the war.[449]

Death tolls in other cities have similarly been revised down in the years since World War II.[450] In Rotterdam, the bombing raids were called off after surrender negotiations started, but one bomber group missed the signal due to weather conditions and proceeded on their mission. The targets they hit included oil storage tankers, normal targets of military value.[451] Additionally, there was no concern on the part of the British in starting the bombing against Germany. It was only after the war that justification was deemed necessary, and for that, the examples of Warsaw and Rotterdam were exaggeratedly used. Churchill referred to the "Massacre of Rotterdam" and it being reduced to a blazing ruin, which was false: Only the southern portion of the city was damaged.[452] After the war, when plans were being made to convene war crimes trials, the suggestion was made that Luftwaffe airmen be put on trial for crimes against humanity for their attacks on these cities. It was decided not to pursue these accusations for fear that accusations could be made against Allied air crews for the same reasons.[453] The line demarking the boundary between justified and unjustified is, indeed, very fine.

An interesting aspect to German terror bombing is that, even with the examples of Guernica, Warsaw, and Rotterdam, after the British instituted the policy of strategic bombing in May 1940, Germany did not initially retaliate. This is in keeping with the fact that Germany viewed its air force as more of a tactical weapon instead of a strategic weapon. Hitler viewed the initial attacks as inadvertent, the result of an individual losing composure or acting on his own accord.[454] Germany instituted the Blitz, the bombing of London, the night after the RAF bombed Berlin on August 25, 1940, in retaliation of the accidental bombing of London when Luftwaffe bombers missed their targets of aircraft factories and oil refineries east of London. In this case, it appears that Churchill was looking for an excuse to begin the bombing of Germany. Even knowing that this event was probably an accident, given the miniscule number of planes involved (twelve) and minimal damage incurred (four casualties), he used the incident as an opportunity to fully unleash the doctrine of unrestricted area bombing.[455] Likewise, the initial bombing of German cities evoked an extreme response from the German public for revenge: "From Danzig it is reported that the *complete destruction of London* is demanded as a tangible lesson to the war mongers and as a reprisal for the bombing of German cities."[456]

However, all these opinions are written from the perspective of the passage of time and using historical research not available when the decisions were made. When presented with a new weapon that would potentially end the war and save American lives, President Truman approved its use. While Truman initially felt that the Soviet invasion of Japan was a necessary step in obtaining Japanese surrender, further knowledge of the potential of the bomb led him to believe that its use alone would negate the need for either Soviet or US invasions.[457] One should also remember that while the bombing of Hiroshima and Nagasaki resulted in the deaths of thousands of Japanese civilians, the death toll during the Rape of Nanjing, by one count, surpassed the combined total for these two atomic bombs, and the actual overall damage caused by these two bombs was significantly less than that caused by the firebombing: The atomic bombs killed approximately one-seventh the number of people as did the incendiaries.[458] General Curtis E. LeMay, in charge of the firebombing missions over Japan, stated, "The assumption seems to be that it is much more wicked to kill people with a nuclear bomb, than to kill people by busting their heads with rocks."[459]

As the results of the firebombing missions against Japan became evident, the USAAF fully embraced the tactic as a way to speedily end the war, and the attacks were extended to other cities.[460] General Henry "Hap" Arnold, USAAF chief, believed that doing so was destroying the Japanese ability to wage war and thereby saving American lives.[461] General LeMay said exactly that: "You possess a natural inborn repugnance against killing people; yet you know you're going to have to do it. . . . And you must train yourself grimly to adopt a philosophical attitude with regard to those losses. If you're going to fight you're going to have some people killed. . . . You have to pay a price in warfare, and part of the price is human life."[462] Dropping incendiaries on Japanese cities killed thousands of civilians but doubtless saved hundreds of thousands of American lives:

> No matter how you slice it, you're going to kill an awful lot of civilians. Thousands and thousands. But, if you don't destroy the Japanese industry, we're going to have to invade Japan. And how many Americans will be killed in an invasion of Japan? Some say a million. We're at war with Japan. We were attacked by Japan. Do you want to kill Japanese, or would you rather have Americans killed?[463]

While such remarks may call into question US racism toward the Japanese in instituting these policies, such attitudes are not indicated in the official documents of the times. However, officials also knew that because of the underlying racism, such policies and actions would probably be overlooked in general.[464]

PROPAGANDA

In the Western world, the word *propaganda* has come to carry with it a negative connotation of the government forcing lies or extreme views on the people in the hopes that they will eventually believe them. In getting the masses to believe the message presented, the governments use the ideas of "exclusion" and "threat" to portray the perceived threat as legitimate.[465] Exclusion includes portraying the unwanted population as less than human—animals, demons, vermin, or disease.[466] Likewise the idea of said population being a threat was conveyed by communicating that the targeted groups were going to either take over the country or exterminate its inhabitants.[467] The issue with propaganda is knowing where to draw the line. How much is too much or too little information? Or how often do we broach the same topic?

Propaganda in World War II portrayed the fight as one between good and evil, that the United States and the Allies were good and virtuous, fighting the forces of evil in their attempt to take over the world. The main Axis leaders, Hitler, Mussolini, and Tojo, were all portrayed as either evil caricatures or buffoons. In this aspect, the government tended to not dictate how to portray the leaders but occasionally suggested an angle to take. One such angle was that the children might need to be "reeducated" in order to minimize the effects of the brainwashing by the Axis powers. Two days later, Dr. Seuss had a cartoon showing a child getting his thoughts cleaned out by the use of bellows.[468]

Propaganda in the days following Pearl Harbor had to be focused on Germany in keeping with the Allies' goal of "Germany first." The inclination of the public was to deal with Japan first, as they had attacked the United States, not Germany. This was especially so with the news of what was happening in Nanjing that had come out of China in the late 1930s. When the Japanese were targeted, they tended to utilize the same tropes: sneaky, grotesque, blindly fanatical, and malevolent. And even though it was what we would today call misinformation, Japanese Americans were portrayed as waiting for the appropriate moment to perform the acts of sabotage they had been trained to do.

Slogans, easy to remember with alliteration, proved very effective: "Loose lips sink ships." Equally effective were sayings that emphasized the need for security and to not spread rumors: "A careless word, a needless loss." Filmmaker Alfred Hitchcock even made a photographic dramatization of the perils of rumormongering for *Life* magazine.[469]

Propagandists had to walk a fine line between how much information to disseminate or not. Too little, to protect national security and avoid traumatizing the public, would lead to the populace demanding to know more of what was happening. Victories were touted, while defeats were minimized. Pearl Harbor was acknowledged as a significant loss, but for security reasons, President Roosevelt said that he could not give any additional information.[470] As is often the case, weaknesses were downplayed, while strengths were extolled. Hitler used the same technique in the days leading up to the start of the war:

When showing off the new air force, he made sure to emphasize how impressive it was, knowing that it was still in its infancy but not wanting the world to know it. Though the deceased were typically not shown, so as not to dwell on the negative, Roosevelt did finally allow it so that the public would not grow overly confident about the war efforts.[471]

Heroes played an important role. In spite of his incompetency in handling the defense of the Philippines, MacArthur was portrayed as a hero, not because he was, but because the public needed him to be so. The Sullivan brothers were portrayed as the heroes they were for having given their lives on the USS *Juneau*, but so, too, were the brothers' parents as they made visits to armament factories and shipyards, encouraging increased production.[472]

Decreased consumer consumption was encouraged. Rubber was in short supply, so it was rationed. The need for rationing was explained to the public to encourage cheerful compliance. Actions as simple as not asking for an extra cup of coffee or pat of butter were seen as ways to support the war effort. All consumer goods were rationed, to the point that the automobile manufacturers ceased making cars for 1943 through 1945. When the new models came out in 1946, they were the same as the 1942 models, as all design and manufacturing efforts had gone into manufacturing military vehicles.[473] The simple action of taking a day off or changing jobs was frowned upon. "The Enemy Laughs When You Loaf" and "Don't Be a Job Hopper" were commonly seen on posters.

However, the Japanese viewed propaganda in a very positive light. In Germany, the Soviet Union, and the United States, propaganda was a one-way street, originating in the highest levels of the government and flowing downward into the general public. In contrast, Japanese propaganda was a two-way street, flowing not only from the government down to the general public but also from the public up to the highest levels of government. As Barak Kushner writes, "The ultimate goal of Japanese wartime propaganda was to connect the home front to the battlefront and structure a stable Japanese society that could help win the war."[474] Unlike Nazi Germany, there didn't exist a centralized Ministry of Propaganda. Instead, multiple entities worked together to disseminate the desired ideology to the people: The Cabinet Board of Information, the Ministry of Foreign Affairs, military propaganda platoons tasked with dealings in China, and the Special Higher Police were all just a portion of the propaganda apparatus in Japan.[475] Kushner notes, "Japanese society believed they had a role in producing propaganda in coordination with the government and military. The government did not simply mandate what society's actions should be."[476] Messages like the following from a Japanese intellectual directed to the outside world were common: "Japan is at peace, we have bathtubs. We have paved roads. We also have stamps. You can find and eat Western food. There are no poisonous insects here. Cholera does not run rampant."[477]

This is not to say that the government exercised no control over propaganda, because it did. One message conveyed by the government was that sex was fine but only if you followed societal norms. Masturbation was frowned on ("Think about what you are doing."), while sex with a comfort woman was deemed acceptable.[478] It also didn't mean that the messages presented were accurate. As is the case in any propaganda, much of it was in error, but the accuracy of the content was not the point. Propaganda does not work on a rational level; it appeals to emotion. Gregory S. Gordon sums it up: "Regardless of its mass dissemination, propaganda works on an individual and human level."[479] In this vein, Americans were portrayed as weak and decadent, without the will for a prolonged conflict involving personal sacrifice.[480] "Better stock up with fifty years' supply of clothes. It's going to be a long war, and money won't do you any good when there is nothing to buy" was one example of such messages.[481]

And just as Hitler and Tojo were objects of American propaganda, Roosevelt was a frequent target of the Japanese, who portrayed him as a tyrannical dictator who falsified the war news in order to deceive the American populace, a "Don Quixote of the present century living in his dreams," a "paralytic cripple with a warped brain."[482] Likewise, MacArthur was a coward for fleeing the Philippines, a "Braggart and Crybaby" who also falsified war results and

Figure 27. The Nazis conducted two major anti-Semitic campaigns during the first half of 1943. This one was issued probably during the first of them in February, just after the defeat at Stalingrad. The text: "The slogan for 1943: Unstoppably onward until final victory!" IMAGE AND CAPTION USED COURTESY OF DR. RANDALL L. BYTWERK, "GERMAN PROPAGANDA ARCHIVE," CALVIN UNIVERSITY, ACCESSED MAY 22, 2025, HTTPS://RESEARCH.CALVIN.EDU/GERMAN-PROPAGANDA-ARCHIVE/POSTERS/1943AA.JPG.

"[concocted] victory news."[483] However, the propaganda was presented in such a way that it was able to pivot 180 degrees and go from portraying the United States as Satan incarnate to the most faithful friend ever once the war was over.

Nazi Germany and Joseph Goebbels were masters of propaganda: "National Socialism was the most prolific rhetorical movement of the twentieth century."[484] Both Hitler and Goebbels were master orators, able to appeal to millions of citizens, getting them to believe something when the exact opposite was blatantly apparent. Goebbels's "Total War" speech of February 18, 1943 is, as Bytwerk points out, probably the single most famous Nazi speech.[485] It was not just a speech; it was a media event. There was no escaping it. It was broadcast in its entirety on national radio twice, printed in all the major newspapers, and reprinted into a booklet with circulation in the millions, and portions of the speech were shown on the newsreel of February 24 before every motion picture shown.[486]

Coming on the heels of the defeat at Stalingrad, it was the first time that the government tacitly acknowledged that defeat might occur. By the end of the war, Germany was so desperate for manpower that they began conscripting males from the age of sixteen to sixty for participation in the Volkssturm, a compulsory home defense militia. These conscripts were to fight the advancing Allied forces and shatter the Allied morale as well as increase the morale of the conscripts and prepare the civilian population for total war mobilization. The messages portrayed were the same as put forth by the United States. The enemy was a nefarious villain, and the people had to do their part in order to win the war (see figure 27).

CHAPTER 6

POST-WORLD WAR II, 1945-PRESENT

"If liberty means anything at all, it means the right to tell people what they do not want to hear."

—George Orwell

Sun Tzu said, "To subdue the enemy without fighting is the acme of skill."[1] After reading this chapter, one may very well walk away wondering if we were successful. In some ways, we were, but in many, we weren't.

As chapter 1 shows, war need not involve large armies amassed against each other in formal battle. War, or, to use a more encompassing term, *conflict*, can occur with any number of combatants. Indeed, as time went on, the use of large-scale forces, such as in World War II, Korea, and Vietnam, gave way to small, more precision-like operations, such as the United States in Grenada in 1983 and Somalia in 1992, or to the war on terror in the twenty-first century. Most of these incidents were racially or religiously motivated and did not necessarily constitute war per se, but they were still armed conflicts resulting in large numbers of deaths. Regardless, wherever there is conflict, there is the possibility of noncombatant involvement.

Asymmetric warfare is defined as warfare between opposing forces that differ greatly in military power; it typically involves the use of unconventional weapons and tactics (such as those associated with guerrilla warfare and terrorist attacks).[2] While the term was first used in 1975, the principle is as old as time. Asymmetric warfare is characterized by the following traits:

Disparity in Power: At its core, asymmetric warfare hinges on a significant imbalance in military strength. The weaker side, lacking the resources and

conventional military might of a state actor, must innovate and adapt unconventional strategies to level the playing field.

Unconventional Tactics: Guerrilla warfare, ambushes, sabotage, and terrorism are quintessential tactics of asymmetric warfare. These methods aim
to maximize damage to the stronger force while minimizing losses for the
weaker side, showcasing the ingenuity and resourcefulness of asymmetric
actors.

Blurred Lines: Asymmetric warfare often blurs the distinction between combatants and civilians. Insurgents may operate within civilian populations,
making it challenging for the stronger force to discern between military
targets and innocent bystanders, complicating the ethical and strategic di-
mensions of conflict.

Terrorism: In the arsenal of weaker actors facing off against stronger adversaries, terrorism emerges as a potent tool for disrupting stability and exerting
influence. Characterized by its asymmetrical nature, terrorism encompasses
a spectrum of violent acts aimed at civilian populations or noncombatant
targets, leveraging fear and intimidation to achieve political, ideological,
or strategic objectives.[3]

The timing of the term's origin is appropriate, given that US involvement in
Vietnam ended that year with the fall of Saigon. Vietnam was a classic David-
and-Goliath encounter. Twenty years earlier, the United States became, along
with the Soviet Union, the dominant world power, with a military that could
vanquish any foe—any foe that was willing to fight a traditional war, that is.
With the same tactics the Americans had utilized two hundred years earlier,
the Vietnamese achieved the same results as the colonists had at that time.

The conflicts that have followed have all been nontraditional encounters.
When the Twin Towers were hit by planes on September 11, 2001, the United
States was once again ready to go to war. But against whom? Al-Qaeda was a
non–state-sponsored entity with no formal headquarters, no established bases.
This is the face of war today. And as always, whenever there is a war, there are
victims, both intended and unintended. What happens to these unintended
victims brings myriad quandaries, as governing bodies attempt to modernize
the rules of law to cope with new circumstances.

These unintended victims, the civilians caught in the cross-fire of insurgents, the human shields whom cowards use to protect themselves from
attack—these are the concern of this book. What follows is a brief overview
of some conflicts since the end of World War II and their impact on civilians.

MUTUALLY ASSURED DESTRUCTION (MAD)
According to the editorial team at Total Military Insight, "Mutually assured
destruction, often abbreviated as MAD, is a doctrine of military strategy characterized by the idea that full-scale use of nuclear weapons by opposing sides

would result in the annihilation of both."[4] With the explosion of the atomic bomb over Hiroshima on August 6, 1945, the world entered the nuclear age. The United States would not be the only country to have nuclear weapons in its arsenal. On August 29, 1949, the Soviet Union detonated their first atomic bomb, and with that, the world entered the nuclear arms race, the result being the adoption of the doctrine of MAD. And while not resulting in a war and the use of these weapons, it very nearly did.

General Curtis LeMay fully embraced the Douhetian theory of total war. He believed a war with the Soviet Union to be inevitable and repeatedly recommended a total preemptive strike against that country to prevent them from striking first. He believed that the number of US casualties would be far fewer than if the Soviet Union struck first. Targeting the Soviet nuclear weapons systems would force them to use conventional weapons to retaliate, causing them to question the feasibility of retaliating. In his view, winning consisted of "achieving a condition wherein the enemy cannot impose his will on us, but we can impose our will on him."[5] The only way to achieve this was through a "successful strategic air *offensive*."[6] The repercussions of such a strategy, had it been implemented, can only be imagined. Fortunately, Presidents Eisenhower and Kennedy did not adopt his recommendations.

When Kennedy was confronted with a situation in which nuclear weapons might have been needed, the Cuban missile crisis of October 1962, he was given six options on how to proceed: do nothing, negotiate, invade, institute a blockade of Cuba, bomb missile bases, or use nuclear weapons.[7] He chose the blockade of Cuba, with Soviet Premier Nikita Khrushchev eventually backing down.

ALGERIA, 1945, 1954–1962

On May 8, 1945, the Algerians celebrated the end of World War II in Europe. However, in the Algerian towns of Sétif and Guelma, the celebrations soon turned to protests. The French police responded by firing on the protestors. Having been under French rule for decades, the Algerians wanted independence from France. The French wanted nothing to do with that and strove to keep the Algerians under their thumb. After back-and-forth sparring since 1943, the French moved to quash the independence movement.

On April 25, 1945, Messali Hadj, leader of the Algerian People's Party (PPA), who were advocating for independence, and proclaimed leader of the Algerian people, was abducted and placed under house arrest.[8] The French, and Europeans in general, feared having to treat the Algerians as equals. Being in the minority fueled French fears: The town of Guelma housed about 4,000 French and 16,500 Muslim Algerians.[9] To help suppress their fears, the French started forming militias to counter the PPA. The militias began rounding up PPA members, which led to the Algerians to take revenge on them. On May 8, a demonstrator in Guelma was killed. By the end of the month, between

1,500 and 2,000 Muslims had died in this town.[10] In Sétif, French civilians were stabbed or shot by the Muslims.

The escalation of hostilities exacerbated the problem, leading to the police, with help from the European civilians in the country, to begin mass executions. The bodies were buried in mass graves and burned in the lime kilns at Heliopolis.[11] The Algerian death toll ranged from a minimum of 1,000 by French estimates to around 45,000 or even 80,000 according to the Arabs.[12] There were only 102 French casualties.[13] However, of those 102, many had their throats cut, genitals removed, bellies slit open and eviscerated, and hands and feet cut off. Regardless, the rebellion was quashed, at least for a few years.

But the rebellion wasn't destroyed. On November 1, 1954, the Front de Libération Nationale (National Liberation Front; FLN) started the War of Independence by calling on all Algerians to revolt against French rule. While the Algerians primarily used guerrilla warfare against the French, the French continued to call in more troops. By April 1956, there were nearly a half-million troops in the country.[14] While both sides committed atrocities, the French crimes became more well known. Henri Alleg published the book *La Question* giving detailed accounts of French war crimes during the Battle of Algiers. After selling more than 65,000 copies, the French government banned it.[15]

Among those helping the French were French settlers known as pied-noirs. Additionally, there were Algerians who helped the French. Known as harkis, they helped France for numerous reasons.[16] After the war, the pied-noirs fled to France, while the vast majority of harkis were left in Algeria. Here, they became the targets of retaliation for their role in aiding the French.

The French had their own militant groups seeking to settle their grudges. The Organisation Armée Secrete (OAS) was a group of French Army members who refused to recognize an independent Algeria. They proceeded to slaughter those who fought for Algerian independence, including several attempts to assassinate French President Charles de Gaulle, who was leading the push for independence.

The estimate of Muslims killed by the French as they put down the riots of 1945 is at least 15,000. Numbers from the 1954–1962 violence range from 90,000 to 300,000. The World Peace Foundation further lists civilian deaths at 55,000 to 60,000; harkis deaths at 30,000; and European civilians at 2,788.[17]

The problems of trying to establish a true democracy continue today. For example, in elections for president, incumbent president Abulmadjid Tebboune was declared the winner on September 9, 2024, having garnered 94.7 percent of the vote, followed by Islamist Abdelaali Hassani Cherif, with 3.2 percent, and socialist Youcef Aouchiche, with 2.2 percent.[18] Tebboune was elected in 2019, replacing President Abdelaziz Bouteflika, who had been in office for twenty years. Having run on a pro-democracy platform, Tebboune, upon assuming power, proceeded with the all-too-familiar routine of cracking down on antigovernment protests and using the military to ensure his reelection would be a mere formality by disqualifying nearly every challenger.[19]

GERMANY, 1945-1950

The aftermath of the Second World War saw Germany in shambles. It was occupied by troops from the victorious Allies, and the country was divided into East and West Germany, with the Soviet Union controlling East Germany and the United States, Britain, and France controlling West Germany. Berlin was similarly divided. However, as this book demonstrates, armies are wont to immoral acts.

The civilian and military survivors, the Jewish Holocaust survivors, and the displaced persons from Eastern Europe feared retribution and revenge. And they had good reason to fear revenge. The Soviets especially knew no limits in their lust for vengeance against the atrocities inflicted on them by the Nazis. No woman was safe from the threat of rape. Nobody knows for sure how many victims there actually were, nor will they. Estimates of two million women raped have been given, but estimates like this are merely guesses and are not reliable. The women had to deal not only with the shame associated with rape but also with the fact that immediately following the war and for many years after, Germans were viewed as the conquered enemy to whom no remorse or sympathy was to be shown. The topic wasn't to be broached in East Germany due to Soviet censorship. It wasn't until these women were in their eighties that they began to speak out about what happened.

Rape is not an outlet for pent-up sexual desires. It is used to dominate and humiliate victims. And when soldiers commit this act, it is often as a sign to the populace that their government cannot protect them anymore. This is why military rapes often occur in public.[20] One study found that the average age of the women at the time they were raped during the occupation was 16.7 years.[21] One woman told her story of being raped repeatedly for two weeks when she was only fifteen. When she was reunited with her mother several months later, her mother treated her coldly and told her never to talk about what had happened to her, although it was acceptable to write about it, which she did.[22]

Jews worried about retribution due to the preferential treatment they received after the war in deference to the atrocities heaped on them. The number of Jews in postwar Germany swelled from 20,000 in 1945 to 250,000 in 1947.[23] Besides the rapes, citizens had to deal with threats posed by all the new people coming into the country. For example, marauding foreigners, labeled Poles, Russians, and Ukrainians, roamed the streets. In Ansbach in the American zone, gangs of Poles roamed the streets shooting the population with machine guns.[24] In another incident, thirty Poles in October 1945 broke into the house of a gentleman, murdered him and his war-disabled son, raped the daughter and a farm laborer, and robbed the house.

Even the Americans were not friendly. While they might have been there to help restore order and rebuild the country, they were told that they were not liberators but victors and occupiers. Orders given to the American soldiers stated as much:

4. Basic Objectives of Military Government in Germany:

a. It should be brought home to the Germans that Germany's ruthless warfare and the fanatical Nazi resistance have destroyed the German economy and made chaos and suffering inevitable and that the Germans cannot escape responsibility for what they have brought upon themselves.

b. Germany will not be occupied for the purpose of liberation but as a defeated enemy nation. Your aim is not oppression but to occupy Germany for the purpose of realizing certain important Allied objectives. In the conduct of your occupation and administration you should be just but firm and aloof. You will strongly discourage fraternization with the German officials and population.

c. The principal Allied objective is to prevent Germany from ever again becoming a threat to the peace of the world. Essential steps in the accomplishment of this objective are the elimination of Nazism and militarism in all their forms, the immediate apprehension of war criminals for punishment, the industrial disarmament and demilitarization of Germany, with continuing control over Germany's capacity to make war, and the preparation for an eventual reconstruction of German political life on a democratic basis.[25]

A survey of American soldiers in April 1945 reported that 76 percent "hated" the German civilians, and 71 percent assigned blame for the war to "all or most Germans."[26] The conduct of the GIs in the beginning of the occupation expressed these sentiments. Serious crimes, including homicide, rape, and robbery, increased from 3.7 per 10,000 troops in August 1945 to 11.1 per 10,000 in January 1946.[27] So while the Americans might not have been as malicious as the Soviets, they were hardly blameless in the postwar terror instilled in the German population.

CRIMEAN TATARS, 1945–PRESENT

After the end of World War II, little changed for the Tatars. They were forbidden from returning to their homes and homelands. In 1967, they were completely exonerated of all crimes. However, they were still not allowed to return to their home; Crimea became a Slavic territory. In 1968, three hundred families were allowed to return to Crimea. This was done to report to the world that the Soviet Union was allowing the Tatars to return to their homes. However, with only three hundred allowed to return, that claim was clearly propaganda.[28] To continue to enforce the ban on returning, the old propaganda about their treasonous activities during the war was revived.

With the downfall of the former Soviet Union in the late 1980s came a return of the Tatars to Crimea, to the point that in 2015, there were approximately 230,000 to 300,000 Tatars residing in Crimea, comprising 10.6 percent of the Crimean population.[29] Beginning in 2014, Russia once again occupied Crimea, reannexing it to Russia in March, against Crimea's desires. With this annexation came the abuses against the Tatars. A draft resolution from November 2016 by the UN General Assembly's Human Rights Committee condemning Russia's occupation listed "arbitrary detentions, torture and other cruel, inhumane or degrading treatment, and . . . discriminatory legislation" as some of the abuses against the Crimeans.[30] Additionally, the resolution called for the release of all Ukrainians who were illegally detained, the reopening of cultural and religious institutions, and the revocation by Russia of their decision to label the Mejlis of the Crimean Tartar People (the self-governing body of the Crimean Tatars) a banned extremist organization.[31]

The Office of the UN High Commissioner for Human Rights further noted the "arrests, ill-treatment, torture and intimidation perpetrated against political opponents and minorities, as well as the denial of basic human rights to those who do not accept the forced imposition of Russian legislation and citizenship on the Peninsula."[32] Human Rights Watch expanded on these accusations by noting that the Russian authorities have "subjected members of Crimean Tatar community and their supporters, including journalists, bloggers, activists, and others to harassment, intimidation, threats, intrusive and unlawful searches of their homes, physical attacks, and enforced disappearances" as well as outlawing media and organizations that criticized Russia's actions in Crimea.[33] One specific instance that the head of the UN Human Rights Monitoring Mission in Ukraine, Fiona Frazer, gave was that of a Tatar man who was arrested by the Russian Federal Security Service (FSB, successor to the KGB) and held incommunicado until the next day, when he was left at a bus station in Simferopol in critical condition with signs of having been beaten and tortured, including by use of electric shock, in order to incriminate himself and others.[34]

VIETNAM, 1946–1975

Typically, when people think of the US involvement in Vietnam, they think of 1968–1975. However, the actual conflict began in 1946. Vietnam had been a French possession prior to World War II, but toward the end of the war, France lost possession of Indochina to the Japanese. Even though France regained possession after the war, the Vietnamese, under Ho Chi Minh, declared themselves independent of France and soon began a guerrilla war against it. Soon after, President Truman announced that the United States would help any nation who was threatened by communism, in what became known as the Truman Doctrine, and began giving assistance to France.

The first US soldiers arrived in the country shortly thereafter, and by 1963, there were 16,000 US advisors (the term given to US troops in the conflict) in the country.[35] One of the reasons that the conflict became so protracted was the US insistence on waging limited warfare. They didn't fight to win, as they had in World War II; they fought to keep from losing, which is what happened.

They sprayed defoliants like Agent Orange over rural South Vietnam to try to kill the vegetation that gave cover to the North Vietnamese guerrillas. This was chemical warfare, directed at vegetation but with the unforeseen conse-quences of causing numerous physical ailments to the US troops, which they would deal with for decades after the war. There were no real objectives other than the ambiguous "stop the spread of communism."

On January 30, 1968, about 85,000 Viet Cong and North Vietnamese troops launched the Tet Offensive. One of the most hard-fought battles was for the city of Hué. The battle raged for nearly a month, leaving 216 US troops, 384 South Vietnamese, and more than 5,000 Communists dead. The battle led to the disappearance (i.e., slaughter) of five thousand South Vietnamese civil-ians who were deemed to have come from social categories that were not to be a part of the new North Vietnamese society. Additionally, US air and artillery support destroyed half of the city, leaving more than 100,000 inhabitants home-less.[36] This was an object lesson in how not to win the hearts and affection of the locals, the result being a tactical loss but strategic victory for the North.

My Lai
My Lai is a subhamlet of the Tu Chung hamlet, Son My Village, Son Tinh District, Quang Ngai Province, in the former Republic of South Vietnam.[37] The area was well known for insurgent activity. As such, it had been declared a "free-fire zone," where "all civilians were automatically suspected of being Viet Cong or Viet Cong sympathizers." Subsequently, US forces did not need to get approval from Saigon or local officials before staging bombing missions and ar-tillery attacks.[38] A search-and-destroy mission was scheduled for March 16–18, 1968, to find and eliminate Viet Cong forces in the area. Reports stated that all the villagers would have left for the market no later than 7:00 a.m., leaving only Viet Cong in the area. Charlie Company, First Battalion, Twentieth In-fantry Regiment, Eleventh Infantry Brigade, with Second Lieutenant William L. Calley serving as first platoon leader, was tasked with clearing the area of the Viet Cong. Upon arriving, the company was met with silence. There were no Viet Cong present, only women, children, and older men preparing their breakfasts. The villagers were rounded up and guarded while the hamlet was searched. The search revealed only a few weapons.

The company was already on edge as they undertook the mission. They had lost more than forty men in three months and were convinced that the offending guerrillas were in the village.[39] Additionally, the company was under orders from Colonel Oran Henderson to "Go in there aggressively. Wipe them out for good."[40] Furthermore, when they asked for clarification on whether to kill the civilians, Captain Ernest Medina replied, "Get rid of them. . . . No, I want them killed. Fire when I say fire."[41] After all, for all the soldiers knew, mothers holding babies could have concealed grenades to be used when the soldiers were not expecting it.[42] At about 7:30 on the morning of March 16, 1968, the carnage began (see figures 28–30).

Soldiers threw grenades pell-mell into houses full of women, children, and elders; livestock and crops were destroyed, as was the village water supply when a person and then a grenade was thrown into the well; villagers cowering in their huts were shot and bayoneted; women were raped; huts were burned; villagers found hiding in ditches were removed and shot; tongues were cut out; and people were scalped. No men of military age were found. The village had been razed. Initial reports listed 128 Viet Cong having been killed.[43] The final death toll was 504, including 182 women, 17 of whom were pregnant, and 173 children, which included 56 infants.[44] The incident was covered up until the story broke a year later.

The final question came down to following orders and knowing whether an order is illegal or not. Fourteen men were charged in the incident, including Calley, Medina, and Henderson. All were acquitted, except for Calley, who was convicted of murder of twenty-two Vietnamese civilians. Calley claimed to have only been following orders, which the other defendants were able to successfully use as a defense. That same argument, when used in the Nuremberg Trials some twenty-five years earlier, was dismissed, as the soldiers should have known better.

Figure 28. Some villagers huddle together with black bags covering their heads. Son My, South Vietnam, March 16, 1968. Source: "The My Lai Massacre: 33 Disturbing Photos of the War Crime the U.S. Got Away With," ed. John Kuroski, All That's Interesting, updated May 20, 2021, https://allthatsinteresting .com/my-lai-massacre-photos#6.

Figure 29. The road out of My Lai, littered with dead bodies. Son My, South Vietnam, March 16, 1968. Source: "The My Lai Massacre: 33 Disturbing Photos of the War Crime the U.S. Got Away With," ed. John Kuroski, All That's Interesting, updated May 20, 2021, https://allthatsinteresting.com/my-lai-massacre-photos#9.

Figure 30. A man and his son lie dead in the dirt. Son My, South Vietnam, March 16, 1968. Source: "The My Lai Massacre: 33 Disturbing Photos of the War Crime the U.S. Got Away With," ed. John Kuroski, All That's Interesting, updated May 20, 2021, https://allthatsinteresting.com/my-lai-massacre-photos#20.

One soldier, as his defense, never argued the fact that the incident occurred. Instead, he reasoned that only the army could give him legal orders; therefore, in following orders, he was not guilty of assault with intent to kill. While the order was ruled illegal, the question came down to the ability of a soldier to decide if it is illegal or not.[45] Many viewed Calley and his conviction a scapegoat for all his superiors, who should have known that such an order was illegal, which in retrospect appears was the case. Other people believed the same, enough that President Richard M. Nixon ordered him placed under house arrest pending his appeal. He was paroled after serving about three and a half years.

The tragedy in all of this is not just the fact that such a heinous crime was committed, which is tragedy enough. Instead, it was later found out that this was not an isolated incident but was the only incident to be uncovered. A massacre occurred in the village of My Khe as well as an operation code-named Speedy Express, in which thousands of Vietnamese civilians were killed in the Mekong Delta area.[46]

Chemical Weapons

In the late 1940s and into the 1950s, Britain was involved in a war with Malaya as they sought to obtain their independence from Britain. As part of their war efforts, the British employed American-developed herbicides in attempts to deny the guerrillas cover as well as deprive them of crops. The effects of this warfare were communicated to a small group of American scientists. With the commencement of hostilities in Southeast Asia, more impetus was given for the development of effective herbicides. Of 26,000 chemicals evaluated for their potential as effective defoliants, 6 were chosen for the job.[47]

The six defoliants were given code names of colors of the rainbow—Agents Green, Pink, Purple, White, Blue, and Orange, based on the color painted onto the drums in which the agents were delivered.[48] Agent Orange was the most potent of the group and ended up being the most-used defoliant during the war. Beginning in 1961, over the course of the Vietnam War, more than 20 million gallons of herbicides were dropped over Vietnam, Laos, and Cambodia during what was known as Operation Ranch Hand. Of those gallons, more than 13 million gallons were Agent Orange.[49]

The operation was judged a success. Nearly 12,000 square miles of Vietnamese jungles were sprayed with the defoliant.[50] In some areas where the agent was used, American officers noted that the ambush rate had dropped by 90 percent.[51] While results were typically seen in a matter of hours, some areas were hit several times to get the desired results. There was just one, not-so-small problem with the agent: It was as poisonous to humans as it was to plants.

In the manufacture of Agent Orange, which is composed of two chemicals, one chemical produces in small but varying quantities a substance known as dioxin. The amount of dioxin produced was so small that it was deemed of no concern. The military repeatedly said how safe the substance was. However, dioxin is one of the most poisonous substances ever created, at least as toxic as nerve gas and known to cause birth defects.[52] Additionally, it can stay in the soil for one hundred years or more.[53] When used repeatedly, the minuscule amounts accumulate, and it doesn't break down, so the effects begin to show.

The total amount of dioxin produced from those 13 million gallons varies from 240 pounds to 375 pounds.[54] Given that a few ounces placed in the water supply would have been enough to kill the entire population of New York City, there is no doubt of its lethality. It has been linked to type 2 diabetes; immune system dysfunction; nerve disorders; muscular dysfunction; hormone disruption; heart disease; liver disease; Hodgkin's disease; lymphomas; and cancers affecting the bladder, prostate, and respiratory system. And the effects weren't limited to those exposed. Horrendous birth defects, including spina bifida, missing limbs, blindness, deafness, and cranial abnormalities, started appearing in not just the Vietnamese who had be exposed to it from Operation Ranch Hand but also Americans who had come into contact with it. In 1984, a lawsuit against the manufacturers of the chemical in Agent Orange was settled out of court for $180 million ($545 million in 2024), and the US Department of Veterans Affairs awarded additional compensation to about 1,800 veterans.

All of this raises the question of its legality under the 1925 Geneva Protocol. The US government claimed that what they were doing was perfectly legitimate. And as it had not ratified the protocol at that time, they were not bound by it. However, "no State on ratifying the 1925 Geneva Gas Protocol has made a reservation or declaration of interpretation to the effect that the Protocol does not apply to herbicides."[55] The United Nations passed the Environmental Modification Convention, which in articles 1 and 2 states,

> Each State Party to this convention undertakes not to engage in military or any other hostile use of environmental modification techniques having widespread, long-lasting or severe effects as the means of destruction, damage or injury to any other State Party. As used in article I, the term "environmental modification techniques" refers to any technique for changing—through the deliberate manipulation of natural processes—the dynamics, composition or structure of the Earth, including its biota, lithosphere, hydrosphere and atmosphere, or of outer space.[56]

As mentioned in chapter 4, the protocol did not cover the use of chemicals in a purely domestic situation. Both Britain and the United States were producing tear gas, but Britain found out that the one in use could be circumvented by simply closing the eyes. Something that affected multiple systems was needed. The new agent, code-named CS, caused burning eyes, itching skin, running nose, coughing, and vomiting. The United States immediately turned it into an offensive weapon in Vietnam, filling grenades with it and spraying it from aircraft and helicopters.

The British could legitimately say that they had only used CS in domestic incidents. The United States had no such excuse, so when used, the troops were instructed to refer to it as "tear gas," not just "gas." An American spokesman reported later that the "purpose of the gas attack was to force the Viet Cong troops to the surface, where they would be more vulnerable to the fragmentation effects of the bomb bursts."[57] The United States used thousands of tons of CS in Vietnam.

Domestically, research continued into chemical and biological agents, including nerve gases. Production of these agents continued unabated. Cluster bombs filled with mustard gas and phosgene, more stocks of nerve gas, and ordnance dumps spread throughout the United States in Colorado at Rocky Mountain Arsenal and Utah at Tooele, as well as Arkansas, Indiana, Alabama, Kentucky, Oregon, and Maryland, in addition to Okinawa and West Germany. The United States had enough gas for a twelve-month campaign.[58]

ISRAEL AND PALESTINE, 1948-PRESENT

Israel has been under near-constant attack by the Arab nations since its formation in 1948. The Jewish people have been under attack for longer than that, dating to biblical times. The modern state of Israel began to take form in the late nineteenth century. Pogroms targeting Jews increased in intensity and frequency, resulting in a large influx of Jews to the area of Palestine. Theodor Herzl, a journalist from Austria-Hungary, was the first to propose a new Jewish state in Palestine, then ruled by the Ottoman Empire. His vision of a homeland where Jews could experience state-controlled self-determination became known as Zionism and quickly gained a following.[59] Continued pogroms with the concomitant death of numerous Jews and destruction of their property led to the exodus of tens of thousands from Russia and Eastern Europe to Palestine.

With the loss of World War I by the Ottoman Empire (as part of the Central Powers), the land of Palestine was given to Britain as a mandate. As part of that mandate, Britain was to establish a national home for Jews, ensuring that doing so did not cause contention with the non-Jews residing there. This mandate had been preceded and made possible by the Balfour Declaration of 1917, in which British Foreign Secretary Arthur James Balfour expressed Britain's support for the creation of a Jewish national home in Palestine.[60] The declaration was strongly opposed by the Palestinian Arabs.[61]

Not surprisingly, contention did arise. In what became known as the Hebron Massacre, Palestinian Arabs attacked Jews, with at least sixty-six Jews dead and an undetermined number of Arabs.[62] This incident was followed by the Arab Revolt of 1936, in which Palestinian Arabs demanded sovereignty and the end to Jewish immigration.[63] While the Arabs may have been the most vocal anti-Semites, the vast majority of the world followed their path. With the White Paper of 1939, Britain effectively stopped Jewish immigration into Palestine. Hitler had proposed forcibly removing Jews to Madagascar, which ultimately did not happen. As had happened so many times before, nobody wanted Jews. They weren't accepted where they were living, namely Europe, nor were they allowed to immigrate to any other countries, such as

Palestine. See chapter 4's coverage of the M.S. *St. Louis*. The stage was set for the Holocaust.

On September 9, 1940, the Italian air force bombed Tel Aviv. The attack killed 137 people, all in residential areas. The seemingly obvious target, the port of Jaffa, was untouched, although the Italians claimed that they had hit it.[64] As mentioned in chapter 5, the alliance between Nazi Germany and Palestine is evident in the animosity the Muslim community in general has toward the Jews. Walter Schellenberg, head of SD-Ausland (*Sicherheitsdienst*, or Nazi Party foreign intelligence), wrote in one of his reports on the mood in Palestine when the Palestinians learned of Rommel's success in North Africa in 1942: "The extraordinarily pro-German attitude of the Arabs is due primarily to the fact that they 'hope Hitler will come' to drive out the Jews. Field Marshal Rommel has become a legendary figure. Thus, all the Arabs today long for the arrival of the Germans; they continually ask when the Germans will come and are downright distressed that they have no weapons."[65]

With the defeat of the Axis powers in 1945 came the plans for a distinct Jewish homeland. On November 29, 1947, the United Nations voted to divide Palestine into two states: one Jewish and one Arab. The Arabs rejected this and began a civil war, with a siege of Jerusalem from December 1947 until the first truce was brokered on June 11, 1948. Strict rationing of food was instituted, and starvation was avoided on June 14 by the establishment of a road, allowing meager amounts of supplies to enter the city. That one road was the only route for supplying Jerusalem until the Valor Road opened several months later. During the siege, the Arabs ambushed a large group en route to a hospital. They killed seventy-seven doctors, nurses, university scientists, and administrative staff.[66]

The 1948 Arab-Israeli War (War of Independence) started within hours of the formal establishment of the state of Israel on May 14, 1948. As a new country, Israel was severely lacking in defensive forces. Things were bleak enough that the Jews wondered if a new Holocaust was going to occur.[67] In an attempt to prevent the Arab militiamen from returning to their villages, the main Jewish militia, the Haganah, decided that they needed to destroy the village wells, as had been occurring throughout the skirmishes. However, the Haganah lacked sufficient forces to occupy each village, and many of the villages had been acquired from outside the boundaries the United Nations initially mandated, so there was question of whether they would be allowed to hold the villages.[68] In an act of desperation, Israeli leaders with the Haganah formulated Operation Cast Thy Bread, the use of biological warfare in an attempt to poison targeted wells, rendering them useless and preventing the return of the Arabs.

The most serious and potent use of these agents was the use of typhoid, or typhoid and dysentery, against the Kabri aqueduct used by the Arab town of Acre. Starting in the latter part of April 1948, Acre experienced an epidemic of typhoid and dysentery. On May 8, ninety cases among the villagers and sixty-five cases involving the British soldiers stationed there were reported. The Ittihad Hospital reported fifty-eight cases of typhoid and twelve of dysentery between May 25 and July 18, with five fatalities from typhoid.[69] Israeli officials credited this "epidemic" with facilitating the ease with which the town was later taken by the Haganah.[70] In a separate incident, two Israeli agents were caught poisoning a couple of wells in Gaza. They were interrogated, tortured, tried for their crime, and subsequently executed for their actions.[71] These were the only incidents and had little, if any, bearing on the outcome of the war.

The War of Independence was followed by the Sinai War of 1956; the Six-Day War of 1967; the War of Attrition of 1967; the Yom Kippur War of 1973; the two Lebanon Wars of 1982 and 2006; the two Intifadas of 1987 and 2000; the three Gaza Wars of 2008, 2014, and 2021; Operation Pillar of Defense of 2012; and the Israel-Palestine crisis of 2021. Each of these wars came as a result of Arab forces attempting to destroy Israel.[72]

On October 7, 2023, the terrorist group Hamas launched an attack on Israel that included the firing of thousands of rockets into Israel and the infiltration of Israel by land, sea, and air.[73] One of the targets was a Supernova music festival attended by approximately three thousand people. Totals for the carnage were 1,200 dead, mostly civilians, and 253 abducted.[74] Additional attacks have been vowed by Hamas until Israel is "annihilated." Hamas has stated that international law allows them to fight Israel because they claim Israel "invaded" their country; they conveniently have left out that the same international law prohibits them from attacking Israel within its own legitimate and legally recognized borders and from attacking innocent civilians.[75] Hamas leaders have said that they do not want peace with Israel but a permanent war and that they have no intention of governing Gaza or helping the Palestinians residing therein:

> "I hope that the state of war with Israel will become permanent on all the borders, and that the Arab world will stand with us," Taher El-Nounou, a Hamas media adviser, told *The Times.*
>
> "This battle was not because we wanted fuel or laborers," he added. "It did not seek to improve the situation in Gaza. This battle is to completely overthrow the situation."[76]

Hamas and Hezbollah have taken to using human shields by placing their headquarters and weapons caches in hospitals and schools. On March 31, 2024, a weapons cache was found in the maternity ward of Gaza's Shifa Hospital.[77] This was after the Israel Defense Forces (IDF) were fired on from inside the hospital, including the emergency room, the maternity ward, and the burn unit. IDF was trying to allow normal hospital functions but was thwarted by the Islamic terrorists in Hamas.[78] In May, IDF killed fifteen terrorists, ten of whom were from Hamas, inside a "war room" established in a school run by United Nations Relief and Works Agency (UNRWA).[79] This led to allegations that UNRWA was a front for terrorist activities, as tunnels under the UNRWA contained equipment used for intelligence and communications.[80] UNRWA used Hamas to escort humanitarian aid trucks entering the Gaza Strip, so Hamas controlled the shipments. Humanitarian aid was going to Hamas and not the residents of Gaza, all with the tacit help of the United Nations.[81] Hamas and Islamic jihad terrorists were also using the UNRWA headquarters as a command center.[82] And the United Nations was forced to admit that some of UNRWA's employees may have participated in the attack.[83]

Furthermore, these acts show that schools run by the United Nations are being used to indoctrinate children into wanting the death of the Jews. A copy of Hitler's book *Mein Kampf* was found in a child's bedroom at a Hamas terror base in the Gaza Strip. It had been marked and annotated, showing signs that it had been used.[84] Schools inculcated children into hating Jews from a very young age.[85] Additional documents were found showing that Hamas had maps to target schools and other locations where children would be, instructing the savages to "kills [sic] as many as possible" and "capture hostages."[86]

On October 1, 2024, Iran joined in the attack on Israel, launching scores of missiles at the country.[87] On October 16, 2024, the mastermind behind the October 7 Massacre, Yahya Sinwar, was killed during action in Gaza.[88] He apparently was trying to flee into Egypt; he had passports and cash on his body. One of the passports belonged to a UNRWA employee who may have been one of Sinwar's bodyguards.[89] This successful operation was conducted despite US Secretary of Defense Lloyd Austin threatening Israel with an arms embargo unless they continued to work with UNRWA, and President Joe Biden, with Vice President Kamala Harris, threatening Israel with an arms embargo unless they suspended Gaza operations and "surged" aid into the area.[90] Then, after demanding the cessation of Gaza operations upon threat of an embargo, President Biden, in a move often seen in US foreign relations wherein it tries to play both sides of the game (refer to chapter 5 and the United States' dealings with Unit 731), stated,

> Shortly after the October 7 massacres, I directed Special Operations personnel and our intelligence professionals to work side-by-side with their Israeli counterparts to help locate and track Sinwar and other Hamas leaders hiding in Gaza.
>
> With our intelligence help, the IDF relentlessly pursued Hamas's leaders, flushing them out of their hiding places and forcing them onto the run.[91]

As the war continued, the casualties continued to climb. According to the United Nations Office for the Coordination of Humanitarian Affairs (OCHA), as of September 23, 2024, casualties in the Gaza Strip included at least 41,431 Palestinians dead and another 95,818 injured, while Israeli totals for that same period (October 7, 2023–September 23, 2024) stood at 1,546 dead and 2,287 soldiers wounded.[92] Given actions like those of Hamas and their supporters, we are clearly headed toward another Holocaust.

CHINA, 1949–PRESENT

China's persecution of its citizens can only be described as genocide. Their efforts to suppress the past by forcibly "reeducating" those who strive to keep the memories of each incident alive are blatant persecution of those individuals within its borders, stretching into areas whose authority is disputed, such as Tibet and Taiwan. While governments in general emphasize those aspects they wish to maintain as official and quiet the others, often under the guise of misinformation, it is especially heinous in China.

With the end of World War II, the Communist Chinese Party (CCP), under Mao Zedong, and the Kuomintang (KMT), under Generalissimo Chiang Kai-shek, resumed a civil war that had been suspended while they joined forces against the Japanese. The end of this civil war resulted in the KMT fleeing to Taiwan, while the CCP assumed authority over mainland China. The cost of the war in human lives was astronomical: 1.5 million Communists dead and 600,000 Nationalist troops dead, with approximately three times that defecting to the Communists and another seven million troops captured by the Communists. About five million civilians died from combat, disease, and famine.[93] However, the purge in the years following dwarfed any other despot in history.

With the start of Mao's Great Leap Forward, famine, disease, and violent death to those suspected of hoarding grain or trying to escape the famines by moving to the cities led to an estimated 20 to 45 million dead.[94] One man, whose son stole a handful of grain, was forced to bury him alive as punishment. The father died of grief a few days later. The only other person in history who comes close to Mao's death totals is Genghis Khan (see chapter 1). He is thought to have also killed some 40 million in his conquests, accounting for *11 percent* of the world's population at that time.[95] Unfortunately, the deaths and oppression didn't stop with either Genghis Khan or Mao.

Tibet, 1950–Present

On October 29, 1950, Communist China, having won its civil war against the Nationalists the year before, invaded Tibet over disputes regarding its sovereignty.[96] Under the ruse of the need to destroy inequitable feudalism and enact modernization and social justice, the 40,000 Chinese troops easily overpowered the 4,000 Tibetan troops.[97] Initially promising not to change the political structure of Tibet, the Communists soon reneged and began the occupation of Tibet.[98] Since then, the Chinese continually tried to erase the Tibetan identity by forging them into Chinese subjects. Forced relocation of entire villages

was common. To get the villages to "voluntarily" relocate, Chinese forces proceeded door to door, engaged in "publicity work."[99] These visits were intrusive
and used the threats of terminating essential services to their homes if they did
not comply. Relocation entailed moving from remote rural areas to semiurban
areas, where farming and herding was impossible. This left the villagers without work and dependent on the state.[100] These actions continue to this day.

In March 1959, when Tibet tried to revolt over fears of an attempt to kid-
nap the Dalai Lama, the Chinese aimed artillery at Norbulinka Palace, home
of the Dalai Lama. On March 21, 1959, shelling of the palace began, with tens
of thousands of Tibetans slaughtered in the process. Afterward, the Chinese
executed the guards of the Dalai Lama and destroyed the major monasteries of
the capital of Lhasa as well as their thousands of inhabitants.[101]

These acts have established the pattern by which the Chinese attempt
to maintain control over areas where their jurisdiction is of questionable authority. The Chinese government routinely detains human rights defenders;
tightens control over society, media, and the internet; and employs mass surveillance technology in an effort to maintain domination over the Indigenous
population.[102]

Tiananmen Square

In June 1989, Tiananmen Square, one of the largest public squares in the
world, located in central Beijing, was occupied by university students and many
other Chinese in an effort to expand the foreign ideas and values as well as the
affluence that the nation was undergoing.[103] The protests started on April 22,
1989, the day of the funeral of Hu Yaobang, who had previously been the CCP
general secretary until 1987, when he was forced out by the CCP due to his
encouragement of democratic reforms. These protests continued, constantly
growing in size, until the last two weeks of May, when martial law was declared.

Beijing residents took to the streets to prevent the troops from reaching
the square. Finally, on the night of June 3–4, troops accompanied by heavily
armed forces reached the square. By June 5, the CCP was in firm control of
the square, except for one lone protestor, dubbed "Tank Man," who defiantly
stood in front of an advancing tank column (see figure 31). By this time, the
CCP had tried to block all communication with the outside world with propaganda. The photographer, Jeff Widener, had American exchange student Kirk
Martsen help him by smuggling the roll of film in his pants out of the hotel
where Widener was staying.[104] The protest was brutally put down, with estimates of several hundred to 10,000 dead; up to 10,000 arrested; and several
dozen executed.[105]

Thirty-five years after the protest and subsequent massacre, the CCP refused to acknowledge that the protest ever occurred. Widener's photos are
banned in China, as are public commemorations of the event. Hong Kong was
the only Chinese province allowed to have any kind of commemoration, but

Figure 31. A man blocks a line of tanks on Beijing's Tiananmen Square on June 5, 1989. Photo by Jeff Widener/AP. Source: Kyle Almond, "The Story Behind the Iconic 'Tank Man' Photo," CNN, accessed May 22, 2025, https://www.cnn.com/interactive/2019/05/world/tiananmen-square-tank-man-cnnphotos/.

Figure 32. The front page of *Christian Times* on June 2, 2024, showing the manner in which they were forced to protest the Tiananmen Square Massacre of June 4–5, 1989. Source: Christian Times, accessed May 22, 2025, https://christiantimes.org.hk/News/174535/1918.jpg.

that was banned in 2020 in an effort to stop the spread of the COVID-19 virus. However, with the virus basically under control, the CCP still refused to lift the ban. Those who protest are sentenced to jail.[106] The weekly Hong Kong periodical *Christian Times*, to protest in a manner that wouldn't warrant arrest and seizure of the paper, printed a nearly blank front page on its June 2, 2024, issue to signify the prohibition of any mention of the protests (see figure 32).[107] Additionally, the Communist government still fills Hong Kong's Victoria Park with pro-regime festivals every June 4, "seeking to erase the memory of the great candlelight vigils that were once held there."[108]

Uyghurs, 2014–Present

The Uyghurs are a Turkic-speaking people native to northwest China. Following the pattern established in Tibet, the Chinese have systematically tried to eliminate all trace of the Uyghur culture. More than one million Uyghurs have been forcibly detained in "reeducation camps." Up to a half-million people have been forced to pick cotton in one of the Uyghur areas. Forced sterilization of women in order to minimize the population and removing children from their families are common. As of 2020, there were an estimated 380 reeducation camps in the area.[109] In 2017 and 2018, almost 23,000 residents of one county were in a camp or in prison, more than 12 percent of the adult population. If extrapolated out to the entire Xinjiang Province, where they reside, there would be more than 1.2 million people detained.[110] Escaped prisoners of the camps report regular physical, mental, and sexual torture, along with mass rape and sexual abuse.

Xinjiang is run like Tibet, with police, checkpoints, and surveillance cameras monitoring nearly every move the people make. Police use a mobile app to monitor how much electricity people are using and how many times they use their front doors.[111] The Chinese say they are engaging in counterterror activities and boast of how peaceful the area is as a result. Consequently, because of this operation, initiated in 2014 under the name "Strike Hard Campaign Against Violent Terrorism," the Chinese government has been accused of "imprisonment or other deprivation of liberty in violation of international law; persecution of an identifiable ethnic or religious group; enforced disappearance; torture; murder; and alleged inhumane acts intentionally causing great suffering or serious injury to mental or physical health, notably forced labor and sexual violence."[112] Human Rights Watch has a list of the alleged crimes.[113]

KOREA, 1950

When the conflict first started in June 1950, the US Air Force called for the firebombing of industrial centers. Initially, the US government said no to this suggestion. However, once China entered the fray, the United States decided that it, too, needed to up the stakes, so the firebombing of North Korea began.[114]

North Korea became one of the most heavily bombed countries in history. The United States dropped more bombs on North Korea than it dropped in the entire Pacific theater during World War II.[115] These bombings targeted both civilians and military. Whole cities were destroyed, with thousands of civilians dead and more left destitute.[116] When the United States ran out of urban targets, they began bombing hydroelectric and irrigation dams.[117] It was said that if two bricks were still on top of each other, they bombed them.[118]

The devastation was described in a diplomatic cable sent to the United Nations in January 1951:

> On January 3 at 10:30 am, an armade of 82 flying fortresses loosed their death-dealing load on the city of Pyongyang. . . .
>
> Hundreds of tons of bombs and incendiary compound were simultaneously dropped throughout the city, causing annihilating fires. In order to prevent the extinction of these fires, the trans-Atlantic barbarians bombed the city with delayed-action high-explosive bombs which exploded at intervals throughout for a whole day, making it impossible for the people to come out onto the streets. The entire city has now been burning, enveloped in flames, for two days. By the second day 7,812 civilians' houses had been burnt down. The Americans were well aware that there were no military objectives left in Pyongyang. . . .
>
> The number of inhabitants of Pyongyang killed by bomb splinters, burnt alive and suffocated by smoke is incalculable, since no computation is possible. Some fifty thousand inhabitants remain in the city, which before the war had a population of five hundred thousand.[119]

Situations like this help account for North Korea's extreme animosity toward the United States. The United States originally stated that they would only bomb military targets. However, with the entry of China into the conflict, all of North Korea soon became a military target. Regardless, the similarity between Japan's refusal to apologize for their role in World War II and the United States' refusal to apologize for their role in Korea is apparent with little effort. Mike Bassett stated, "The US will never admit what it did in Korea."[120]

The end of the war saw an unknown number of dead and wounded. The US military listed 33,629 battle deaths, with 20,617 dead from other causes. South Korean military deaths have been estimated in excess of 400,000, while the South Korean Ministry of Defense listed 281,257 dead and missing. Communists dead were estimated to be about a half-million. Total dead were estimated at more than three million Korean civilians, with 20 percent of the population of North Korea dead.[121] Korean civilians dead accounted for upward of 70 percent of total casualties.[122]

The civilians were seen as the enemy, either potential or actual, by both Korean sides in the war, which led to the large number of civilians dead.[123] South Koreans captured by North Korea were viewed as traitors and treated as such. Therefore, it should come as no surprise that war crimes were committed by both North and South Korea. In a report by the South Korean Truth and Reconciliation Commission, the list of verifiable war crimes was published. The report found that crimes were indeed committed by both sides as well as the United States.[124] However, perhaps surprisingly, the vast majority of civilian killings by US forces were deemed militarily necessary.[125]

In one atrocity listed in the report, South Korean police, in an attempt to identify Communist sympathizers, dressed as a North Korean soldier, entered the village of Naju, and killed whoever greeted them with Communist flags.[126] Many of the atrocities committed by South Korea came when they crossed into North Korea. Japanese sources put the total of executed or kidnapped at 150,000.[127] American soldiers were documented to have "affixed Chinese skulls to spikes on the forward sponsons of their tanks."[128] No Gun Ri, in late July 1950, saw three to four hundred refugees, including women and children strafed, by US planes and shot by members of the Seventh Cavalry.[129]

Additionally, prisoners of war were frequently tortured. In one report, the list of alleged crimes against American prisoners of war was laid out:

> The evidence before the subcommittee conclusively proves that American prisoners of war who were not deliberately murdered at the time of capture or shortly after capture, were beaten, wounded, starved, and tortured; molested, displayed, and humiliated before the civilian populace and/or forced to march long distances without benefit of adequate food, water, shelter, clothing, or medical care to Communist prison camps, and there to experience further acts of human indignities.[130]

The report continued, listing eight separate instances of prisoner torture, along with forced marches and treatment in Communist prison camps.[131]

While the conflict ended, the barbarity in the North never ceased. North Korea is recognized as one of the most repressive regimes in the world. Their crimes include extrajudicial executions, rape, forced abortions, jail without trial, torture, and starvation rations that force prisoners to search for insects or rodents for food. There are an estimated 200,000 prison camp internees.[132]

The Korean War is often referred to as the "forgotten war." While those who served there deserve to be recognized, perhaps it is time for both sides to remember, to acknowledge their roles in the war, and to turn the armistice into a peace treaty.

BANGLADESH, 1971–1972

On March 25, 1971, West Pakistan (now Pakistan) began Operation Search-light, the military subjugation of Bengalis, specifically Bengali Hindus, living in East Pakistan (now Bangladesh). The Bengalis had been calling for inde-pendence from West Pakistan due to the dominance of West Pakistan in most aspects of daily life. However, in Pakistan's first honest election after years of military rule, the ruling party in East Pakistan was victorious.[133] After having been the dominant party in Pakistan, West Pakistan was unwilling to share power with East Pakistan and launched Operation Searchlight.

As is so typical of armed conflict, the invading forces during the opera-tion targeted civilians, shooting students in college dormitories, raping women, and instilling a general fear into the Bengalis. Houses were burned, and the infrastructure suffered massive destruction.[134] The Bengalis were specifically targeted during the operation, leading to this act being considered genocide.[135] The United States knew of the genocide of the East Pakistanis but blatantly sided with West Pakistan.[136] Estimates vary, but as a result, between 300,000 and 3 million Bengalis were killed; around 2 million to 3 million Bengali women were raped, with approximately 25,000 being forcibly impregnated; and 10 million became refugees in India.[137]

UGANDA, 1971–1979, 1980–1986, 2024

Uganda gained its independence in 1962. From 1962 to 1971, it enjoyed a rel-atively peaceful existence. In 1971, Idi Amin staged a coup and took control of the country. He ruled from 1971 to 1979, eight horrific, blood-filled years. Lord Owen, British foreign secretary during this time, stated, "His regime goes down in the scale of Pol Pot as one of the worst of all African regimes."[138] He report-edly fed his opponents to his pet crocodiles. His goal was to rid the country of the Lango and Acholi ethnic groups. His junta was responsible for the death of 300,000.[139]

Since his downfall, Uganda's history has been one despotic leader after an-other, stoking racial tensions between differing tribes in order to stay in power. As of 2005, Uganda's leader, Joseph Kony, had enslaved more than 20,000 chil-dren, most under the age of thirteen.[140] Along with the death of the people comes the near destruction of northern Uganda's economic base of agriculture.

The current insurrectionist is still Kony. He has been charged by the Inter-national Criminal Court with crimes against humanity and war crimes, resulting in 2 counts of murder for the deaths of 405 civilians. The case lists thirty-six counts of intentionally directing attacks against the civilian population, murder, attempted murder, torture, severe abuse and mistreatment, enslavement, pillag-ing, destruction of property, persecution, forced marriage, rape, using children to participate actively in hostilities, sexual slavery, and forced pregnancy.[141] The charges go into great detail about the events leading to these charges, which stem only from actions from July 1, 2002, to December 31, 2005.

Uganda remains relatively stable, despite the despotic nature of its ruler. Nevertheless, tensions are again mounting with many of the same telltale signs. On July 24, 2024, forty-two youths were arrested on charges of "being 'idle and disorderly' and being a 'common nuisance.'"[142] They were protesting the corruption of the government, whose members, including President Yoweri Meseveni and his sycophantic cronies, have been accused of embezzlement and misappropriation of funds. Protestors have posted photos of themselves on social media "so they could be easily identified if Museveni's troops decided to kill or abduct them."[143] Museveni previously called the demonstrators "thieves and parasites" and had members of Parliament of the political opposition party detained while placing their headquarters "under siege."[144] Clearly, it would not take much to push the country over the edge into renewed genocide.

CAMBODIA (KHMER ROUGE), 1975–1979

The Khmer Rouge were a small band of Communists operating in Cambodia near Vietnam. With the US bombings of Cambodia and incursion by US ground troops during the Vietnam War, the Khmer Rouge was able to recruit new members. In 1970, the military overthrew the ruling royal family. The ensuing civil war saw the Khmer Rouge, with backing from North Vietnam and the Viet Cong, fighting against the Khmer Republic, backed by the United States and South Vietnam. With the fall of the capital, Phnom Penh, the United States left the country, allowing the Khmer Rouge leader Pol Pot to take control. He proceeded to remake Cambodia into an agrarian Communist country. In doing so, he opened the gates of hell on the country.

As is typical of Communist regimes, intellectuals were the enemy. Feeling that the country had been contaminated by Western influence, cities were emptied, and the population was marched into the country to work the fields. Thousands of Cambodians were murdered simply for wearing glasses or speaking a foreign language.[145] Ethnic Vietnamese, Thai, Chinese, and Cham Muslims, along with Cambodian Christians and Buddhist monks, were the other targets. The Cham Muslims lost half of their population during this time. All civil and political rights, private property, money, religious practices, minority languages, and foreign clothing were abolished.[146] The executions and persecutions finally came to an end when Vietnam invaded the country in 1978.

Starvation and disease, as well as the effects of overwork, soon set in. The Tuol Sleng Prison in Phnom Penh became a mass murder center, with an estimated 17,000 men, women, and children jailed there over the course of the regime's rule.[147] Additionally, rural areas were set up as execution sites, later known as the Killing Fields. Not even members of the Khmer Rouge were safe from harm, as several thousand were murdered on suspicion of being traitors and spies. Merely questioning the new regime led to death. A warning from the Khmer Rouge stated simply, "To keep you is no gain; to lose you is no loss."[148] As she recounted her trauma, Teeda Butt Mam said this same sentiment was told:

I was fifteen years old when the Khmer Rouge came to power in April 1975. I can still remember how overwhelmed with joy I was that the war had finally ended. It did not matter who won. I and many Cambodians wanted peace at any price. The civil war had tired us out, and we could not make much sense out of killing our own brothers and sisters for a cause that was not ours. We were ready to support our new government to rebuild our country. We wanted to bring back that slow-paced, simple life we grew up with and loved dearly. At the time we didn't realize how high the price was that we had to pay for the Khmer Rouge's peace.

The Khmer Rouge were very clever and brutal. Their tactics were effective because most of us refused to believe their malicious intentions. Their goal was to liberate us. They risked their own lives and gave up their families for "justice" and "equality."

Even after our warmest welcome, the first word from the Khmer Rouge was a lie wrapped around a deep anger and hatred of the kind of society they felt Cambodia was becoming. They told us that Americans were going to bomb the cities. They forced millions of residents of Phnom Penh and other cities out of their homes. They separated us from our friends and neighbors to keep us off balance, to prevent us from forming any alliance to stand up and win back our rights. They ripped off our homes and our possessions. They did this intentionally, without mercy.

They were willing to pay any cost, any lost lives for their mission. Innocent children, old women, and sick patients from hospital beds were included. Along the way, many innocent Cambodians were dying of starvation, disease, loss of loved ones, confusion, and execution.

We were seduced into returning to our hometowns in the villages so they could reveal our true identities. Then the genocide began. First, it was the men.

They took my father. They told my family that my father needed to be reeducated. Brainwashed. But my father's fate is unknown to this day. We can only imagine what happened to him. This is true for almost all Cambodian widows and orphans. We live in fear of finding out what atrocities were committed against our fathers, husbands, and brothers. What could they have done that deserved a tortured death?

Later the Khmer Rouge killed the wives and children of the executed men in order to avoid revenge. They encouraged children to find fault with their own parents and spy on them. They openly showed their intention to destroy the family structure that once held love, faith, comfort, happiness, and companionship. They took young children from their homes to live in a commune so that they could indoctrinate them.

Parents lost their children. Families were separated. We were not allowed to cry or show any grief when they took away our loved ones. A man would be killed if he lost an ox he was assigned to tend. A woman would be killed if she was too tired to work. Human life wasn't even worth a bullet. They clubbed the back of our necks and pushed us down to smother us and let us die in a deep hole with hundreds of other bodies.

They told us we were void. We were less than a grain of rice in a large pile. The Khmer Rouge said that the Communist revolution could be successful with only two people. Our lives had no significance to their great Communist nation, and they told us, "To keep you is no benefit, to destroy you is no loss."[149]

In all, an estimated two million people were murdered during the reign of Pol Pot, a number even the Khmer Rouge acknowledges as accurate (see figure 33). Before the Khmer Rouge came to power, the population had been 7.5 to 8 million.

Figure 33. A charnel house at Tonle Bati, Cambodia, with collected skulls and bones from a mass grave of Khmer Rouge victims at Tonle Bati, some thirty kilometers south of Phnom Penh. SOURCE: WESLEY RAHN, ED., "WILL FINAL KHMER ROUGE RULING END DARK CHAPTER?" DEUTSCHE WELLE, SEPTEMBER 26, 2022, HTTPS://WWW.DW.COM/EN/CAMBODIA-WILL-THE-FINAL-KHMER-ROUGE-RULING-CLOSE-A-DARK-CHAPTER/A-63243079.

IRAN AND IRAQ, 1980–1988

In 1979, the shah of Iran, who had been installed by the United States, was overthrown by a popular uprising. This was followed by the takeover of the US embassy in Tehran on November 4, 1979. Of the sixty-six staff members inside who were taken hostage, fifty-two were held for 444 days, until January 20, 1981. While captive, the hostages were kept tied up and handcuffed; paraded in front of angry crowds; placed in solitary confinement; subjected to Russian roulette; and threatened with execution, having their feet boiled in oil, and having their eyes cut out.[150] While the hostages were released, the revolution led to the installation of the Ayatollah Khomeini as the supreme leader of Iran. This, along with other reasons, such as religious differences, led to the invasion of Iran by Saddam Hussein of Iraq, and the Iran-Iraq War from September 22, 1980, to August 20, 1988.

As part of the war against Iran, Hussein openly used chemical weapons against not just the Iranian military but his own citizens as well.[151] By Iraq's own admission, "it consumed almost 19,500 chemical bombs, over 54,000 chemical artillery shells and 27,000 short-range chemical rockets between 1983 and 1988. Iraq declared it consumed about 1,800 tons of mustard gas, 140 tons of Tabun, and over 600 tons of Sarin."[152] The use of these chemicals resulted in nearly 5,000 dead and more than 100,000 ill Iranians.[153] Thirty years later, 56,000 Iranians were still dealing with the effects of these attacks.[154]

In the northeast of Iraq lies the town of Halabja. Following its capture forty-eight hours earlier by Iranian forces, Iraq attacked the town on March 16, 1988, with chemical agents. The attack was to deal with not only the Iranian army but also the pro-Iranian Kurdish forces who lived there. The attack killed as many as 5,000 and injured 7,000 Iraqis, with other estimates of Kurds killed at 50,000 to 100,000.[155] Additionally, in 1988, under orders from Khomeini, the Iranians executed their political prisoners throughout the country. The estimated total number of those executed ranges from 2,000–5,000 to 30,000 in thirty-two cities.[156]

BOSNIA AND HERZEGOVINA, 1991, AND KOSOVO, 1998

The Balkans of southeast Europe has been an area of contention for nearly as long as can be remembered. Consisting of Serbs, Croats, Muslims, Slovenes, Albanians, Macedonians, and Montenegrins, it became Yugoslavia after the end of World War I. During World War II, the country was led by Josip Broz Tito, who managed to keep the country at peace for nearly forty years.

In May 1991, Croatia and Slovenia announced their separation from the Serbian portion of the country. A year later, Bosnia-Herzegovina followed suit. Bosnia-Herzegovina's ethnic makeup consisted of 44 percent Muslims, 31 percent Serbs, and 17 percent Croats. Bosnian Serbs, with the Serbian armed forces, proceeded to seize control of most of Bosnia. The Serbs were against independence. With support from the military of neighboring Serbia and refusing to recognize the country's independence, the ethnic cleansing of the country began.

Muslims and Croats were driven from homes by the Serbs and subsequently expelled from the country, tortured, and murdered. Toward the end of first part of the war, in 1995, the Croatian army attacked areas in Croatia under Serbian control, and more atrocities were committed. Then in 1998, Kosovo declared its independence, and ethnic Albanians began targeting ethnic Serbians with the same heinous crimes. It was during this time that Slobodan Milošević, president of Serbia and of the Federal Republic of Yugoslavia, became infamous for his role in perpetrating these war crimes. He became the first sitting head of state to be charged with war crimes.

In order to propagate the cleansing of the non-Serbs, the propaganda apparatus was initiated. The Bosnian Serbs seized control of the Bosnian television transmitters in the summer of 1991, and any non-Serb employees were fired. Furthermore, radio stations and newspapers experienced the same fate. The propaganda was successful, turning formerly peaceful neighbors against each other and inciting the Bosnian Serbs to become murderers.[157] As Noel Malcolm explains, "It was as if all the TV stations in the USA had been taken over by the Ku Klux Klan."[158]

The main message being transmitted was that of a single Serb state, combining Bosnia, under the leadership of Radovan Karadžić, with the remnants of Yugoslavia, led by Milošević. What started out as hate speech turned into speech encouraging genocide. The Bosnian Serbs were told that the Bosnian Muslims (Bosniaks) were planning genocide against the Serbs and that the Serbs needed to eliminate the Muslims before they themselves were exterminated.

In March 1995, the town of Srebrenica was attacked in an attempt to rid the area of Bosniaks and annex the area to Serbia. Karadžić ordered the military to "create an unbearable situation of total insecurity with no hope of further survival or life for the inhabitants of Srebrenica."[159] During the night of July 11, 10,000 Bosniak men left the town to try to reach safety. The Serbs were able to stop them and, using UN equipment and falsely promising safety to the men if they surrendered, execute many of them. Many more were removed to Bosniak territory. On the night of July 13, mass executions began and continued through July 16. Execution sites included football fields, meadows, a warehouse, a factory, a school, a dirt road, and a cultural center.[160]

As many as 200,000 people were killed, although this number was revised to about 100,000, with more than 2 million people displaced.[161] All told, between 7,000 and 8,000 people were killed in Srebrenica alone.[162] Milošević was arrested in 2001 and charged with genocide and crimes against humanity. He died in prison in 2006 before his trial could be concluded. Karadžić went into hiding in 1997 and was not arrested until July 2008. He was convicted of the same crimes as Milošević and sentenced to life in prison.[163]

During this time, Vojislav Šešelj was advocating for the same acts to occur in Serbia. He was accused of summoning Serbian fighters to eliminate Croats, Muslims, and many other non-Serbian residents of large parts of Croatia. He reportedly called for the "eyes of Croats to be 'gouged out with rusty spoons.'"[164] He was charged and acquitted of nine war crimes and crimes against humanity by the UN war crimes court on March 31, 2016.[165] Nonetheless, on April 11, 2018, the acquittal was overturned by another UN war crimes court. He was sentenced to ten years in prison, but because he had spent twelve years in pretrial detention, he was not required to return to prison.[166]

RWANDA, 1994, 1996–1997

Oftentimes, people wonder, What if? What if France and Britain had acted more forcefully against Hitler prior to the invasion of Poland? Or even, What if somebody had killed Hitler as a child? Rwanda is another such case study of What if?

Rwanda, comprising 84 percent Hutu and 15 percent Tutsi, gained its independence in 1962.[167] Racism once again reared its ugly head. The Hutus hated the Tutsi due to, among other things, having been disenfranchised by Belgium when Rwanda was one of their colonies.[168] When the Hutus assumed official government power, they quickly targeted the Tutsi for violence. Between December 21, 1963, and January 12, 1964, some 10,000 Tutsis were slaughtered by government troops. Another 160,000 fled to neighboring countries.[169] This first wave of genocide subsided, but even during this "quiet" phase, Tutsis were still targeted.

On September 30, 1990, a group of exiles who had fled to Uganda formed the Rwanda Patriotic Front (Front Patriotique Rwandais; RPF) and invaded Rwanda. Attempts were made to broker a cease-fire, but the Rwandan government refused to negotiate with the RPF. Nevertheless, hostilities ended on November 1, 1990, after troops from Belgium, France, and the Congo were sent to assist the Rwandan government.[170]

During this same year, an extremist newspaper, *Kangura*, was founded by Sassan Ngeze. The paper has been described as "one of the most virulent voices of hate."[171] Shortly after the abovementioned deployment, *Kangura* printed in its pages "The Hutu Ten Commandments." These "commandments" provide a valuable insight into the belief system of not just Hutus but also most racist groups:

1. Every Hutu should know that a Tutsi woman, whoever she is, works for the interest of her Tutsi ethnic group. As a result, we shall consider a traitor any Hutu who
   - marries a Tutsi woman
   - befriends a Tutsi woman
   - employs a Tutsi woman as a secretary or a concubine.

2. Every Hutu should know that our Hutu daughters are more suitable and conscientious in their role as woman, wife and mother of the family. Are they not beautiful, good secretaries and more honest?

3. Hutu women, be vigilant and try to bring your husbands, brothers and sons back to reason.

4. Every Hutu should know that every Tutsi is dishonest in business. His only aim is the supremacy of his ethnic group. As a result, any Hutu who does the following is a traitor:
   • makes a partnership with Tutsi in business
   • invests his money or the government's money in a Tutsi enterprise
   • lends or borrows money from a Tutsi
   • gives favours to Tutsi in business (obtaining import licenses, bank loans, construction sites, public markets, etc.).

5. All strategic positions, political, administrative, economic, military and security should be entrusted only to Hutu.

6. The education sector (school pupils, students, teachers) must be majority Hutu.

7. The Rwandan Armed Forces should be exclusively Hutu. The experience of the October 1990 war has taught us a lesson. No member of the military shall marry a Tutsi.

8. The Hutu should stop having mercy on the Tutsi.

9. The Hutu, wherever they are, must have unity and solidarity and be concerned with the fate of their Hutu brothers.
   • The Hutu inside and outside Rwanda must constantly look for friends and allies for the Hutu cause, starting with their Hutu brothers.
   • They must constantly counteract Tutsi propaganda.
   • The Hutu must be firm and vigilant against their common Tutsi enemy.

10. The Social Revolution of 1959, the Referendum of 1961, and the Hutu Ideology, must be taught to every Hutu at every level. Every Hutu must spread this ideology widely. Any Hutu who persecutes his brother Hutu for having read, spread, and taught this ideology is a traitor.[172]

After this tract was published, President Habyarimana and his Rwandan government were forced to share their control with the RPF and with UN peacekeeping forces monitoring the implementation of this new government.[173] Radical Hutus refused to accept this situation and undertook to stymie the treaty that established the government, accusing Habyarimana of being a sellout by not continuing the complete banishment of Tutsis from any type of recognition.

Using radio, newspapers, and similar media, they unleashed their tirade against the Tutsis. By now, the formula for these attacks was well known. The

Tutsis were portrayed as "cockroaches that have to be crushed."[174] Misinformation was aired over Radio Rwanda, claiming that the Tutsis had put together a hit list for murder.[175] A broadcast on April 3, 1994, denounced a doctor, resulting in him being burned alive in front of his house three days later.[176] The fuse was ready for the bomb to go off.

The spark that lit the fuse was the downing of the plane carrying President Habyarimana by a surface-to-air missile that hit as it was landing. Blaming the Tutsi, although nobody knows for sure who fired the missile, the Hutu, led by a militia known as the Interahamwe, began their purge. Along with the Tutsi, any moderate Hutus who might have opposed the oncoming genocide were immediately murdered. Tutsis fled to areas that previously had been safe refuges, such as schools and churches. However, this time, these sites became the sites of major massacres. Entire families were murdered, many by being slashed to death by machetes.[177]

The carnage lasted for just more than one hundred days, finally ending when the RPF overthrew the Hutu government. All told, hundreds of thousands Rwandan Hutus participated in the slaughter, with as many as one million dead, mainly Tutsis. More than one million Hutus, including those who participated in the genocide, fled to neighboring countries, leading to more violence. More than five million total died in all the violence associated with the genocide (see figure 34).[178]

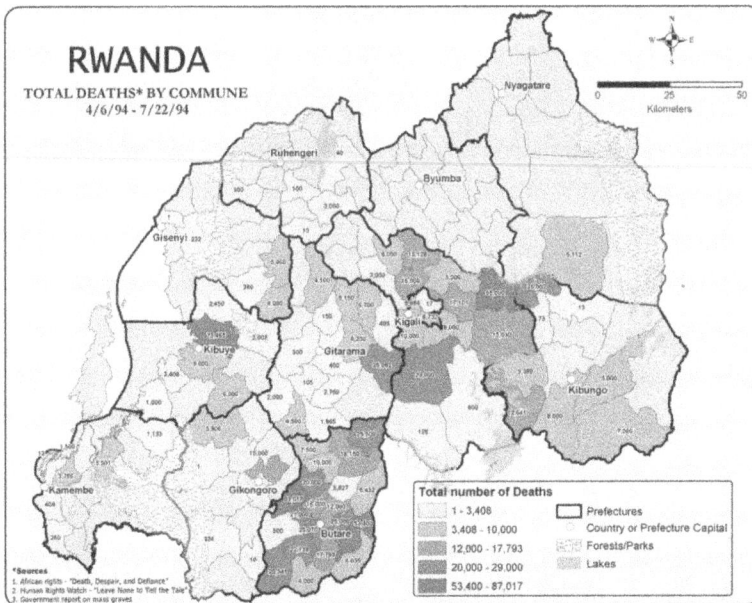

Figure 34. Total deaths by commune in Rwanda. Source: University of Minnesota, College of Liberal Arts, "Rwanda," accessed May 22, 2025, https://cla.umn.edu/chgs/holocaust-genocide-education/resource-guides/rwanda.

Where the What if? comes into play is in the prelude to the genocide. On January 11, 1994, General Romeo Dallaire, who was force commander of the UN peacekeeping mission in Rwanda, sent what came to be known as the "Genocide Fax" to the UN headquarters in New York City. In it he stressed that Hutu extremists were stockpiling and distributing arms to the Interahamwe militias. Items 6 through 9 of the fax stated,

6. Principal aim of Interhamwe in the past was to protect Kigali from RPF. Since UNAMIR mandate he has been ordered to register all Tutsi in Kigali. He suspects it is for their extermination. Example he gave was that in 20 minutes his personnel could kill up to 1000 Tutsis.
7. Informant states he disagrees with anti-Tutsi extermination. He supports opposition to RPF but cannot support killing of innocent persons. He also stated that he believes the president does not have full control over all elements of his old party/faction.
8. Informant is prepared to provide location of major weapons cache with at least 135 weapons. He already has distributed 110 weapons including 35 with ammunition and can give us details of their location. Type of weapons are G3 and AK47 provided by [informant] RGF. He was ready to go to the arms cache tonight—if we gave him the following guarantee. He requests that he and his family (his wife and four children) be placed under *our* protection.
9. It is our intention to take action within the next 36 hours with a possible H HR of Wednesday at dawn (local). Informant states that hostilities may commence again if political deadlock ends. Violence could take place day of the ceremonies or the day after; therefore Wednesday will give greatest chance of success and also be most timely to provide significant input to on-going political negotiations.[179]

While there was some concern about the sudden change of heart of this informant and the possibility of disinformation, Dallaire was willing to give him the benefit of the doubt, at least with reservations.[180]

The UN response to this fax was less than helpful. The very first point of the reply fax was, "We have carefully reviewed the situation in the light of your MIR-79. We cannot agree to the operation contemplated in paragraph 7 of your cable, as it clearly goes beyond the mandate entrusted to UNAMIR under resolution 872 (1993)."[181] The final point of this fax was, "If you have major problems with the guidance provided above, you may consult us further. We wish to stress, however, that the overriding consideration is the need to *avoid entering into a course of action that might lead to the use of force and unanticipated repercussions.* Regards."[182] In a case of morbid irony, it was the *inaction* by the United Nations that led to the "unanticipated repercussions" that became the Rwandan genocide. Hence the question, What if? What if the United Nations had acted on Dallaire's initial fax? Would the genocide have been avoided? Or would it simply have been postponed to a later date? We will obviously never know.

These governing institutions had a horrible track record with such interventions. As to the United States, the government knew the potential for such brutalities before they happened. US President Bill Clinton knew of the situation when he took office in 1993. One report dating to Clinton's predecessor, President George H. W. Bush, stated, "Hutu extremists with links to Rwanda's ruling party were believed to be advocating the extermination of ethnic Tutsis."[183] The UN Office on Genocide Prevention and the Responsibility to Protect stated that genocides are a process, not a spontaneous incident. They are too well organized to happen spontaneously.[184]

After the incident, the United States actively warned against labeling what had happened a genocide. As Collin Woldt writes, "Simply put, the United States was presented with the facts and chose to ignore them. Perhaps it is best summarized by Greg Stanton, a professor at George Mason University and the president of Genocide Watch: 'When President Clinton said after the Rwandan genocide, "We really didn't know," I'll be direct. He was lying.'"[185] In fact, the history of US involvement in such situations is abysmal. However, it is difficult to imagine a situation where US involvement could have resulted in a *worse* situation than the one that unfolded.

DEMOCRATIC REPUBLIC OF CONGO, 1998–2009

In August 1998, rebels, with the support of Rwanda, Uganda, and Burundi, started a war with the Congolese (Democratic Republic of Congo, DRC), who were helped by Zimbabwe, Angola, and Namibia. However, in 1999, Rwanda and Uganda began fighting each other. The remaining factions splintered into smaller rebel groups, drawn primarily along ethnic lines. The DRC, with its various allies, began targeting Tutsi, Rwandan, and Banyamulenge groups, all of Rwandan ancestry.

This targeting included arbitrary arrests; disappearances; mass executions; rape and other sexual violence, including the killing of pregnant women and the rape of men; the use of weapons with the intent to maim the victims; cannibalism; and the devastation of crops, health care system, and other civilian entities.[186] The genocide was carried out not only by the differing militias but also by women and children. In one incident, 996 members of the Hema tribe in Drodro, DRC, were murdered with guns and machetes by the Lendu tribe.[187] Targeting members of differing tribes became not just a means to an end but an end in and of itself.[188]

The carnage lessened in 2003, but still there were millions dead from the turmoil in the country, the vast majority civilian deaths. The total number of dead by all causes, not just killing, was reported by the International Rescue Committee to be up to 5.4 million between August 1998 and April 2007, including child soldiers who had been forcibly removed from their families and pressed into service with the various factions.[189] That total had passed 6 million, with an additional 7.8 million displaced, by April 2025.[190]

Christian Malanga, a naturalized US citizen, formed a new political party to oust President Felix Tshisekedi for allegedly running a dictatorship. The coup on May 19, 2024, failed when a hit squad arrived at the home of one target only to find him not home. They subsequently got lost trying to find the house of the next intended victim.[191] Prosecutors in the DRC said they would seek the death penalty for fifty people who were supposedly involved in the coup.

A jailbreak at Makala Central Prison in July 2024 resulted in the death of 129, including 24 inmates, who were shot by prison guards; most of the others were killed in a stampede. Human rights activists and political opposition leaders claim the official government statistic is too low. The institution is designed to hold 1,500 prisoners but routinely holds more than 12,000, with death by starvation common. DRC Justice Minister Constant Mutamba claimed the riot was a deliberate act of sabotage and promised a "stern response" for those who planned the riot.[192]

IRAQ, 2003

On March 20, 2003, the United States, along with coalition forces, again invaded Iraq. The formal war was over on May 1, 2003. However, the occupation of the country continued until December 18, 2011. One major difference between this war and the previous one is that in this one, both Iraq and the United States were committing war crimes.

Some estimates put the civilian death toll by 2007 at 650,000.[193] In the actual battle, estimates of fatalities were around 9,200 combatants and approximately 3,750 noncombatants, for a total of about 12,950 fatalities.[194] In the same report, Iraqi fatalities in just Baghdad are listed at 2,224 to 3,531 combatant dead and 1,990 to 2,347 noncombatant dead.[195] In addition, 3 million Iraqis of a population of 32 million were refugees by the end of the war.[196]

The United States was first accused of war crimes shortly after the war began. The accusations included dropping cluster munitions in civilian-occupied areas and blanket air strikes, killing civilians.[197] The name Abu Ghraib became synonymous with the atrocities perpetrated by the United States at this time. In this camp, which had been used by the Iraqi government as a prison where its perceived enemies were tortured, detainees were also routinely tortured by the United States. Such methods included sexual abuse, secret detention and detainee transfers, enforced disappearance, killing protestors, sleep deprivation, forced nudity, lack of adequate food and water, mock executions, threats of and actual rape, and using private contractors to kill and injure civilians (see figure 35).[198] Unfortunately, these are just some of the torture and humiliation techniques used at Abu Ghraib. In 2009, *The International Initiative to Prosecute US Genocide in Iraq* unsuccessfully filed charges against US presidents and UK prime ministers for the acts that occurred in that country.[199]

Figure 35. Sabrina Harman poses for a photo behind naked Iraqi detainees forced to form a human pyramid, while Charles Graner watches, November 7, 2003, 11:50 p.m. This photo from Abu Ghraib prison in Iraq was part of the evidence used against US soldiers accused of abusing and humiliating inmates. SOURCE: WIKIPEDIA, "ABU GHRAIB TORTURE AND PRISONER ABUSE," ACCESSED MAY 22, 2025, HTTPS://EN.WIKIPEDIA.ORG/WIKI/ABU_GHRAIB_TORTURE_AND_PRISONER_ABUSE#/MEDIA/ FILE:ABU_GHRAIB_48.JPG.

SUDAN, 2003

After suffering from famine in the 1980s, Sudan broke into civil war in 2003. The rebels were trying to gain economic and political power for the people, but the government responded by utilizing gang rapes, mass murder, torture, and mutilation of the non-Arab population.[200] The villages were raided by the government troops, often attacking from opposite sides of the village to prevent escape. Everything valuable was stolen, and the villagers were tortured, raped, and enslaved. After this, the villages were burned. The animals were taken, and the village wells were contaminated by throwing corpses into them.[201] Estimates put the number of dead between 300,000 and 500,000, with 2.5 million refugees and more than 2,000 villages destroyed.[202]

The conflict has continued intermittently since then, with militias being reorganized into the Rapid Support Forces (RSF). The RSF have been accused of ethnic cleansing of non-Arabs in the region.[203] In 2019, President Omar al-Bashir was overthrown in a coup d'état, and a joint military-civilian government was formed. It, too, was overthrown in October 2021. Now, however, the coup partners have broken into rival factions, with the avowed goal of fighting until the other is dead.[204]

The fighting has become so severe that Doctors Without Borders abandoned one of the refugee camps they staffed, where about 800,000 civilians were caught in the middle of the war. Both sides are specifically targeting the civilians to inflict homelessness and starvation on them.[205] Both sides have launched air strikes and artillery at hospitals, targeting the patients and doctors still inside.[206] So far, there are more than 15,000 dead, with 7–9 million more displaced. With the current famine, there are estimates that as many as two million more could die from starvation.[207] Unfortunately, with the war between Russia and Ukraine raging, little attention has been given to the country, including from the United States, whose main source of help has been President Biden telling both sides to "stop blocking aid to the Sudanese people," which has not solved the issue.[208]

KENYA, 2007
Again, as a result of a disputed election, violence erupted in Kenya in 2007. This time, according to the *New York Times*, violence started merely fifteen minutes after the winner of the election was declared. It was disputed, as Kenya's election commission ignored evidence of vote rigging in order to keep the existing government in power.[209] Martial law was declared, with the banning of all live media broadcasts.

The war was, once again, between rival tribes, this time the Kikuyu (President Mwai Kibaki) and the Luo (challenger Raila Odinga). And the crimes against the civilians were the same, namely an estimated nine hundred gang rapes of not just women but also children as young as three, pregnant women, and nursing mothers. To a lesser extent, men and boys were penetrated with guns, sticks, and bottles. They were stripped naked, sexually fondled, humiliated, and circumcised or castrated. Sexual attacks were followed by such physical abuse as stabbing, kicking, cutting with machetes, and beatings with heavy objects. At least 1,133 people were left dead, with more than 600,000 people displaced.[210]

CÔTE D'IVOIRE, 2011
In November 2010, elections were held in Côte d'Ivoire (Ivory Coast), pitting incumbent president Laurent Gbagbo against challenger Alassane Ouattara. The elections were verified as free and fair, and Ouattara was declared the winner. However, Gbagbo refused to concede and was finally arrested in May 2011.[211] In helping him in his bid to stay in power, the radio stations began broadcasting hate speech directed toward Ouattara and his supporters.[212] Pro-Gbagbo forces dragged pro-Ouattara personnel from restaurants, after which family members found their bodies full of bullets in morgues several days later. Persons merely suspected of being pro-Ouattara were beaten to death with bricks, shot at point-blank range, or burned alive. Villages were burned to the ground, and wells were stuffed with bodies.[213] From November to May, militias

from both sides fought and committed atrocities, with hundreds of women and girls raped and hundreds of thousands fleeing to neighboring Liberia.[214] Once Ouattara took office, he began to investigate these war crimes but only against pro-Gbagbo supporters. During 2011–2012, the military under Ouattara's direction continued to commit torture, killings, rapes, and extortion.[215] While Ouattara did institute some reforms, such as a program to disarm the 60,000 or so youth who had participated in the war, again, they were directed primarily toward pro-Gbagbo forces.[216] Estimates to the total number of dead are around three thousand for the conflict.[217]

While the fighting has subsided, human trafficking of primarily children, both into and out of the country, has continued with little abatement.[218] Elections held in 2020 were relatively peaceful due to the intervention of outside parties, but the underlying factors are still in play, rendering the question of long-term stability in doubt.[219]

SYRIA, 2011

Starting in late 2010, pro-democracy protests in Tunisia consequently spread to other Arab countries, becoming what is known as the Arab Spring and leading to the removal of several heads of state. The uprising spread to Syria in 2011, when protests began on January 26. Major protests began on March 15 following the arrest and torture of a group of children on the charge of writing antiregime graffiti. The government's oppressive response to the protests only led to more protests throughout the country, and by June, the country had descended into civil war.[220]

This time, the conflict was concerned with political ideologies. Many protestors belonged to the country's Sunni majority, while the country's president Bashar al-Assad and his family belonged to the minority Alawi sect of Shi'ite Muslims. The Alawi also comprised the majority of the security forces and irregular militias. These groups were responsible for some of the worst violence against the protestors and opponents of the Assad regime.[221]

Assad painted the protestors as Sunni extremists similar to al-Qaeda and who were helping foreign countries in their wars against Syria. The accusations were not totally without merit, as al-Qaeda chief Ayman al-Zawahiri in February 2012 called for Sunnis to join a jihad against the Assad regime.[222] The Sunnis were also portrayed as planning violent retaliations against non-Sunnis.[223] The greater the number of protests, the more violent the government response was, which in turn led to even more protests.

In a report by the Syrian Network for Human Rights, seventy-two torture methods used by the Assad government were recorded. Techniques included burning hostages with various objects; tying hostages to a cross as if being crucified followed by beatings; pulling out finger- and toenails; cutting, beating, and damaging reproductive organs; keeping deceased detainees' bodies in cells; beating prisoners' wounds; and abandoning incontinent and physically immobile detainees in their own body waste without helping clean them.[224] In this same report, between March 2011 and September 2019, at least 14,298 individuals, including 178 children and 63 women, died from torture.[225] Additionally, the article estimated that nearly 1.2 million Syrian citizens have been arrested and detained at some point during Assad's reign.[226]

Countries opposed to Assad's violent responses, including the United States, European Union, Qatar, Turkey, and Saudi Arabia, began to take action by introducing sanctions against Syria. Syria's allies, composed of Iran, Russia, and China, attempted to foil any action taken by the anti-Assad coalition. The country was finally invaded by Turkey in 2019 and was subjected to US-led air strikes.

Again, civilians bore the brunt of this civil war. The Russians were accused of using bunker-busting and incendiary bombs against civilian homes and water supplies.[227] Meanwhile, the United States was accused of similar tactics with their air strikes against the city of Raqqa. The precision air strikes were reportedly not as precise as the Americans claimed. Additionally, 30,000 artillery rounds were fired by the US-led coalition during the Raqqa assault.[228] A UN report placed the civilian deaths between March 2011 and March 2021 at 306,887, with total documented deaths of 350,209.[229] As of this writing, there is no end to the conflict in sight.

A report by the US Department of State cited Amnesty International's estimates for individuals who had disappeared since 2011 of between 10,000 and 120,000.[230] And torture was ubiquitous:

> Authorities continued to use 42 methods of torture, documented by the [Syrian Network for Human Rights], including eight common positions involving tying the prisoners' hands and beating their bodies with wires or sticks, in particular in genital areas. Other reported methods of physical torture included: removal of nails and hair; stabbing and cutting off body parts, including ears and genitals; beating the bare soles of feet (falaqua); burning with acid or cigarettes; applying electric shocks; denying medical care; and hanging. Multiple human rights organizations reported other forms of torture, including forcing objects into the rectum and vagina, hyperextending the spine, and putting the victim into the frame of a wheel and whipping exposed body parts.[231]

MYANMAR, 2012

As President Thein Sein tried to guide Myanmar (formerly Burma) from a military dictatorship to a democracy, a national reconciliation between the Burmese Buddhist majority and the other ethnic groups in the country was made a priority. One ethnic group, however, was not a part of this understanding: the Muslim ethnic group the Rohingya, which Myanmar does not recognize as citizens. These Muslims were not allowed to own land or travel outside their small region of Rakhine, bordering Bangladesh. Muslims were dragged from buses and murdered, and one of the country's leading weekly newspapers printed vengeful comments about the Muslims, including from one reader who wrote, "Terrorist is terrorist. Just kill them."[232] A Buddhist monk was quoted as accusing "Muslim men of repeatedly raping Buddhist women, of using their wealth to lure Buddhist women into marriage, then imprisoning them at home."[233] About two hundred were killed, and several thousand were displaced.[234] The country has remained unstable, suffering from another military coup in February 2021, with the military jailing the governing party members and supporters. Protests resulted in a brutal military suppression, with entire villages destroyed, shooting at protestors and into people's homes, and massacring civilians and opposition fighters. More than 8,000 people were arrested, and 1,500 were killed, with many more displaced.[235]

RUSSIA AND UKRAINE, 2014-PRESENT

The Russo-Ukraine War began in February 2014, with Russia subtly invading and annexing Crimea. In late February, Russian troops not bearing any insignia, began occupying Crimea, seizing the Crimean Parliament and government buildings, followed by airports and communication centers. All the while, Russian president Putin denied the forces were Russian.[236] On March 1, the Russian Parliament approved the use of armed forces in Crimea, with its annexation announced on March 18, 2014.[237]

On February 24, 2022, Russia invaded Ukraine. Since then, there have been multiple reports from various organizations about the war crimes being perpetrated. The UN Office of the High Commissioner for Human Rights reported at least 1,035 civilians had been killed, with another 1,650 injured, on March 25, 2022, after only one month of fighting.[238] Wide-area explosives, such as missiles and rockets, were used near populated areas, private houses, residential apartment buildings, medical and educational edifices, and water and electricity systems. On March 3, 2022, forty-seven civilians in two schools and multiple apartment buildings were killed by Russian air strikes, while on March 9, seventeen civilians were injured in Mariupol Hospital No. 3 when it, too, was destroyed by Russian air strikes.[239]

A coroner from the Kyiv suburb of Bucha reported collecting more than one hundred bodies during the attempt by Russian troops to take Kyiv, while another woman said, "They shot everyone they saw. They shot the gas pipe, too, and her [a deceased friend of the woman] mother was in the house."[240] As of February 23, 2024, Ukraine is investigating more than 122,000 suspected war crimes by Russia. Trials have already begun, mostly in absentia, with eighty convictions in Ukraine courts.[241] As of August 2024, the war continues, with beatings, electric shock, broken bones, knocked-out teeth, lack of food, and gangrenous limbs common during the conflict.[242]

Chapter 7

CONCLUSIONS

"Therefore, let him who desires peace prepare for war."

—Publius Flavius Vegetius Renatus

Henry David Thoreau was quite correct when he said, "The savage in man is never quite eradicated."[1] Going back to the question asked in the introduction, Is mankind inherently good or evil? This question has consumed humanity for most of its intellectual history. After reading this book, the seemingly obvious answer to that question is that man is inherently evil. Despite the myriad attempts at controlling our basic nature through laws, treaties, and the like, man continues to wage war in one form or another on one another. Those of a more optimistic viewpoint believe that, with just a few more laws, we can control the beast inside all of us and achieve utopia on earth. However, pessimists point out that most, if not all, treaties have failed, and for all our attempts to the contrary, "bullies" still exist.

Organized religion and philosophical beliefs dwell on this topic extensively. Mengzi (also known as Mencius, 372–289 BC) and Xunzi (310–238 BC) are two great Confucian thinkers. Mengzi believed in the innate goodness of man, identifying four main areas ("sprouts") that form our moral goodness: ren, or benevolence; righteousness, or the ability to feel shame when indicated; ritual proprietary, having a heart that feels courtesy; and wisdom, having a heart that feels approval/disapproval and has a sense of right and wrong.[2] He further stated, "Having these four sprouts within oneself, if one knows to fill them all out, it will be like a fire starting up or a spring breaking through! If one can merely fill them all out, they will be sufficient to care for all within the Four Seas. If one merely fails to fill them out, they will be insufficient to serve one's parents."[3]

Conversely, Xunzi said that Mengzi was wrong, that human nature is innately bad, and this nature is only overcome through striving to surmount it:

> People's nature is such that they are born with a fondness for profit
> in them. If they follow along with this, then struggle and conten-
> tion will arise, and yielding and deference will perish therein. They
> are born with feelings of hate and dislike in them. If they follow
> along with these, then cruelty and villainy will arise, and loyalty
> and trustworthiness will perish therein. They are born with desires
> of the eyes and ears, a fondness for beautiful sights and sounds. If
> they follow along with these, then lasciviousness and chaos will
> arise, and ritual and righteousness, proper form and order, will per-
> ish therein.[4]

Both, however, see the need to cultivate the good in us: Mengzi so that we don't fall into evil, and Xunzi so that we become good.[5]

Buddhism sees man as basically good. Buddha said, "Monks, this mind is originally radiant and clear, but because passing corruptions and defilements come and obscure it, it doesn't show its radiance."[6] The evil that humans ex-hibit is a result of confusion and ignorance: ignorance of the basic nature of ourselves, which causes us to suffer and leads us to mistreat ourselves and oth-ers. This ignorance is temporary, and ignorance and confusion are not innate qualities of individuals. This includes the Hitlers and Pol Pots of the world.[7]

Christianity, particularly Catholicism, teaches the fall of Adam. This fall from grace caused all subsequent humans to be carnal and devilish. The Apos-tle Paul stated in 1 Corinthians 2:14 (KJV), "But the natural man receiveth not the things of the Spirit of God: for they are foolishness unto him: neither can he know them, because they are spiritually discerned." He further wrote in Romans 8:5–10 (KJV):

> For they that are after the flesh do mind the things of the flesh; but
> they that are after the Spirit the things of the Spirit.
>   For to be carnally minded is death; but to be spiritually minded
> is life and peace.
>   Because the carnal mind is enmity against God: for it is not
> subject to the law of God, neither indeed can be.
>   So then they that are in the flesh cannot please God.
>   But ye are not in the flesh, but in the Spirit, if so be that the
> Spirit of God dwell in you. Now if any man have not the Spirit of
> Christ, he is none of his.
>   And if Christ be in you, the body is dead because of sin; but the
> Spirit is life because of righteousness.

From these scriptures, we see that Christianity believes that the "natural" man is an enemy to God. It is only through conscious effort and following the ways of God that we can become good. Of course, some have an easier time doing this than others.

Philosophers tend to see both sides of the argument, as well. Those who espouse the goodness of man point out that few of us, using an example given by Mengzi, would be able to pass a child drowning in a well without stopping to help save the child. Martin and Strudler assert, "All men have a mind which cannot bear to see the sufferings of others."[8] They also point to experiments done on rhesus monkeys that show they would rather starve to death than obtain food if it meant inflicting pain on another monkey.[9]

This same article takes the opposing viewpoint that we are all murderers by not contributing to worthwhile organizations, when doing so would save the life of an individual. It references a psychology experiment in which

> subjects coming out of a phone booth encounter a woman who has just dropped some papers, which are scattered across the ground. Some subjects help pick up the paper; others do not. The difference between the two groups is largely fixed by whether a subject found a coin that the experimenter had planted in the phone booth. Those who find the coin are happy and tend to help; people who don't find a coin tend not to help. Other experiments get similar results even when the stakes are much higher. In some situations, most people will knowingly and voluntarily expose others to death for no good reason. People cannot be trusted.[10]

Will genocide continue in the world? Probably. Could it happen in the United States? James E. Waller believes so: "*Recent political transitions in governance, combined with an escalation in long-term social fragmentation trends, have increased our risk for genocide in the US.*"[11] However, let's examine this topic a bit more.

Waller wrote this in 2017. The polarization has only increased, as former president Donald Trump is again president. Waller gives numerous examples of what he considers to be the inherent genocidal tendencies of the United States, including the treatment of Native Americans (see chapter 1); the history of slavery, including a request in 1951 "that the General Assembly of the United Nations find and declare by resolution that the Government of the United States is guilty of the crime of Genocide against the Negro People of the United States"; and the US involvement in Vietnam and, most recently, in Iraq (chapter 6).[12] He then proceeds with a litany of acts (or "crimes," if you will), which he accuses Trump of committing as evidence that the "President truly appears to be leading a master class in transforming the United States into a dictatorship."[13] Comparisons to Hitler by him as well as others abound. So, too, do examples of supposed diktats. Christopher Browning, a noted Holocaust scholar, laments,

> Our democracy is based on majority rule tempered by minority rights. I had always assumed that the major threat to our democracy,

if one arrived, would come through a "tyranny of the majority" that cast aside or subverted the constitutional protections of the minority. What we have seen between 2010 and 2016, however, is not the emergence of a tyranny of the majority, but an increasingly irreversible capture of our elected institutions by a focused and un-inhibited minority.[14]

However, in listing these grievances, Waller overlooks several inconvenient facts: One, as alluded to by Browning and articulated in Waller, is that Hillary Clinton won the popular vote but lost the 2016 election to the minority who voted for Trump.[15] What he fails to mention is that the US presidency has never been determined by the popular vote. The Electoral College, as set forth in article II of the US Constitution, was established for this very reason—to prevent a tyranny of the majority and allow the minority a fair chance at elect-ing their choice for president.[16]

As to how this could happen, Ballotpedia explains,

> The split in the 2016 presidential can be attributed to three main factors. First, Democratic nominee Hillary Clinton received signif-icant support in populous blue states like New York and California, where together she received more than 13 million votes. Large mar-gins in these states inflated Clinton's popular vote totals while doing little to offset her deficit in the Electoral College.
>
> Similarly, Clinton performed well in several red states that she ended up losing. In Texas, for example, she received 3.8 million votes. Barack Obama's (D) final vote count in Texas in 2012 was 3.3 million. Clinton also added significantly to her popular vote to-tals in Arizona and Georgia. Donald Trump (R) won both of those states.
>
> The third factor was several close races in battleground states with large populations. For example, Clinton and Trump were sep-arated by margins of less than one percent in Wisconsin, Michigan, and Pennsylvania.[17]

With this explanation, the way the presidential election is held can be exem-plified using a sporting analogy, the NFL's Super Bowl and the MLB's World Series. The popular vote is analogous to the Super Bowl: It is one game (elec-tion), winner take all. The Electoral College is comparable to the World Se-ries: A team can score one hundred runs over three games, while the other team scores only four runs over the four other games. However, if during those four games, the opposing team fails to score any runs, then they lose those games and consequently the World Series. It doesn't matter the total number of runs scored; what matters is the number of games won. Both are valid ways to determine the winner of a contest. The Founding Fathers chose to go with the baseball model instead of the football model.[18]

The 2016 election was not the first time this had happened; it also happened in 2000, 1888, and 1876.[19] Waller's observation that the results of that election "[causes] many in our country to question whether the Trump administration is a legitimate actor representative of the people as a whole" demonstrates a severe lack of knowledge of US history and political science.[20] He quotes Corey Saylor of the Council of American-Islamic Relations (CAIR) as saying that Trump had created a "toxic environment" for US Muslims but fails to recognize that CAIR is an unindicted co-conspirator in providing aid to the terrorist group Hamas.[21] I cover Hamas and CAIR more later this chapter.

Waller continues,

> In the Trump administration, global citizenship now has taken a back seat to a prevailing nationalist sentiment. . . . "America First" separates us from, and prioritizes us over, the global community. Moreover, such nationalism can become a global threat if it undermines our commitment to international norms such as R2P (Responsibility to Protect) and deprioritizes foreign aid to needy countries in terms of health, agriculture, banking, security, etc.[22]

The issue with this is that nations still exist, and what might be in the best interests of one nation does not mean it is in the best interests of other or all nations. This is a prime example of why the United States is composed of individual states instead of one giant entity to deal with local and regional issues as well as national issues. What is in the best interests of California or New York is not necessarily what is in the best interests of Alaska or Wyoming. Neither is what is best for Germany or France necessarily best for the United States or Canada. In a quote attributed to President Ronald Reagan, "A nation without borders is not a nation."[23]

Waller writes further, "If the Trump administration eventually goads Iran, North Korea, or China into a war, it will only be a state versus state conflict for a short time."[24] Not only did Trump *not* goad those countries into war, but he was also instrumental in helping achieve a modicum of peace in the Middle East, through the signing of the Abaraham Accords between Israel and the United Arab Emirates and Bahrain on September 15, 2020.[25]

Race relations, which were supposed to have been healed by the election of President Barak Obama, the first biracial president of the United States, have instead worsened.[26] And all of these writings fail to touch on Second Amendment rights and gun control issues or the issue of abortion rights, the rulings on which, by the Supreme Court of the United States (SCOTUS), have the political Left in such a state of apoplexy that President Joe Biden announced sweeping, radical changes to that body (SCOTUS) that alarmists claim would "destroy its status as a separate branch of government shielded from political pressure."[27]

Further proposed changes include removing presidential immunity for acts made while in office as president, immunity that every president since Washington has used in order to be able to perform the responsibilities of the office without having to worry about the repercussions of those acts once out of office. Both of these proposals are blatantly unconstitutional and are so radical that even the left-leaning *Washington Post* called them "dead-on-arrival" and noted that "Chief Justice John G. Roberts Jr.'s opinion in *Trump v. United States* explained that a law that criminalizes the president's performance of core constitutional duties—such as firing a subordinate—is not valid."[28]

The preceding analysis is not meant as a political speech directed toward either side of the US political spectrum. It merely serves to drive home the fact that the risk of the United States devolving into a genocidal state similar to those of the third world is, in fact, a definite possibility, but the threat comes from both sides of the US political system and not just one or the other.

The pessimist would look at the contents of this book and the preceding examination and believe that all we have learned in thousands of years of warfare is how to kill more efficiently. The optimist would say that we need to continue to try for lasting peace; what if the next time would have made the difference, but we failed to make the effort? However, given man's track record, we would be foolish to assume that the next treaty will actually be the one that works. Again, Waller writes, "We would be naïve, though, to believe that we have, or could ever have, mitigated all the risk of genocide. It would be a disingenuous and dangerous denial of our history to believe that our past, present, or future somehow shelters us from the risk of genocide."[29] And all this debate has, so far, failed to address the quagmire that is the Middle East.

Hamas is a terrorist organization whose members have pledged to destroy Israel and all Jews and have professed that they will stop at nothing to do so, as evidenced by the slaughter of 1,400 Israelis in the October 2023 attack covered in chapter 6.[30] In order to facilitate their ability to execute this malicious plan, they have built a vast tunnel system running through Gaza. These tunnels are wired for electricity, have room for stashed weapons, and have a rail system.[31] In their obsession to destroy Israel, Hamas members routinely use human shields, by placing weapons caches inside civilian buildings, such as hospitals and schools, and denying aid meant for the Palestinian civilians in the area.[32] They openly admit that their goal is to create a permanent state of war with Israel, as was previously pointed out: "'I hope that the state of war with Israel will become permanent on all the borders, and that the Arab world will stand with us,' Taher El-Nounou, a Hamas media adviser, told The [New York] Times."[33]

Their combatants regularly pose as civilians in order to blend in and not draw suspicion as they draw up plans for operations.[34] They routinely torture their captives, including rape and other sexual violence as well as executions, all of which they film.[35] Bodies of babies whose heads had been cut off were found.[36] Even more, they have infiltrated the United Nations in order to facilitate their reign of terror.[37]

CAIR called Israel's hostage rescue mission a "horrific massacre," which hearkens to the saying attributed to Joseph Goebbels: "If you repeat a lie often enough, people will believe it and you will even come to believe it yourself."[38] And perhaps the most egregious insult one could give to a country that has just been attacked: When Israel attempted to fight back, to defend its borders, to protect its citizens, the world condemned them while ignoring the atrocities committed by Hamas.[39] They are offered farcical ideas on how to prosecute the war.[40] They are placed on blacklists by institutions that monitor organizations that harm children, while plans to defend themselves against the perpetrators are rebuked by the president of the United States.[41]

US Attorney General Merrick B. Garland, upon unsealing charges against six senior leaders of Hamas of terrorism, murder conspiracy, and sanctions evasion, stated,

> On October 7th, Hamas terrorists, led by these defendants, murdered nearly 1200 people, including over 40 Americans, and kidnapped hundreds of civilians. This weekend, we learned that Hamas murdered an additional six people they had kidnapped and held captive for nearly a year, including Hersh Goldberg-Polin, a 23 year old Israeli American. We are investigating Hersh's murder, and each and every one of Hamas' brutal murders of Americans, as an act of terrorism. The charges unsealed today are just one part of our effort to target every aspect of Hamas' operations. These actions will not be our last.[42]

After the public was made aware of the grizzly horrors of the Holocaust, a collective cry of "Never again" went up. However, "Never again" has turned into "Again and again." Anti-Semitism is on the rise worldwide.[43] Public support for Hamas and their actions against Israel is widespread. College campuses erupted in violence as pro-Palestinian protests occupied school buildings. How does one deal with these situations? Are we on the path to another Holocaust? Already, we are seeing Stars of David painted on the doors and houses of Jews, evoking memories of a Germany from the not-so-distant past.[44] Are we hurdling toward another Kristallnacht?[45]

George Washington, in his fifth annual address to Congress, stated, "There is a rank due to the United States among nations which will be withheld, if not absolutely lost, by the reputation of weakness. If we desire to avoid insult, we must be able to repel it; if we desire to secure peace, one of the most powerful instruments of our rising prosperity, it must be known that we are at all times ready for war."[46] President Ronald Reagan based his entire presidency on the idea of "peace through strength," stating, "We know only too well that war comes not when the forces of freedom are strong, but when they are weak. It is then that tyrants are tempted."[47]

Bullies have existed since the dawn of time, whether it be the kid down the street who steals your lunch money or Adolf Hitler, Joseph Stalin, Pol Pot, or any other tyrant throughout history exploiting weaker persons or countries. The pessimists believe these bullies will never cease to exist, regardless of what we do to stop them. Indeed, this is the classic question of "nurture versus nature."

In the conclusion of his book, *War Before Civilization*, Lawrence H. Keeley writes, "After exploring war before civilization in search of something less terrible than the wars we know, we merely arrive where we started with an all-too-familiar catalog of deaths, rapes, pillage, destruction, and terror. This is a brutal reality that modern Westerners seem very loath to accept."[48] While we may not be killing as many proportionately as in prehistoric times, we are still killing. The basic nature of mankind remains unchanged as we continue to deal with the seven deadly sins that were first put forth by Pope Gregory I (the Great) in the sixth century: pride (excessive love of one's own excellence); greed (immoderate love or desire for riches and earthly possessions); lust (disordered desire for or inordinate enjoyment of sexual pleasure); envy (resentment or sadness at another's good fortune or excellence, with an often-insatiable desire to have it for oneself); gluttony (excess in eating and drinking); wrath (strong feeling of hatred or resentment with a desire for vengeance); and sloth (culpable lack of physical or spiritual effort).[49] All seven sins serve as motivation for war, in both ancient and modern times. It would seem, given this sentiment, that for all intents and purposes, the only thing that man really has learned in the past several millennia is to kill more effectively.

In summary, perhaps the best reason for the use of total war is found in the *United States Strategic Bombing Survey* (USSBS), which predates Reagan's quote by some forty years: "Prevention of war will not come from neglect of strength or lack of foresight or alertness on our part. Those who contemplate evil and aggression find encouragement in such neglect."[50] As long as human beings exist, there will be evil persons willing to do whatever they deem necessary to obtain their desires. We must be prepared to oppose such persons whenever and wherever they arise.

While prevention of events is preferable to reacting to events, acting too early runs the risk of appearing to be the totalitarian regime that we are opposing. Likewise, insufficient use of force against the oppressor only encourages more aggression by the oppressor. General Curtis LeMay stated that he felt that using insufficient force was more immoral than using excessive force: "If you use less force, you kill off more of humanity in the long run, because you are merely protracting the struggle."[51] Regardless of how we act, we need to remember what George Orwell stated: "The choice before human beings is not

. . . between good and evil but between two evils. You can let the Nazis rule the world; that is evil; or you can overthrow them by war, which is also evil. . . . Whichever you choose, you will not come out with clean hands."[52]

Perhaps in the end, it's not so much where we start as it is where we finish. Mengzi and Xunzi started with different assumptions but finished with a morally good individual as a result of active ethical development. An old Cherokee tale tells of a grandfather teaching his grandson a life lesson:

> "A fight is going on inside me," he says to the boy. "It is a terrible fight and it is between two wolves. One is evil—he is anger, envy, sorrow, regret, greed, arrogance, self-pity, guilt, resentment, inferiority, lies, false pride, superiority, and ego." He continued, "The other is good—he is joy, peace, love, hope, serenity, humility, kindness, benevolence, empathy, generosity, truth, compassion and faith.
>
> "The same fight is going on inside you—and inside every other person, too."
>
> The grandson thought about it for a minute and then asked his grandfather, "Which wolf will win?"
>
> The old Cherokee simply replied, "The one you feed."[53]

NOTES

INTRODUCTION

1. Adrian Gregory, *A War of Peoples, 1914–1919* (Oxford University Press, 2014), 1.
2. Gregory, *War of Peoples*, 7.
3. Jeremy Black, *The Age of Total War, 1860–1945* (Praeger Security International, 2006), 1–11.
4. Raymond Aron, in Roy Pierce, "Political Power, Technology, and Total War: Two French Views," *Journal of Conflict Resolution* 2, no. 4 (December 1958): 321, http://www.jstor.org/stable/172889; Bertrand de Jouvenel, in Pierce, "Political Power," 321.
5. Talbot Imlay, "Total War," *Journal of Strategic Studies* 30, no. 3 (June 2007): 551, http://www.tandfonline.com/doi/abs/10.1080/01402390701343516; Lance Janda, "Shutting the Gates of Mercy: The American Origins of Total War, 1860–1880," *Journal of Military History* 59, no. 1 (January 1995): 8, https://doi .org/10.2307/2944362.
6. Stig Förster, in Imlay, "Total War," 555.
7. Quincy Wright, "The Nature of Conflict," *Western Political Quarterly* 4, no. 2 (June 1951): 204 (emphasis added), http://www.jstor.org/stable/443101.
8. Jan Philipp Reemtsma, "The Concept of the War of Annihilation: Clausewitz, Ludendorff, Hitler," in *War of Extermination: The German Military in World War II, 1941–1944*, ed. Hannes Heer and Klaus Naumann (Berghahn Books, 2000), 13.
9. See Black, *Age of Total War*, 2.
10. Lawrence H. Keeley, *War Before Civilization* (Oxford University Press, 1996), 176.
11. Daniel Rothbart, Karina V. Korostelina, and Mohammed D. Cherkaoui, eds., *Civilians and Modern War: Armed Conflict and the Ideology of Violence* (Routledge, 2012), 12.
12. Rothbart, Korostelina, and Cherkaoui, *Civilians and Modern War*, 13.
13. Richard E. Rubenstein, "The Role of Civilians in American War Ideology," in *Civilians and Modern War: Armed Conflict and the Ideology of Violence*, ed.

Daniel Rothbart, Karina V. Korostelina, and Mohammed D. Cherkaoui (Routledge, 2012), 21–50.

14. United States Holocaust Memorial Museum, "Definitions: Types of Mass Atrocities," accessed July 11, 2024, https://www.ushmm.org/genocide-prevention/learn-about-genocide-and-other-mass-atrocities/definitions.
15. United States Holocaust Memorial Museum, "Definitions."
16. World Without Genocide, "Eight Stages of Genocide," accessed July 11, 2024, https://worldwithoutgenocide.org/genocides-and-conflicts/background-and-overview-information/eight-stages-of-genocide.
17. James E. Waller, "It Can Happen Here: Assessing the Risk of Genocide in the US," Center for Develoment of International Law, February 24, 2017, 38, https://worldwithoutgenocide.org/wp-content/uploads/2017/03/Waller-Assessing-the-Risk-of-Genocide.pdf.

CHAPTER 1

1. Jeremy Black, *The Age of Total War, 1860–1945* (Praeger Security International, 2006), 6–7; Lawrence H. Keeley, *War Before Civilization* (Oxford University Press, 1996), 26.
2. For multiple examples of these annihilations, see Keeley, *War Before Civilization*, 59–69.
3. Keeley, *War Before Civilization*, 29. Note that the term *homicide* in this case, along with the terms *feuding* and *vendetta*, refers to warfare among smaller societal units, such as tribes and bands. *Warfare* and related terms are reserved for large units, such as modern states.
4. Black, *Age of Total War*, 5.
5. Georg W. F. Hegel, in Kristof K. P. Vanhoutte, "'Oh God! What a Lovely War': Giorgio Agamben's Clausewitzian Theory of Total/Global (Civil) War," *Sotsiologicheskoe Obozrenie/Russian Sociological Review* 14, no. 4 (2015): 29, https://doi.org/10.17323/1728-192x-2015-4-28-43.
6. Vanhoutte, "Oh God!" 32.
7. Black, *Age of Total War*, 5–6.
8. Black, *Age of Total War*, 15–16; History.com Editors, "Ethnic Cleansing," History, updated April 10, 2025, http://www.history.com/topics/ethnic-cleansing; Kate Raphael, "Mongol Siege Warfare on the Banks of the Euphrates and the Question of Gunpowder (1260–1312)," *Journal of the Royal Asiatic Society* 19, no. 3 (July 2009): 355–70, https://www.jstor.org/stable/27756073.
9. George R. Milner, "Warfare in Prehistoric and Early Historic Eastern North America," *Journal of Archaeological Research* 7, no. 2 (June 1999): 106, https://doi.org/10.1007/s10814-005-0001-x.
10. Milner, "Warfare," 107.
11. R. Brian Ferguson, "War Is Not Part of Human Nature," *Scientific American* 319, no. 3 (September 1, 2018): 76, https://www.scientificamerican.com/article/war-is-not-part-of-human-nature/.
12. Jean-Jacques Rousseau, in Keeley, *War Before Civilization*, 6.
13. Russell Means and Marvin Wolf, in Richard J. Chacon and Rubén G. Mendoza, eds., *North American Indigenous Warfare and Ritual Violence* (University of Arizona Press, 2013).
14. Marta Mirazón Lahr, in Keeley, *War Before Civilization*, 48.

15. University of Cambridge: Research, "Evidence of a Prehistoric Massacre Extends the History of Warfare," January 20, 2016, https://www.cam.ac.uk/research/news/evidence-of-a-prehistoric-massacre-extends-the-history-of-warfare.
16. Keeley, *War Before Civilization*, 27–28.
17. Keeley, *War Before Civilization*, 28.
18. Ernest S. Burch Jr., "Traditional Native Warfare in Western Alaska," in *North American Indigenous Warfare and Ritual Violence*, ed. Richard J. Chacon and Rubén G. Mendoza, 11–29 (University of Arizona Press, 2013), 11.
19. Burch, "Traditional Native Warfare," 19, 27.
20. Burch, "Traditional Native Warfare," 19.
21. Burch, "Traditional Native Warfare," 22.
22. Burch, "Traditional Native Warfare," 22, 25.
23. J. J. K. Simon, in Jerry Melbye and Scott I. Fairgrieve, "A Massacre and Possible Cannibalism in the Canadian Arctic: New Evidence from the Saunaktuk Site (NgTn-1)," *Arctic Anthropology* 31, no. 2 (January 1994): 57, https://www.researchgate.net/publication/288244980_A_massacre_and_possible_cannibalism_in_the_Canadian_Arctic_New_evidence_from_the_Saunaktuk_site_NgTn-1.
24. Melbye and Fairgrieve, "Massacre and Possible Cannibalism," 58.
25. Chacon and Mendoza, *North American Indigenous Warfare*, 37.
26. Nicolas Jeremie, in Chacon and Mendoza, *North American Indigenous Warfare* (spelling in the original).
27. Chacon and Mendoza, *North American Indigenous Warfare*, 38.
28. Chacon and Mendoza, *North American Indigenous Warfare*, 40.
29. Andrew Graham, in Chacon and Mendoza, *North American Indigenous Warfare*, 40–41.
30. Keeley, *War Before Civilization*, 53.
31. Keeley, *War Before Civilization*, 52.
32. Keeley, *War Before Civilization*, 55.
33. Keeley, *War Before Civilization*, 90, figure 6.2; 91.
34. Keeley, *War Before Civilization*, 195, table 6.2.
35. M. Meggitt, in Keeley, *War Before Civilization*, 90.
36. Keeley, *War Before Civilization*, 91.
37. Keeley, *War Before Civilization*, 175.
38. Felix Gilbert, "Machiavelli: The Renaissance of the Art of War," in *Makers of Modern Strategy: From Machiavelli to the Nuclear Age*, ed. Peter Paret, Gordon A. Craig, and Felix Gilbert, 11–31 (Princeton University Press, 1986), 13–14.
39. Milner, "Warfare," 108.
40. Raphael, "Mongol Siege Warfare," 363.
41. Raphael, "Mongol Siege Warfare," 359, 362.
42. Cristina Andrei and Decebal Nedu, "The Campaign of Marcus Atilius Regulus in Africa: Military Operations by Sea and by Land (256–255 B.C.)," *Constanta Maritime University Annals* 13 (2010), https://www.academia.edu/43696687/THE_CAMPAIGN_OF_MARCUS_ATILIUS_REGULUS_IN_AFRICA_MILITARY_OPERATIONS_BY_SEA_AND_BY_LAND_256_255_B_C.
43. Bret Mulligan, "Nepos: Life of Hannibal, the First Punic War," Dickinson College Commentaries, 2013, https://dcc.dickinson.edu/nepos-hannibal/first-punic-war.
44. William D. Rubenstein, in Gregory S. Gordon, *Atrocity Speech Law: Foundation, Fragmentation, Fruition* (Oxford University Press, 2017), 31.
45. Gordon, *Atrocity Speech Law*, 31–32.

46. Ben Kiernan, in Gordon, *Atrocity Speech Law*, 32.

47. V. Barras and G. Greub, "History of Biological Warfare and Bioterrorism," *Clinical Microbiology and Infection* 20, no. 6 (June 2014): 498, https://doi.org/10.1111/1469-0691.12706; Zenobia S. Homan, "Unconventional Warfare in the Ancient Near East," *Social Sciences and Humanities Open* 8, no. 1 (2023): 5, https://doi.org/10.1016/j.ssaho.2023.100501.

48. Homan, "Unconventional Warfare," 6.

49. Homan, "Unconventional Warfare," 7–8.

50. Sarah Everts, "A Brief History of Chemical War," *Distillations Magazine* (May 12, 2015), https://www.sciencehistory.org/stories/magazine/a-brief-history-of-chemical-war/.

51. George Forrest, "The First Sacred War," *Bulletin de Correspondance Hellénique* 80, no. 1 (1956): 33–34, https://doi.org/10.3406/bch.1956.2408.

52. Barras and Greub, "History of Biological Warfare," 498.

53. Everts, "Brief History of Chemical War."

54. Thucydides, "The Siege of Plataea," in *History of the Peloponnesian War*, trans. Richard Crawley, 1910, https://www.livius.org/sources/content/thucydides-historian/siege-of-plataea/.

55. The Editors of Encyclopaedia Britannica, "Greek Fire," Britannica, updated January 31, 2025, https://www.britannica.com/technology/Greek-fire; Royal Museums Greenwich, "What Was Greek Fire?," accessed April 27, 2025, https://www.rmg.co.uk/stories/topics/greek-fire; Mark Cartwright, "Greek Fire," World History Encyclopedia, November 14, 2017, https://www.worldhistory.org/Greek_Fire/.

56. Gabriele de Mussis, in Mark Wheelis, "Biological Warfare at the 1346 Siege of Caffa," *Emerging Infectious Diseases* 8, no. 9 (September 2002): 972–73, https://doi.org/10.3201/eid0809.010536. I encourage you to read this article to learn why Wheelis believes that this siege could not have been the source of the European Black Death. Stefan Riedel, "Biological Warfare and Bioterrorism: A Historical Review," *Baylor University Medical Center Proceedings* 17, no. 4 (2004): 400, https://doi.org/10.1080/08998280.2004.11928002.

57. Sun Tzu, *The Art of War*, trans. Samuel B. Griffith (Oxford University, 1971), 73.

58. Tzu, *Art of War*, 74.

59. Tzu, *Art of War*, 77.

60. Joseph Kleist, "The Battle of Thermopylae: Principles of War on the Ancient Battlefield," *Studia Antiqua* 6, no. 1 (Spring 2008): 76, https://scholarsarchive.byu.edu/cgi/viewcontent.cgi?article=1091&context=studiaantiqua.

61. Kleist, "Battle of Thermopylae," 84.

62. Kleist, "Battle of Thermopylae," 77, 79–82.

63. Talbot Imlay, "Total War," *Journal of Strategic Studies* 30, no. 3 (June 2007): 561–62, 565, http://www.tandfonline.com/doi/abs/10.1080/01402390701343516.

64. See Thomas Head, "The Development of the Peace of God in Aquitaine (970–1005)," *Speculum* 74, no. 3 (July 1999): 656–86, https://doi.org/10.2307/2886764.

65. For a review of both the circumstances leading up to these two conferences and the subsequent results, see Head, "Development of the Peace"; New Catholic Encyclopedia, "Peace of God," Encyclopedia.com, accessed April 26, 2025, https://www.encyclopedia.com/religion/encyclopedias-almanacs-transcripts-and-maps/peace-god; Thomas Gergen, "The Peace of God and Its Legal Practice in the Eleventh Century," *Cuadernos de Historia del Derecho* 9 (July 2002): 11–27, https://

www.researchgate.net/publication/39283239_The_Peace_of_God_its_legal_prac
tice_in_the_Eleventh_Century.

66. R. R. Palmer, "Frederick the Great, Guibert, Bülow: From Dynastic to National War," in *Makers of Modern Strategy: From Machiavelli to the Nuclear Age*, ed. Peter Paret, Gordon A. Craig, and Felix Gilbert, 91–120 (Princeton, NJ: Princeton University Press, 1986).

67. Russell F. Weigley, *The Age of Battles: The Quest for Decisive Warfare from Breitenfeld to Waterloo* (Indiana University Press, 1991), 46, 69–71.

68. Elizabeth A. Fenn, "Biological Warfare in Eighteenth-Century North America: Beyond Jeffery Amherst," *Journal of American History* 86, no. 4 (March 2000): 1573, https://doi.org/10.2307/2567577.

69. Palmer, "Frederick the Great," 94–95.

70. Russell F. Weigley, "American Strategy from Its Beginning Through the First World War," in *Makers of Modern Strategy: From Machiavelli to the Nuclear Age*, edited by Peter Paret, Gordon A. Craig, and Felix Gilbert, 408–43 (Princeton University Press, 1986), 409.

71. Fenn, "Biological Warfare," 1564–65.

72. Harold B. Gill Jr., "Colonial Germ Warfare," *Colonial Williamsburg Journal* 26, no. 1 (Spring 2004): 18–23, https://research.colonialwilliamsburg.org/Foundation/journal/Spring04/warfare.cfm.

73. Pierre-François-Xavier de Charlevoix, in Fenn, "Biological Warfare," 1565.

74. Fenn, "Biological Warfare."

75. Fenn, "Biological Warfare."

76. Fenn, "Biological Warfare," 1565–66.

77. Erica Charters, "Military Medicine and the Ethics of War: British Colonial Warfare During the Seven Years War (1756–63)," *Canadian Bulletin of Medical History* 27, no. 2 (Fall 2010): 274, https://doi.org/10.3138/cbmh.27.2.273 (spelling in the original).

78. Fenn, "Biological Warfare," 1567.

79. Matthew Wills, "How Commonly Was Smallpox Used as a Biological Weapon," *JSTOR Daily*, April 4, 2021, https://daily.jstor.org/how-commonly-was-smallpox-used-as-a-biological-weapon/.

80. Wills, "How Commonly Was Smallpox."

81. Correspondence between Jeffrey Amherst and Henry Bouquet, in Fenn, "Biological Warfare," 1555–1557. See also Gill, "Colonial Germ Warfare" (spelling in the original).

82. William Trent, in Fenn, "Biological Warfare," 1554 (spelling in the original).

83. Fenn, "Biological Warfare," 1557.

84. Martin van Creveld, "Historical Development," under "tactics," Britannica, n.d., updated September 14, 2023, https://www.britannica.com/topic/tactics/Historical-development.

85. van Creveld, "Historical Development."

86. American Battlefield Trust, "Revolutionary War Strategy," accessed April 23, 2025, https://www.battlefields.org/learn/articles/revolutionary-war-strategy.

87. van Creveld, "Historical Development."

88. American Battlefield Trust, "Revolutionary War Strategy."

89. Vincent J. Cirillo, "Two Faces of Death: Fatalities from Disease and Combat in America's Principal Wars, 1775 to Present." *Perspectives in Biology and Medicine* 51, no. 1 (Winter 2008): 125, https://doi.org/10.1353/pbm.2008.0005.

90. American Battlefield Trust, "Revolutionary War Strategy."
91. Fenn, "Biological Warfare."
92. Fenn, "Biological Warfare," 1574.
93. Amherst, in Fenn, "Biological Warfare" (spelling in the original).
94. Niccolò Machiavelli, in Gill, "Colonial Germ Warfare."
95. Inoculation was not the same as immunization. Inoculation involved purposely introducing the disease into the patient's body through an abraded area. This would cause the patient to develop the disease but in a much milder form than if they had been infected normally. The reason for this is unknown. Regardless, this was a major reason for the colonists winning: They didn't have to worry about disease outbreaks (except during the invasion of Canada in 1775). See Cirillo, "Two Faces of Death."
96. Fenn, "Biological Warfare," 1567–1568.
97. Fenn, "Biological Warfare," 1568 (spelling in original).
98. Fenn, "Biological Warfare."
99. Fenn, "Biological Warfare," 1569–70.
100. See Fenn, "Biological Warfare," 1570–73, for these examples.
101. Robert G. Parkinson, "Print, the Press, and the American Revolution," in Oxford Research Encyclopedia of American History (Oxford University Press September 3, 2015), https://doi.org/10.1093/acrefore/9780199329175.013.9.
102. Parkinson, "Print, the Press."
103. Parkinson, "Print, the Press."
104. Parkinson, "Print, the Press."
105. Parkinson, "Print, the Press."
106. Parkinson, "Print, the Press."
107. Niccolò Machiavelli, The Prince, trans. George Bull (Penguin Books, 2003), 10–11.
108. Machiavelli, Prince, 18.
109. Machiavelli, Prince, 18.
110. Gilbert, "Machiavelli," 29.
111. Palmer, "Frederick the Great," 107.
112. Palmer, "Frederick the Great," 109.
113. François-René de Chateaubriand, in Christopher Vere, "The Napoleonic Wars and the Birth of Modern Warfare," Intelligence and National Security 24, no. 3 (2009): 466, http://dx.doi.org/10.1080/02684520903135065.
114. Peter Paret, "Napoleon and the Revolution in War," in Makers of Modern Strategy from Machiavelli to the Nuclear Age, ed. Peter Paret, Gordon A. Craig, and Felix Gilbert, 123–42 (Princeton University Press, 1986), 129, 136, 141.
115. Napoleon Bonaparte, Napoleon on the Art of War, ed. and trans. Jay Luvaas (Touchstone, 2001), 92; Paret, "Napoleon and the Revolution," 125.
116. Bonaparte, Art of War, 116.
117. Weigley, "American Strategy," 330.
118. Weigley, "American Strategy," 10.
119. Berke Gursoy, "The Eagle's Rise: Napoleon Bonaparte's First Campaign and the Birth of Napoleonic Warfare," Armstrong Undergraduate Journal of History 9, no. 1 (April 2019): 18, https://doi.org/10.20429/aujh.2019.090102.
120. Gursoy, "Eagle's Rise," 19–20.
121. Gursoy, "Eagle's Rise," 22–24.

122. John Lynch, "The Lessons of Walcheren Fever, 1809," *Military Medicine* 174, no. 3 (March 2009): 315, https://academic.oup.com/milmed/article/174/3/315/4333688.
123. Lynch, "Lessons of Walcheren Fever."
124. Martin R. Howard, "Walcheren 1809: A Medical Catastrophe," *BMJ* 319, no. 7725 (December 18, 1999): 1643, https://www.bmj.com/content/319/7225/1642.
125. Howard, "Walcheren 1809."
126. Howard, "Walcheren 1809."
127. Howard, "Walcheren 1809."
128. Howard, "Walcheren 1809."
129. Howard, "Walcheren 1809," 1644.
130. Howard, "Walcheren 1809," 1645.
131. Matthew D. Turner and Jason Sapp, "Failure to Plan: The Disease That Cost an American Empire," *Military Medicine* 188, nos. 7–8 (July–August 2023): 171, https://doi.org/10.1093/milmed/usad161.
132. Turner and Sapp, "Failure to Plan," 172.
133. John S. Marr and John T. Cathey, "The 1802 Saint-Domingue Yellow Fever Epidemic and the Louisiana Purchase," *Journal of Public Health Management and Practice* 19, no. 1 (January–February 2013): 78–79, https://doi.org/10.1097/PHH.0b013e318252eea8.
134. Marr and Cathey, "Saint-Domingue Yellow Fever Epidemic," 81.
135. Kelly Gambino-Shirley, "Napoleon's Missed Opportunities to Maintain Combat Forces Through Medical Innovations and Battling the Hidden Enemy," Air Command and Staff College, December 7, 2011, https://apps.dtic.mil/sti/citations/AD1019074.
136. Antoine Henri de Jomini, *The Art of War* (Greenhill Books, 1996), 25.
137. Jomini, *Art of War*, 31–32.
138. Jomini, *Art of War*, 32.
139. Jomini, *Art of War*, 33.
140. Jomini, *Art of War*, 138, 141.
141. Jomini, *Art of War*, 150.
142. Carl von Clausewitz, *On War*, ed. and trans. Michael Howard and Peter Paret (Princeton University Press, 1989), 87.
143. von Clausewitz, *On War*, 580.
144. von Clausewitz, *On War*, 586–92.
145. von Clausewitz, *On War*, 592–93.
146. von Clausewitz, *On War*, 90.
147. von Clausewitz, *On War*, 208.
148. von Clausewitz, *On War*, 108.
149. von Clausewitz, *On War*, 583, 610.
150. von Clausewitz, *On War*, 220.
151. Gambino-Shirley, "Napoleon's Missed Opportunities"; Dino E. Buenviaje and James F. Willis, "Laws of Warfare: World War I," in *World at War: Understanding Conflict and Society* (ABC-CLIO, 2024).
152. History.com Editors, "American Indian Wars: Timeline," History, updated April 15, 2025, https://www.history.com/topics/native-american-history/american-indian-wars-timeline.
153. History.com Editors, "American Indian Wars: Timeline."

154. National Geographic Society, "The Indian Removal Act and the Trail of Tears," *National Geographic*, updated October 1, 2024, https://education.nationalgeo graphic.org/resource/indian-removal-act-and-trail-tears/.

155. National Geographic Society, "Indian Removal Act."

156. Bill of Rights Institute, "A Deep Stain on the American Character: John Marshall and Justice for Native Americans—Handout A: Narrative," accessed April 23, 2025, https://billofrightsinstitute.org/activities/a-deep-stain-on-the-american -character-john-marshall-and-justice-for-native-americans-handout-a-narrative.

157. Bill of Rights Institute, "Deep Stain."

158. Richard E. Rubenstein, "The Role of Civilians in American War Ideology," in *Civilians and Modern War: Armed Conflict and the Ideology of Violence*, ed. Daniel Rothbart, Karina V. Korostelina, and Mohammed D. Cherkaoui, 21–50. (Routledge, 2012). See pp. 27–32 for the source of this information as well as Rubenstein's full coverage of the First Seminole War and his justification of this war as self-defense.

159. Bill of Rights Institute, "Deep Stain."

160. Andrew Jackson, "President Andrew Jackson's Message to Congress 'On Indian Removal' (1830)," National Archives, December 6, 1830, https://www.archives .gov/milestone-documents/jacksons-message-to-congress-on-indian-removal.

161. Bill of Rights Institute, "Deep Stain."

162. Bill of Rights Institute, "Deep Stain"; National Geographic Society, "Indian Removal Act."

163. In National Geographic Society, "Indian Removal Act."

164. History.com Editors, "American-Indian Wars," History, updated February 27, 2025, https://www.history.com/articles/american-indian-wars; History.com Editors, "American Indian Wars: Timeline."

165. Richard W. Stewart, ed., "The Mexican War and After," in *American Military History: The United States Army and the Forging of a Nation, 1775–1917*, vol. 1, 175–96 (Center of Military History, United States Army, 2005), 190.

166. Richard W. Stewart, ed., "Winning the West: The Army in the Indian Wars, 1865–1890," in *American Military History: The United States Army and the Forging of a Nation, 1775–1917*, vol. 1, 321–40 (Center of Military History, United States Army, 2005).

167. Hereafter, the organization is referred to as LDS or the Church, and its members, as Saints, in an effort to respect current church requests. See BYU Department of History, "The Mountain Meadows Massacre," BYU College of Family, Home, and Social Sciences, 2024, https://history.byu.edu/mountainmeadowsmassacre.

168. William P. MacKinnon, "Prelude to Civil War: The Utah War's Impact and Legacy," in *Civil War Saints*, ed. Kenneth L. Alford (Religious Studies Center, Brigham Young University, 2012), 2.

169. MacKinnon, "Prelude to Civil War."

170. Harry Searles, "Utah War Summary," American History Central, updated February 2, 2024, https://www.americanhistorycentral.com/entries/utah-war/; Harry Searles, "Utah War Facts," American History Central, updated June 10, 2024, https://www.americanhistorycentral.com/entries/utah-war-facts/.

171. Richard Poll, "The Utah War," History to Go, 1994, https://historytogo.utah.gov/ utah-war/.

172. Poll, "Utah War."

173. Poll, "Utah War"; Searles, "Utah War Summary."

174. BYU Department of History, "Mountain Meadows Massacre."
175. History.com Editors, "120 Emigrants Murdered at the Mountain Meadows Massacre," updated January 31, 2025, https://www.history.com/this-day-in-history/mormons-and-paiutes-murder-120-emigrants-at-mountain-meadows.
176. Gilbert King, "The Aftermath of Mountain Meadows," *Smithsonian Magazine*, February 29, 2012, https://www.smithsonianmag.com/history/the-aftermath-of-mountain-meadows-110735627/.
177. Richard E. Turley Jr., "The Mountain Meadows Massacre," Church of Jesus Christ of Latter-Day Saints, September 2007, https://www.churchofjesuschrist.org/study/ensign/2007/09/the-mountain-meadows-massacre?lang=eng.
178. History.com Editors, "120 Emigrants Murdered."
179. American Experience, "The Mountain Meadows Massacre," PBS, accessed April 27, 2024, https://www.pbs.org/wgbh/americanexperience/features/mormons-massacre/.
180. MacKinnon, "Prelude to Civil War."
181. Searles, "Utah War Summary."
182. Brandon Rottinghaus and Justin S. Vaughn, *Official Results of the 2024 Presidential Greatness Project Expert Survey*, 2024, https://qzg.wvf.mybluehost.me/wp-content/uploads/2024/05/Presidential-Greatness-White-Paper-2024.pdf.

CHAPTER 2

1. George B. McClellan to Hill Carter, Esq., in Lance Janda, "Shutting the Gates of Mercy: The American Origins of Total War, 1860–1880," *Journal of Military History* 59, no. 1 (January 1995): 12, https://doi.org/10.2307/2944362.
2. William T. Sherman, *Sherman's Civil War: Selected Correspondence of William T. Sherman, 1860–1865*, ed. Brooks D. Simpson and Jean V. Berlin (University of North Carolina Press, 1999), 128–29.
3. John W. Brinsfield, "The Military Ethics of General William T. Sherman: A Reassessment," *Parameters* 12, no. 1 (July 4, 1982): 36–48, https://doi.org/10.55540/0031-1723.1280.
4. Sherman, *Sherman's Civil War*, 266 (emphasis in original).
5. Sherman, in Janda, "Shutting the Gates," 13.
6. Sherman, in Christopher S. Hoffman, *Major General William T. Sherman's Total War in the Savannah and Carolina Campaigns*" (School of Advance Military Studies, US Army Command and General Staff College, 2018), 3, https://apps.dtic.mil/sti/pdfs/AD1071091.pdf.
7. William T. Sherman, *Memoirs of Gen. W. T. Sherman*, vol. 2, 4th ed. (Charles L. Webster, 1891), 111.
8. Sherman, in John Bennett Walters, *Merchant of Terror: General Sherman and Total War* (Bobbs-Merrill, 1973), 70. Compare Niccolò Machiavelli, *The Prince*, trans. George Bull (Penguin Books, 2003), 54.
9. Hoffman, *Sherman's Total War*, 2 (emphasis added).
10. Janda, "Shutting the Gates," 16–17.
11. Articles of War: An Act for Establishing Rules and Articles for the Government of the Armies of the United States, 9th Cong., approved April 10, 1806, https://military-justice.ca/wp-content/uploads/2019/01/U.S.-ARTICLES-OF-WAR.pdf.
12. Sherman, in Brinsfield, "Military Ethics of Sherman," 41.
13. W. T. Sherman to J. B. Hood, in Sherman, *Memoirs*, 118.

14. J. B. Hood to W. T. Sherman, in Sherman, *Memoirs*, 124.

15. W. T. Sherman to J. M. Calhoun, in Sherman, *Memoirs*, 125–26.

16. Adam Badeau, in Janda, "Shutting the Gates," 12.

17. Philip Sheridan, in National Park Service (NPS), "The Burning," accessed March 25, 2024, https://www.nps.gov/articles/000/the-burning-shenandoah-valley-in-flames.htm.

18. Ulysses S. Grant, in NPS, "Burning."

19. Sheridan and Grant, in NPS, "Burning."

20. NPS, "Burning."

21. Janda, "Shutting the Gates," 10.

22. Sheridan, in NPS, "Burning."

23. Stephen Davis, "Atlanta Campaign," New Georgia Ecyclopedia, updated September 17, 2018, https://www.georgiaencyclopedia.org/articles/history-archaeology/atlanta-campaign/.

24. Davis, "Atlanta Campaign."

25. Davis, "Atlanta Campaign."

26. Sherman, in Stephen Davis, "The Burning of Atlanta: What Really Happened?" *Civil War News*, September 22, 2022, https://www.historicalpublicationsllc.com/civilwarnews/the-burning-of-atlanta-what-really-happened/article_0f304440-34f0-11ed-acfc-7b4f80e30e60.html.

27. Sherman, in Davis, "Burning of Atlanta."

28. History.com Editors, "Union General Sherman's Scorched-Earth March to the Sea Campaign Begins," History, updated January 30, 2025, https://www.history.com/this-day-in-history/the-march-to-the-sea-begins.

29. Gordon Jones, in Davis, "Burning of Atlanta."

30. Hoffman, "Sherman's Total War."

31. Sherman, *Memoirs*, 175. For his complete orders for the march, see 174–76.

32. Sherman, *Memoirs*.

33. Hoffman, "Sherman's Total War," 14.

34. Sherman, *Memoirs*, 182.

35. Sherman, *Sherman's Civil War*, 764.

36. Sherman, in Mark, "General William T. Sherman's Report on the March to the Sea and Capture of Savannah," Iron Brigader, updated August 21, 2018, https://ironbrigader.com/2014/11/22/general-william-t-shermans-report-march-sea-capture-savannah/.

37. Sherman, *Memoirs*, 231.

38. Sherman, *Sherman's Civil War*, 776.

39. Sherman, *Sherman's Civil War*, 776–77.

40. Spencer Glasgow Welch, *A Confederate Surgeon's Letters to His Wife* (Neale, 1911) (spelling in original).

41. Sherman, *Sherman's Civil War*.

42. Emma LeConte, in Erik Sass, "Fall of the South: The Burning of Columbia," Mental Floss, February 18, 2015, https://www.mentalfloss.com/article/61743/fall-south-burning-columbia.

43. Henry W. Slocum, in HistoryNet Staff, "The Burning of Columbia from the Union and Confederate Perspectives," HistoryNet, September 23, 1998, https://www.historynet.com/the-burning-of-columbia-from-the-union-and-confederate-perspectives-october-1998-civil-war-times-feature/.

44. Thomas Osborn, in Sass, "Fall of the South."

45. Sherman, *Memoirs*.
46. Hannah Plummer, in Sass, "Fall of the South."
47. National Park Service (NPS), "History and Legal Status of Prisoners of War," updated October 26, 2022, https://www.nps.gov/ande/learn/historyculture/history-legal-status-pows.htm.
48. NPS, "Prisoners of War."
49. Alan Marsh, "POWs in American History: A Synopsis," National Park Service, 1998, https://www.nps.gov/ande/learn/historyculture/pow_synopsis.htm.
50. Paul A. Chase and James Wood II, "Prisoners of War (POWs) During the American Revolution," Sons of the American Revoluation, accessed August 21, 2024, https://www.sar.org/wp-content/uploads/2020/06/Prisoner-of-War-Briefing-by-Paul-Chase.pdf, 2.
51. Chase and Wood, "POWs During the Revolution," 1.
52. Marsh, "POWs in American History"; Matthew Brenckle, "British and American Naval Prisoners of War During the War of 1812," USS Constitution Museum, 2005, https://ussconstitutionmuseum.org/wp-content/uploads/2020/05/British-and-American-Naval-Prisoners-of-War-During-the-War-of-1812.pdf, 7.
53. Robert H. Kellog, in Gary Flavion, "Civil War Prison Camps," American Battlefield Trust, accessed August 21, 2024, https://www.battlefields.org/learn/articles/civil-war-prison-camps.
54. Marsh, "POWs in American History."
55. Marsh, "POWs in American History."
56. Flavion, "Civil War Prison Camps."
57. Flavion, "Civil War Prison Camps."
58. Flavion, "Civil War Prison Camps."
59. Flavion, "Civil War Prison Camps."
60. Sanitary Commission, in Flavion, "Civil War Prison Camps."
61. Lucius Eugene Chittenden, in Flavion, "Civil War Prison Camps." For a comprehensive list of both Union and Confederate POW camps and data regarding each, see American Civil War, "Confederate Prisoner of War Camps," accessed August 21, 2024, https://www.mycivilwar.com/pow/confederate.html, and American Civil War, "Union Prisoner of War Camps," accessed August 21, 2024, https://www.mycivilwar.com/pow/union.html.
62. Flavion, "Civil War Prison Camps"; National Park Service (NPS), "Search for Prisoners," accessed August 21, 2024, https://www.nps.gov/civilwar/search-prisoners.htm.
63. Robert Scott Davis, "Andersonville Prison," New Georgia Encyclopedia, updated July 15, 2020, https://www.georgiaencyclopedia.org/articles/history-archaeology/andersonville-prison/.
64. Flavion, "Civil War Prison Camps."
65. Davis, "Andersonville Prison."
66. Davis, "Andersonville Prison."
67. Sand Creek Massacre Foundation, "The Massacre," accessed August 21, 2024, https://www.sandcreekmassacrefoundation.org/massacre; American Battlefield Trust, "Sand Creek: Sand Creek Massacre, Chivington Massacre," accessed April 23, 2025, https://www.battlefields.org/learn/civil-war/battles/sand-creek; National Park Service (NPS), "Sand Creek Massacre," updated March 13, 2017, https://www.nps.gov/sand/learn/historyculture/massacre.htm.

68. Guy R Hasegawa, "Proposals for Chemical Weapons During the American Civil War," *Military Medicine* 173, no. 5 (May 2008): 499, https://doi.org/10.7205/MILMED.173.5.499.
69. Hasegawa, "Proposals for Chemical Weapons."
70. Hasegawa, "Proposals for Chemical Weapons," 500.
71. Hasegawa, "Proposals for Chemical Weapons," 500–503.
72. Hasegawa, "Proposals for Chemical Weapons," 504.
73. Hasegawa, "Proposals for Chemical Weapons."
74. Hasegawa, "Proposals for Chemical Weapons."
75. Hasegawa, "Proposals for Chemical Weapons."
76. Jeremy Black, *The Age of Total War, 1860–1945* (Praeger Security International, 2006).
77. Black, *Age of Total War.*
78. Black, *Age of Total War*, 36–37.
79. Black, *Age of Total War*, 38.
80. V. Golovnin, in Yaroslav A. Shulatov, "Russia as a 'Trauma': The Rise and Fall of Japan as a Great Power," *Russia in Global Affairs* 17, no. 4 (October–December 2019): 82, https://doi.org/10.31278/1810-6374-2019-17-4-78-108.
81. Chan Cheng Lin, "Nanjing Massacre and Sook Ching Massacre: Shaping of Chinese Popular Memories in China and Singapore, 1945–2015" (master's thesis, National University of Singapore, December 18, 2015), 23, https://scholarbank.nus.edu.sg/entities/publication/c1d85a55-c675-4093-a31b-a24ecc6d260c.
82. Gerhard Krebs, "World War Zero? Re-Assessing the Global Impact of the Russo-Japanese War 1904–05," *Asia-Pacific Journal: Japan Focus* 10, no. 21 (May 19, 2012): 1–24, https://apjjf.org/2012/10/21/gerhard-krebs/3755/article; John W. Steinberg, "Was the Russo-Japanese War World War Zero?" *Russian Review* 67, no. 1 (January 2008): 1–7, https://www.jstor.org/stable/20620667. The reference of the first Russo-Japanese War as "World War Zero" should not be confused with the climate change initiative spearheaded by John Kerry in 2019 under the same name.
83. History.com Editors, "Russo-Japanese War," History, updated February 27, 2025, https://www.history.com/topics/asian-history/russo-japanese-war.
84. George E. Mowry, "The First Roosevelt,"*American Mercury*, November 1946, 578–84, https://www.unz.com/print/AmMercury-1946nov-00578/.
85. Steinberg, "Russo-Japanese War," 3, 5–7. For a thorough analysis of this entire situation with Japan, consult Richard Storry, *Japan and the Decline of the West in Asia, 1894–1943* (St. Martin's Press, 1979).
86. Richard W. Stewart, ed., "Winning the West: The Army in the Indian Wars, 1865–1890," in *American Military History: The United States Army and the Forging of a Nation, 1775–1917*, vol. 1, 321–40 (Center of Military History, United States Army, 2005), 326.
87. Stewart, "Winning the West," 330.
88. Stewart, "Winning the West," 336–38.
89. William D. Carrigan and Clive Webb, "The Lynching of Persons of Mexican Origin or Descent in the United States, 1848 to 1928," *Journal of Social History* 37, no. 2 (Winter 2003): 413, https://www.jstor.org/stable/3790404.
90. Carrigan and Webb, "Lynching of Persons," 416.
91. Carrigan and Webb, "Lynching of Persons," 418.

92. Richard W. Stewart, ed., "Emergence to World Power 1898–1902," in *American Military History: The United States Army and the Forging of a Nation, 1775–1917,* vol. 1, 341–64 (Center of Military History, United States Army, 2005), 341.

93. Jennifer Llewellyn, Steve Thompson, and Jim Southey, "Militarism as a Cause of World War I," Alpha History, updated January 15, 2025, https://alphahistory.com/worldwar1/militarism/.

94. Edward Grey, in Llewellyn, Thompson, and Southey, "Militarism as a Cause."

95. Joint Resolution to Provide for Annexing the Hawaiian Islands to the United States, H. Res. 259. 55th Cong., 1898, https://www.archives.gov/milestone-documents/joint-resolution-for-annexing-the-hawaiian-islands (emphasis in original).

96. Joint Resolution for Annexing Hawaiian Islands.

97. Francis P. Sempa, "The Geopolitical Vision of Alfred Thayer Mahan," *Diplomat,* December 30, 2014, https://thediplomat.com/2014/12/the-geopolitical-vision-of-alfred-thayer-mahan/.

98. A. T. Mahan, *The Influence of Sea Power upon History, 1660–1783,* 12th ed. (Little, Brown, 1918), https://www.gutenberg.org/ebooks/13529, 82.

99. Sempa, "Geopolitical Vision of Mahan."

100. Stewart, "Emergence to World Power," 343–46. See pages 348–52 for examples of these challenges.

101. Stewart, "Emergence to World Power," 346.

102. Stewart, "Emergence to World Power," 354–55.

103. Stewart, "Emergence to World Power," 357.

104. Stewart, "Emergence to World Power."

105. Stewart, "Emergence to World Power," 358.

106. Stewart, "Emergence to World Power," 359.

107. C. R. Coulter, "Our Policy in the Philippines," *San Francisco Call,* August 1, 1899, https://cdnc.ucr.edu/?a=d&d=SFC18990801.2.111.

108. Stuart Creighton Miller, *Benevolent Assimilation: The American Conquest of the Philippines, 1899–1903* (Yale University Press, 1982), 88.

109. Letters reprinted in History Matters, "American Soldiers in the Philippines Write Home About the War," accessed April 23, 2025, https://historymatters.gmu.edu/d/58/.

110. "President Retires Gen. Jacob H. Smith," *New York Times,* July 17, 1902, https://timesmachine.nytimes.com/timesmachine/1902/07/17/101959147.pdf.

111. Jennie Cohen, "6 Things You May Not Know About the Spanish American War," History, updated February 18, 2025, https://www.history.com/news/6-things-you-may-not-know-about-the-spanish-american-war.

112. Martin Olson, in History Matters, "American Soldiers in the Philippines."

113. Lesley Kennedy, "Did Yellow Journalism Fuel the Outbreak of the Spanish American War?" History, updated February 18, 2025, https://www.history.com/articles/spanish-american-war-yellow-journalism-hearst-pulitzer.

114. SparkNotes, "The Spanish American War (1898–1901)," accessed April 27, 2025, https://www.sparknotes.com/history/american/spanishamerican/context/.

115. U.S. Department of State Archive, "U.S. Invasion and Occupation of Haiti, 1915–34," accessed March 31, 2024, https://2001-2009.state.gov/r/pa/ho/time/wwi/88275.htm.

116. U.S. Department of State Archive, "U.S. Invasion of Haiti."

117. Jean-Philippe Belleau, "Massacres Perpetrated in the 20th Century in Haiti," SciencesPo, April 2, 2008, https://www.sciencespo.fr/mass-violence-war-massacre -resistance/en/document/massacres-perpetrated-20th-century-haiti.html.
118. U.S. Department of State Archive, "U.S. Invasion of Haiti."
119. Belleau, "Massacres Perpetrated in Haiti."
120. Roger Gaillard in Belleau, "Massacres Perpetrated in Haiti."
121. Belleau, "Massacres Perpetrated in Haiti."
122. Belleau, "Massacres Perpetrated in Haiti."
123. Stewart, "Emergence to World Power," 362.
124. National Museum of Australia, "Breaker Morant Executed," updated September 25, 2024, https://www.nma.gov.au/defining-moments/resources/breaker-morant -executed; South African History Online, "Women and Children in White Concentration Camps During the Anglo-Boer War, 1900–1902," updated September 1, 2023, https://www.sahistory.org.za/article/women-and-children-white-concen tration-camps-during-anglo-boer-war-1900-1902.
125. South African History Online, "Women and Children."
126. National Museum of Australia, "Breaker Morant Executed"; Dan Zar, "February 27, 1902: Breaker Morant Executed for War Crimes," History and Headlines, updated February 14, 2020, https://www.historyandheadlines.com/ february-27-1902-breaker-morant-executed-war-crimes/.
127. Ernst Moritz Arndt in Hamza Elshakankiri, "Prussian Militarism and the German Wars of Unification," Armstrong Undergraduate Journal of History 11, no. 2 (October 2021): 41, https://doi.org/10.20429/aujh.2021.110203.
128. Elshakankiri, "Prussian Militarism," 41–42.
129. Jonathan Sperber and Brian Vick, "From Vormärz to Prussian Dominance (1815–1866)," German History in Documents and Images, accessed April 18, 2024, https://germanhistorydocs.org/en/from-vormaerz-to-prussian-domi nance-1815-1866/introduction; Helmuth von Moltke, "Memorandum on a Possible War Between Prussia and Austria (1866)," German History in Documents and Images, accessed April 5, 2024, https://germanhistorydocs.org/en/ from-vormaerz-to-prussian-dominance-1815-1866/helmuth-von-moltke-memora dum-on-the-possible-war-between-prussia-and-austria-1866; Helmuth von Moltke, "Memorandum on the Effect of Improvements in Firearms on Battlefield Tactics (1861)," German History in Documents and Images, accessed April 6, 2024, https://germanhistorydocs.org/en/from-vormaerz-to-prussian-domi nance-1815-1866/helmuth-von-moltke-memorandum-on-the-effect-of-improve ments-in-firearms-on-battlefield-tactics-1861.
130. Celeste Neill, "Otto von Bismarck: Architect of German Unification," History Hit, May 12, 2023, https://www.historyhit.com/1871-unification-germany/.
131. Black, Age of Total War, 39.
132. Black, Age of Total War.
133. For the names of some of these scholars, see Bastian Matteo Scianna, "A Predisposition to Brutality? German Practices Against Civilians and Francs-Tireurs During the Franco-Prussian War 1870–1871 and Their Relevance for the German 'Military Sonderweg' Debate," Small Wars and Insurgencies 30, nos. 4–5 (2019): 970–71, https://doi.org/10.1080/09592318.2019.1638551.
134. Jefferson Davis, in James R. Gilmore, "Our Visit to Richmond," Atlantic Monthly 14, no. 83 (September 1864): 379–80, 382, https://www.gutenberg.org/ ebooks/20350.

CHAPTER 3

1. Michael Geyer, "German Strategy in the Age of Machine Warfare, 1914–1945," in *Makers of Modern Strategy: From Machiavelli to the Nuclear Age*, ed. Peter Paret, Gordon A. Craig, and Felix Gilbert, 527–97 (Princeton University Press, 1986), 548–51.
2. Sun Tzu, *The Art of War*, trans. Samuel B. Griffith (Oxford University, 1971), 85.
3. Giulio Douhet, "The Command of the Air," in *Roots of Strategy: Book 4*, ed. David Jablonsky, 263–408 (Stackpole Books, 1999), 283.
4. Douhet, "Command of the Air," 294.
5. Douhet, "Command of the Air," 394.
6. William Mitchell, "Winged Defense: The Development and Possibilities of Modern Air Power, Economic and Military," in *Roots of Strategy: Book 4*, ed. David Jablonsky, 409–515 (Stackpole Books, 1999), 489.
7. Mitchell, "Winged Defense," 441.
8. Francis P. Sempa, "The Geopolitical Vision of Alfred Thayer Mahan," *Diplomat*, December 30, 2014, https://thediplomat.com/2014/12/the-geopolitical-vision-of-alfred-thayer-mahan/.
9. Sempa, "Geopolitical Vision of Mahan."
10. Sempa, "Geopolitical Vision of Mahan."
11. James R. Holmes and Toshi Yoshihara, *Chinese Naval Strategy in the 21st Century: The Turn to Mahan* (Routledge, 2008).
12. Jeremy Black, *A Century of Conflict: War, 1914–2014* (Oxford University Press, 2015), 19.
13. Imperial War Museum, "Voices of the First World War: Trench Life," accessed April 26, 2025, https://www.iwm.org.uk/history/voices-of-the-first-world-war-trench-life.
14. Black, *Century of Conflict*, 18.
15. Black, *Century of Conflict*, 19–20.
16. Black, *Century of Conflict*, 20.
17. Lawrence H. Keeley, *War Before Civilization* (Oxford University Press, 1996), 54.
18. Adrian Gregory, *A War of Peoples, 1914–1919* (Oxford University Press, 2014), 83. For a brief synopsis of the battle and casualties, see pages 82–85.
19. Black, *Century of Conflict*, 37.
20. Gregory, *War of Peoples*, 90–91.
21. Black, *Century of Conflict*.
22. Black, *Century of Conflict*.
23. Gregory, *War of Peoples*, 28–29.
24. Gregory, *War of Peoples*, 29.
25. Gregory, *War of Peoples*.
26. J. B. Bryce, *Report of the Committee on Alleged German Outrages Appointed by His Britannic Majesty's Government and Presided over by the Right Hon. Viscount Bryce* (Macmillan, 1915), https://archive.org/details/reportofcommitte00grea/page/n3/mode/2up, 14–16.
27. Bryce, *Report of the Committee*, 9.
28. Bryce, *Report of the Committee*, 10.
29. Bryce, *Report of the Committee*, 22.
30. Bryce, *Report of the Committee*, 29.
31. Bryce, *Report of the Committee*, 31; Gregory, *War of Peoples*, 32.
32. Bryce, *Report of the Committee*, 33.

33. Bryce, *Report of the Committee*, 35.
34. Bryce, *Report of the Committee*, 10–39.
35. Bryce, *Report of the Committee*, 39.
36. Bryce, *Report of the Committee*, 40.
37. Gregory, *War of Peoples*, 44.
38. Ronald Grigor Suny, "Armenian Genocide," Britannica, updated April 26, 2025, https://www.britannica.com/event/Armenian-Genocide.
39. Suny, "Armenian Genocide."
40. John Kifner, "Armenian Genocide of 1915: An Overview," *New York Times*, accessed May 4, 2024, https://archive.nytimes.com/www.nytimes.com/ref/times topics/topics_armeniangenocide.html.
41. Suny, "Armenian Genocide."
42. Gregory S. Gordon, *Atrocity Speech Law: Foundation, Fragmentation, Fruition* (Oxford University Press, 2017), 33.
43. Suny, "Armenian Genocide."
44. Suny, "Armenian Genocide." See also Kifner, "Armenian Genocide of 1915."
45. Suny, "Armenian Genocide."
46. Suny, "Armenian Genocide."
47. Suny, "Armenian Genocide."
48. Suny, "Armenian Genocide."
49. Gordon, *Atrocity Speech Law*, 34.
50. Gordon, *Atrocity Speech Law*.
51. Gordon, *Atrocity Speech Law*, 35.
52. Kifner, "Armenian Genocide of 1915."
53. Gregory, *War of Peoples*, 47–48.
54. Gregory, *War of Peoples*, 48.
55. Kifner, "Armenian Genocide of 1915."
56. Gregory, *War of Peoples*.
57. Kifner, "Armenian Genocide of 1915."
58. Gregory, *A War of Peoples*, 47.
59. *The Hague Conventions of 1899 (II) and 1907 (IV) Respecting the Laws and Customs of War on Land*, pamphlet no. 5 (Carnegie Endowment for International Peace, 1915) 9–11, https://ia600406.us.archive.org/28/items/hagueconventions00inte_0/hagueconventions00inte_0.pdf.
60. Heather Jones, "Prisoners of War," International Encyclopedia of the First World War, updated October 8, 2014, https://doi.org/10.15463/ie1418.10475.
61. National WWI Museum and Memorial, "Captured," October 28, 2022–April 30, 2023, https://www.theworldwar.org/captured.
62. Jones, "Prisoners of War."
63. Jones, "Prisoners of War."
64. Jones, "Prisoners of War."
65. Jones, "Prisoners of War."
66. Correspondence with the German Government Respecting the Death by Burning of J. P. Genower, Able Seaman, When Prisoner of War at Brandenburg Camp," parliamentary paper, misc. no. 6, London, 1918, https://dn790003.ca.archive.org/0/items/correspondenceger00grea/correspondenceger00grea.pdf, 2–3.
67. Jones, "Prisoners of War."
68. Jones, "Prisoners of War."
69. Jones, "Prisoners of War."

70. Jones, "Prisoners of War."
71. Jones, "Prisoners of War."
72. Jones, "Prisoners of War."
73. Doina Anca Cretu, "Health, Disease, Mortality; Demographic Effects," International Encyclopedia of the First World War, updated November 17, 2020, https://encyclopedia.1914-1918-online.net/article/health-disease -mortality-demographic-effects/.
74. Cretu, "Health, Disease, Mortality."
75. Cretu, "Health, Disease, Mortality."
76. Jones, "Prisoners of War."
77. Jones, "Prisoners of War."
78. Martin Greener, in Imperial War Museum, "Voices of the First World War: Gas Attack at Ypres," accessed April 26, 2025, https://www.iwm.org.uk/history/ voices-of-the-first-world-war-gas-attack-at-ypres.
79. Archibald James, in Imperial War Museum, "Voices of First World War."
80. K. Lee Lerner, "Biological and Chemical Weapons," in Global Issues in Context (Gale, 2018), https://www.academia.edu/102704126/Biological _and_Chemical_Weapons_Overview.
81. Lerner, "Biological and Chemical Weapons."
82. Theo Emery, Hellfire Boys: The Birth of the U.S. Chemical Warfare Service and the Race for the World's Deadliest Weapons (Little, Brown, 2017), 16.
83. Nobel Prize, "Fritz Haber," accessed April 26, 2025, https://www.nobelprize.org/ prizes/chemistry/1918/haber/biographical/.
84. Jason G. Ramirez and Douglas R. Bacon, "Modern Chemical Warfare Agents: The Anesthesiologist's Perspective," Seminars in Anesthesia, Perioperative Medicine and Pain 22, no. 4 (December 2003): 240, https://doi.org/10.1053/ S0277-0326(03)00041-2.
85. Ramirez and Bacon, "Modern Chemical Warfare Agents."
86. Olivier Lepick, "France's Political and Military Reaction in the Aftermath of the First German Chemical Offensive in April 1915: The Road to Retaliation in Kind," in One Hundred Years of Chemical Warfare: Research, Deployment, Consequences, ed. Breitslav Friedrich, Dieter Hoffmann, Jürgen Renn, Florian Schmaltz, and Martin Worl, 69–76 (Springer International, 2017), 69–76, https://doi .org/10.1007/978-3-319-51664-6_5.
87. Lepick, "France's Political Reaction."
88. Lepick, "France's Political Reaction," 71.
89. Maxine Weygand, in Lepick, "France's Political Reaction" (emphasis added).
90. Lepick, "France's Political Reaction," 73.
91. Jeffery K. Smart, "History of Chemical and Biological Warfare: An American Perspective," in Medical Aspects of Chemical and Biological Warfare, ed. Frederick R. Sidell, Ernest T. Takafuji, and David R. Franz, 9–86 (Borden Institute, Walter Reed Army Medical Center, 1997), https://medcoeckapwstorprd01.blob.core.us govcloudapi.net/pfw-images/borden/chembio/Ch2.pdf.
92. History Skills, "The 4 Most Lethal Chemical Weapons Used in WWI," accessed August 13, 2024, https://www.historyskills.com/classroom/year-9/wwi-gas-attacks/.
93. New World Encyclopedia, "Phosgene," accessed September 8, 2024, https://www .newworldencyclopedia.org/entry/Phosgene; Centers for Disease Control and Prevention (CDC), "Phosgene," September 6, 2024, https://emergency.cdc.gov/ agent/phosgene/basics/facts.asp.

94. Emery, *Hellfire Boys*, 112.
95. Robert Harris and Jeremy Paxman, *A Higher Form of Killing: The Secret History of Chemical and Biological Warfare* (Random House Trade Paperbacks, 2002), 26–28.
96. Centers for Disease Control and Prevention (CDC), "Mustard Gas," September 6, 2024, https://www.cdc.gov/chemical-emergencies/chemical-fact-sheets/mustard-gas.html.
97. Jason G. Ramirez and Douglas R. Bacon, "Modern Chemical Warfare: A History," *Bulletin of Anesthesia History* 22, no. 2 (April 2004): 4–5, https://doi.org/10.1016/s1522-8649(04)50015-x.
98. Ramirez and Bacon, "Modern Chemical Warfare."
99. Sarah Everts, "A Brief History of Chemical War," *Distillations Magazine*, May 12, 2015, https://www.sciencehistory.org/stories/magazine/a-brief-history-of-chemical-war/.
100. Emery, *Hellfire Boys*, 50–56.
101. Amos A. Fries, in Emery, *Hellfire Boys*, 118.
102. Emery, *Hellfire Boys*.
103. Emery, *Hellfire Boys*, 143.
104. Fritz Haber, in Smart, "Chemical and Biological Warfare," 25.
105. Gregory, *War of Peoples*, 43.
106. Gregory, *War of Peoples*, 44.
107. Larry Holzwarth, "The American Submarine Campaign in the Pacific Changed the Tides of WWII," History Collection, October 26, 2020, https://historycollection.com/the-dramatic-american-submarine-campaign-in-the-pacific-changed-the-tides-of-world-war-ii/.
108. Gregory, *War of Peoples*, 58.
109. Black, *Century of Conflict*, 27–28.
110. Gregory, *War of Peoples*, 42–43.
111. Gregory, *War of Peoples*, 100.
112. Black, *Century of Conflict*, 29.
113. Gregory, *War of Peoples*.
114. Black, *Century of Conflict*, 30.
115. Black, *Century of Conflict*, 35.
116. Gregory, *War of Peoples*, 159.
117. Gregory, *War of Peoples*, 178.
118. Military History Matters, "Submarine—The History of Submarine War," March 26, 2011, https://www.military-history.org/feature/submarine-the-history-of-submarine-war.htm; Holzwarth, "American Submarine Campaign."
119. History Matters, "Submarine."
120. Black, *Century of Conflict*, 28.
121. J. M. Spaight, *Aircraft in War* (Macmillan, 1914), https://ia902807.us.archive.org/35/items/aircraftinwar00spai/aircraftinwar00spai.pdf, 3.
122. Spaight, *Aircraft in War*, 14–15.
123. Pamela Feltus, "Aerial Reconnaissance in World War I," U.S. Centennial of Flight Commission, accessed June 6, 2024, https://www.centennialofflight.net/essay/Air_Power/WWI-reconnaissance/AP2.htm; Spaight, *Aircraft in War*.
124. Spaight, *Aircraft in War*, 9.
125. Spaight, *Aircraft in War*, 10.
126. Tami Davis Biddle, *Rhetoric and Reality in Air Warfare* (Princeton University Press, 2002), 21.

NOTES

275

127. Biddle, *Rhetoric and Reality*, 23.

128. Biddle, *Rhetoric and Reality*, 22.

129. Gregory, *War of Peoples*, 124.

130. Jim Haviland, "Gothas: The German Bombers of World War I," *Military Heritage* 5, no. 5 (April 2004), https://warfarehistorynetwork.com/article/gothas-the-german-bombers-of-world-war-i/.

131. Haviland, "Gothas."

132. Gregory, *War of Peoples*, 124–25.

133. Haviland, "Gothas."

134. One interesting side note regarding Germany's strategic bombing involves Britain's royal family. In existence since the Middle Ages under various names, the House of Saxe-Coburg and Gotha added England to its house with the marriage of Prince Albert, son of Duke Ernest Anton of Saxe-Coburg and Gotha, to Queen Victoria of England in 1840. However, sharing the name *Gotha* with a German bomber, as well as being at war with Germany in World War I, proved a little too uncomfortable for the royal family. Looking to change their name from one that sounded very German to something more English, King George V settled on the name *Windsor*, which is still in use today. See Gregory, *War of Peoples*, 125; Jonathan Spangler, "Dukes of Saxe-Coburg and Saxe-Gotha, Families of Two British Consorts," Dukes and Princes, July 2, 2020, https://dukesandprinces.org/2020/07/02/dukes-of-saxe-coburg-and-saxe-gotha-families-of-two-british-consorts/; Their Majesties' Work as Prince of Wales and Duchess of Cornwall, "Saxe-Coburg-Gotha," accessed July 25, 2024, https://www.royal.uk/saxe-coburg-gotha.

135. Aaron Baar, "Global Ad Spending on Track to Top $1T for First Time, WARC Says," MarketingDive, August 28, 2023, https://www.marketingdive.com/news/global-ad-spending-2023-2024-1t-trillion-warc/692010/.

136. Eberhard Demm and Christopher H. Sterling, "Propaganda: World War I," in *World at War: Understanding Conflict and Society* (ABC-CLIO, 2024).

137. Randall L. Bytwerk, *Bending Spines: The Propagandas of Nazi Germany and the German Democratic Republic* (Michigan State University Press, 2004), 4.

138. *Merriam-Webster*, "Propaganda," accessed August 18, 2024, https://www.merriam-webster.com/dictionary/propaganda.

139. Jacques Ellul, in Elisabeth Fondren and John Maxwell Hamilton, "The Universal Laws of Propaganda: World War I and the Origins of Government Manufacture of Opinion," *Journal of Intelligence History* 22, no. 1 (2023): 4, https://doi.org/10.1080/16161262.2022.2036498.

140. Demm and Sterling, "Propaganda."

141. Demm and Sterling, "Propaganda."

142. Gregory, *War of Peoples*, 59.

143. Demm and Sterling, "Propaganda."

144. Demm and Sterling, "Propaganda."

145. Gregory, *War of Peoples*, 42.

146. Robert Niemi, "World War I in Film," *Pop Culture Universe: Icons, Idols, Ideas* (ABC-CLIO, 2024), https://popculture2-abc-clio-com.ezproxy1.apus.edu/Search/Display/1917732; Fondren and Hamilton, "Universal Laws of Propaganda," 16.

147. Fondren and Hamilton, "Universal Laws of Propaganda."

148. Gregory, *War of Peoples*.

149. Demm and Sterling, "Propaganda."

150. Black, *Century of Conflict*, 37.
151. Gregory, *War of Peoples*, 95.
152. Gregory, *War of Peoples*, 94.
153. Gregory, *War of Peoples*, 96.
154. Gregory, *War of Peoples*, 88.
155. Gregory, *War of Peoples*, 88–89. For more on the law of unintended conse-quences, see Tejvan Pettinger, "Law of Unintended Consequences," *Economics Help* (blog), September 17, 2019, https://www.economicshelp.org/blog/2381/eco nomics/law-of-unintended-consequences/, and Rob Norton, "Unintended Con-sequences," Econlib, accessed July 25, 2024, https://www.econlib.org/library/Enc/ UnintendedConsequences.html.
156. Black, *Century of Conflict: War*, 33.
157. Gregory, *War of Peoples*, 68.
158. Gregory, *War of Peoples*.
159. Black, *Century of Conflict*, 33–34.
160. Black, *Century of Conflict*, 34.
161. Gregory, *War of Peoples*, 86.
162. Emery, *Hellfire Boys*, 23.
163. Emery, *Hellfire Boys*, 24.
164. Emery, *Hellfire Boys*, 28–29.
165. Emery, *Hellfire Boys*, 31.
166. Emery, *Hellfire Boys*, 34.
167. Patricia O'Toole, "How the US Government Used Propaganda to Sell Americans on World War I," History, updated January 31, 2025, https://www.history.com/ news/world-war-1-propaganda-woodrow-wilson-fake-news.
168. Livia Gershon, "The US Propaganda Machine of World War I," *JSTOR Daily*, November 17, 2023, https://daily.jstor.org/the-us-propaganda-machine-of -world-war-i/.
169. Demm and Sterling, "Propaganda"; Gershon, "US Propaganda Machine."
170. Fondren and Hamilton, "Universal Laws of Propaganda," 11.
171. O'Toole, "US Government Used Propaganda."
172. Gregory, *War of Peoples*, 142.
173. Thirteen PBS, "The Sedition Act of 1918," accessed April 27, 2025, https://www .thirteen.org/wnet/supremecourt/capitalism/sources_document1.html.
174. Gregory, *War of Peoples*.
175. George Creel, in Fondren and Hamilton, "Universal Laws of Propaganda," 9.
176. Fondren and Hamilton, "Universal Laws of Propaganda," 8.
177. Fondren and Hamilton, "Universal Laws of Propaganda," 9.
178. Fondren and Hamilton, "Universal Laws of Propaganda," 17.
179. Emery, *Hellfire Boys*, 90.
180. Walter Gifford, in Emery, *Hellfire Boys*, 91.
181. Jim Garamone, "April 1917: America Entered the First World War," U.S. Army, April 7, 2017, https://www.army.mil/article/184897/april_1917_america_entered _the_first_world_war.
182. Garamone, "April 1917."
183. Richard W. Stewart, ed., *The United States Army in a Global Era, 1917–2008*, 2nd ed. (Center of Military History, United States Army, 2008), 19.
184. Stewart, *United States Army*, 20.
185. Stewart, *United States Army*, 21.

186. Emery, *Hellfire Boys*, 117–18.
187. Gregory, *War of Peoples*, 135–37.
188. Georges Clemenceau, in Gregory, *War of Peoples*, 136.
189. Black, *Century of Conflict*, 36.
190. Gregory, *War of Peoples*, 163.
191. Gregory, *War of Peoples*.
192. Gregory, *War of Peoples*, 159.
193. Gregory, *War of Peoples*, 160.
194. Gregory, *War of Peoples*, 159–60.
195. Black, *Century of Conflict*, 39.
196. Gregory, *War of Peoples*, 177.

CHAPTER 4

1. Quintus Ennius, in Jakub Jasiński, "Quotes of Ennius," Imperium Romanum, October 28, 2020, https://imperiumromanum.pl/en/roman-art-and-culture/golden-thoughts-of-romans/quotes-of-ennius/.
2. History.com Editors, "Treaty of Versailles," History, updated April 1, 2025, https://www.history.com/topics/world-war-i/treaty-of-versailles-1.
3. Ferdinand Foch, "Ferdinand Foch 1851–1929," in *Oxford Essential Quotations*, 5th ed., ed. Susan Ratcliffe (Oxford University Press, 2017), https://www.oxfordreference.com/display/10.1093/acref/9780191843730.001.0001/q-oro-ed5-00004492?print.
4. See Fritz Fischer, *Germany's Aims in the First World War* (W. W. Norton, 1967).
5. History.com Editors, "Treaty of Versailles."
6. History.com Editors, "Treaty of Versailles."
7. Adrian Gregory, *A War of Peoples, 1914–1919* (Oxford University Press, 2014), 180.
8. Articles 231–247 and Annexes [Treaty of Versailles], World War I Document Archives, accessed April 23, 2025, https://wwi.lib.byu.edu/index.php/Articles_231_-_247_and_Annexes.
9. Gregory, *War of Peoples*.
10. Articles 231–247 and Annexes.
11. ABC News, "Germany Set to Make Final World War I Reparation Payment," September 29, 2010, https://abcnews.go.com/International/germany-makes-final-reparation-payments-world-war/story?id=11755920.
12. Holocaust Encyclopedia, "Treaty of Versailles," United States Holocaust Memorial Museum, accessed April 27, 2025, https://encyclopedia.ushmm.org/content/en/article/treaty-of-versailles.
13. Holocaust Encyclopedia, "Treaty of Versailles."
14. Gregory, *War of Peoples*, 172.
15. Arthur L. Frothingham, *National Security League Handbook of War Facts and Peace Problems*, 4th ed. (National Security League, 1919), 158–59, https://books.google.com/books?id=qVKQAAAAMAAJ&pg=PA1#v=onepage&q&f=true.
16. Frothingham, *National Security League Handbook*, 160.
17. Isabel V. Hull, *Absolute Destruction: Military Culture and the Practices of War in Imperial Germany* (Cornell University Press, 2005), 56.
18. Frothingham, *National Security League Handbook*, 161–62.
19. Gregory, *War of Peoples*, 175.

20. Carlo Kopp, "Chemical and Biological Weapons," *Defence Today* 6, no. 3 (2007): 28–29, https://docslib.org/doc/6265356/chemical-and-biological-weapons-pdf.
21. Kopp, "Chemical and Biological Weapons."
22. Kopp, "Chemical and Biological Weapons."
23. Kopp, "Chemical and Biological Weapons."
24. Kopp, "Chemical and Biological Weapons."
25. Kopp, "Chemical and Biological Weapons," 29.
26. Kopp, "Chemical and Biological Weapons."
27. Treaty of Peace with Germany [Treaty of Versailles], *American Journal of International Law* 13, no. S3 (July 1919): 151–56, https://doi.org/10.2307/2213120. This same article prohibits the manufacture and import of any matériel relating to armored cars and tanks. This was another article that would be broken by Germany.
28. Treaty Relating to the Use of Submarines and Noxious Gases in Warfare, International Humanitarian Law Databases, February 6, 1922, https://ihl-databases.icrc.org/assets/treaties/270-IHL-34-EN.pdf.
29. Stefan Riedel, "Biological Warfare and Bioterrorism: A Historical Review," *Baylor University Medical Center Proceedings* 17, no. 4 (2004): 400–406, https://doi.org/10.1080/08998280.2004.11928002.
30. Holland Committee, in Robert Harris and Jeremy Paxman, *A Higher Form of Killing: The Secret History of Chemical and Biological Warfare* (Random House Trade Paperbacks, 2002), 44.
31. Harris and Paxman, *Higher Form of Killing*.
32. Editors of Encyclopaedia Britannica, "Geneva Gas Protocol," Britannica, updated February 28, 2023, https://www.britannica.com/event/Geneva-Gas-Protocol.
33. Harris and Paxman, *Higher Form of Killing*, 45.
34. Harris and Paxman, *Higher Form of Killing*, 49.
35. Harris and Paxman, *Higher Form of Killing*, 49–50.
36. Riedel, "Biological Warfare and Bioterrorism."
37. Sheldon Harris, *Factories of Death: Japanesse Biological Warfare, 1932–1945, and the American Cover-Up*, rev. ed. (Routledge, 2002), 42.
38. Harris and Paxman, *Higher Form of Killing*, 78; Harris, *Factories of Death*, 78.
39. V. Barras and G. Greub, "History of Biological Warfare and Bioterrorism," *Clinical Microbiology and Infection* 20, no. 6 (June 2014): 500, https://doi.org/10.1111/1469-0691.12706.
40. Susan K. Lewis, "History of Biowarfare," NOVA Online, accessed April 26, 2025, https://www.pbs.org/wgbh/nova/bioterror/hist_nf.html.
41. Harris and Paxman, *Higher Form of Killing*, 49–50.
42. Barras and Greub, "History of Biological Warfare," 500.
43. Barras and Greub, "History of Biological Warfare."
44. Pacific Atrocities Education, "The Development of Unit 731," accessed April 23, 2025, https://www.pacificatrocities.org/the-development-of-unit-731.html.
45. Pacific Atrocities Education, "Plan Kantokuen and Bacteriological Warfare," accessed April 23, 2025, https://www.pacificatrocities.org/plan-kantokuen-and-bacteriological-warfare.html.
46. Convention Relative to the Treatment of Prisoners of and War, International Humanitarian Law Databases, July 27, 1929, https://ihl-databases.icrc.org/assets/treaties/305-IHL-GC-1929-2-EN.pdf.
47. Harris and Paxman, *Higher Form of Killing*, 45–46.

48. Sarah Everts, "A Brief History of Chemical War," *Distillations Magazine*, May 12, 2015, https://www.sciencehistory.org/stories/magazine/a-brief-history-of -chemical-war/.

49. Harris and Paxman, *Higher Form of Killing*, 51–52.

50. Harris and Paxman, *Higher Form of Killing*, 52–54.

51. Harris and Paxman, *Higher Form of Killing*, 54, ephasis in original.

52. Harris and Paxman, *Higher Form of Killing*, 52–53.

53. Harris and Paxman, *Higher Form of Killing*, 53.

54. Harris and Paxman, *Higher Form of Killing*, 53–54.

55. Harris and Paxman, *Higher Form of Killing*, 55.

56. Edmund P. Russell III, "'Speaking of Annihilation': Mobilizing for War Against Human and Insect Enemies, 1914–1945," *Journal of American History* 82, no. 4 (March 1996): 1520, https://doi.org/10.2307/2945309.

57. Harris and Paxman, *Higher Form of Killing*, 56.

58. Sarah Everts, "The Nazi Origins of Deadly Nerve Gases," *C&EN Global Enterprise* 94, no. 41 (October 17, 2016): 26–28, https://doi.org/10.1021/ CEN-09441-SCITECH2.

59. Friedrich Frischknecht, "The History of Biological Warfare: Human Experimenta- tion, Modern Nightmares and Lone Madmen in the Twentieth Century," *EMBO Reports* 4, no. S1 (June 1, 2003): S47, https://doi.org/10.1038/sj.embor.embor849.

60. Harris and Paxman, *Higher Form of Killing*, 102.

61. Shūmei Ōkawa, in Mark Felton, "Why Were the Japanese So Cruel in World War II?," HistoryNet, November 6, 2017, https://www.historynet.com/ a-culture-of-cruelty/.

62. Ōkawa, in Felton, "Why Were Japanese Cruel."

63. Saburo Ienaga, "The Glorification of War in Japanese Education," *International Security* 18, no. 3 (Winter 1993–1994): 115, https://doi.org/10.2307/2539207.

64. Felton, "Why Were Japanese Cruel."

65. Ienaga, "Glorification of War," 122.

66. Ienaga, "Glorification of War," 122n.

67. Felton, "Why Were Japanese Cruel"; Ronald H. Spector, *Eagle Against the Sun* (Vintage Books, 1985), 35.

68. Iris Chang, *The Rape of Nanking: The Forgotten Holocaust of World War II* (Pen- guin Books, 1998), 217.

69. Dan van der Vat, *The Pacific Campaign: The U.S.-Japanese Naval War 1941–1945* (Touchstone, 1991), 44; John Toland, *The Rising Sun: The Decline and Fall of the Japanese Empire, 1936–1945* (Modern Library, 2003), 398.

70. Toland, *Rising Sun*, 301.

71. Chang, *Rape of Nanking*, 217.

72. van der Vat, *Pacific Campaign*, 44.

73. Felton, "Why Were Japanese Cruel."

74. John W. Dower, *War Without Mercy: Race and Power in the Pacific War* (Pantheon Books, 1986), 203, 205.

75. Dower, *War Without Mercy*, 8.

76. Dower, *War Without Mercy*, 25, 27.

77. Dower, *War Without Mercy*, 277–78.

78. Walter Zapotoczny, "The Rape of Nanking: Reasons and Recrimination," 2008, https://www.wzaponline.com/yahoo_site_admin/assets/docs/TheRapeofNanking.2 92125034.pdf.

79. "Alive and Safe, the Brutal Japanese Soldiers Who Butchered 20,000 Allied Seamen in Cold Blood," *The Standard*, April 12, 2012, https://www.standard.co.uk/hp/front/alive-and-safe-the-brutal-japanese-soldiers-who-butchered-20-000-allied-seamen-in-cold-blood-6636703.html.
80. Felton, "Why Were Japanese Cruel."
81. Richard J. Evans, *The Coming of the Third Reich* (Penguin Books, 2003), 25.
82. Evans, *Coming of Third Reich*, 27.
83. Evans, *Coming of Third Reich*.
84. Evans, *Coming of Third Reich*, 31–32; Manfred Messerschmidt, "The Wehrmacht and the Volksgemeinschaft," *Journal of Contemporary History* 18, no. 4 (October 1983): 721, 726, http://www.jstor.org/stable/260309.
85. Omer Bartov, *Mirrors of Destruction: War, Genocide, and Modern Identity* (Oxford University Press, 2000), 34–35, 43.
86. George L. Mosse, *The Crisis of German Ideology: Intellectual Origins of the Third Reich* (Universal Library, 1971), 4.
87. Mosse, *Crisis of German Ideology*.
88. Randall L. Bytwerk, "The Argument for Genocide in Nazi Propaganda," *Quarterly Journal of Speech* 91, no. 1 (February 2005): 42, http://dx.doi.org 10.1080/00335630500157516.
89. Bytwerk, "Argument for Genocide," 43.
90. Bytwerk, "Argument for Genocide," 44–45.
91. Bytwerk, "Argument for Genocide," 45.
92. Bytwerk, "Argument for Genocide," 47.
93. Bytwerk, "Argument for Genocide," 48.
94. J. P. Stern, in Bytwerk, "Argument for Genocide," 55.
95. Wiener Holocaust Library, "Science and Suffering: Victims and Perpetrators of Nazi Human Experimentation," accessed April 27, 2025, https://wienerholocaust library.org/exhibition/science-and-suffering-victims-and-perpetrators-of-nazi-human -experimentation/.
96. Mark Roseman, *The Wannsee Conference and the Final Solution: A Reconsideration* (Picador, 2002), 35.
97. Michael Burleigh, "Euthanasia and the Third Reich," *History Today* 40, no. 2 (1990): 11, https://www.historytoday.com/archive/euthanasia-and-third-reich.
98. Burleigh, "Euthanasia and Third Reich," 12; Diego F. Wyszynski, "Men with White Coats and SS Boots: The Children's Euthanasia Programme During the Third Reich," *Paediatric and Perinatal Epidemiology* 14, no. 4 (October 2000): 298, https://doi.org/10.1046/j.1365-3016.2000.00278.x.
99. Wyszynski, "Men with White Coats."
100. Wyszynski, "Men with White Coats," 297–98.
101. Alexander B. Rossino, *Hitler Strikes Poland: Blitzkrieg, Ideology, and Atrocity* (Univeristy Press of Kansas, 2003), 226.
102. Joseph Goebbels, "The Radio as the Eighth Great Power," German Propaganda Archive, updated January 10, 2024, http://research.calvin.edu/german-propaganda-archive/goeb56.htm.
103. Goebbels, "Radio as Eighth Great Power"; Joseph Goebbels, "Two Speeches on the Tasks of the Reich Ministry for Popular Enlightenment and Propaganda (March 15/March 25, 1933)," ed. and trans. Jeremy Noakes and Geoffrey Pridham, German History in Documents and Images, accessed April 26, 2025, http://germanhistorydocs.ghi-dc.org/docpage.cfm?docpage_id=2431.

104. Eugen Hadamovsky, "The Living Bridge: On the Nature of Radio Warden Activity," German Propaganda Archive, 2006, http://research.calvin.edu/german-propa
ganda-archive/hada3.htm.
105. Hadamovsky, "Living Bridge."
106. Maja Adena, Ruben Enikolopov, Maria Petrova, Veronica Santarosa, and Ekaterina Zhuravshaya, "Radio and the Rise of Nazis in Prewar Germany," *Quarterly
Journal of Economics* (2013): 14, https://doi.org/10.2139/ssrn.2242446.
107. Adena et al., "Radio and Rise of Nazis," 15.
108. Adena et al., "Radio and Rise of Nazis."
109. Robert Gellately, *Backing Hitler: Consent and Coercion in Nazi Germany* (Oxford
University Press, 2001), 185.
110. Gellately, *Backing Hitler*.
111. David Welch, *The Third Reich: Politics and Propaganda* (Routledge, 2008), 42.
112. Welch, *Third Reich*.
113. Hadamovsky, "Living Bridge."
114. Adena et al., "Radio and Rise of Nazis," 25–26, 28–29.
115. Adena et al., "Radio and Rise of Nazis," 31–32.
116. Welch, *Third Reich*, 42.
117. Julius Streicher, "The Poisonous Mushroom," German Propaganda Archive, trans.
Randall L. Bytwerk, 1999, http://research.calvin.edu/german-propaganda-archive/
story2.htm.
118. Streicher, "Poisonous Mushroom"; Julius Streicher, "How to Tell a Jew," trans.
Randall Bytwerk, German Propaganda Archive, 1999, http://research.calvin.edu/
german-propaganda-archive/story3.htm.
119. William L. Shirer, in Rossino, *Hitler Strikes Poland*, 215.
120. Bytwerk, "Argument for Genocide," 42.
121. George Washington, "Washington's Farewell Address to the People of the United
States," United States Senate, September 19, 1796, https://www.senate.gov/
artandhistory/history/resources/pdf/Washingtons_Farewell_Address.pdf.
122. Matthew Specter and Varsha Venkatasubramanian, "'America First': Nationalism,
Nativism, and the Fascism Question, 1880–2020," in *Fascism in America: Past and
Present*, ed. Gavriel D. Rosenfeld and Janet Ward (Cambridge University Press,
2023), 111.
123. Specter and Venkatasubramanian, "America First," 126.
124. Specter and Venkatasubramanian, "America First," 112.
125. Specter and Venkatasubramanian, "America First," 113.
126. Justus D. Doenecke, "Explaining the Antiwar Movement, 1939–1941: The
Next Assignment," *Journal of Libertarian Studies* 8, no. 1 (Winter 1986):
139, https://mises.org/journal-libertarian-studies/explaining-antiwar-move
ment-1939-1941-next-assignment?d7_alias_migrate=1.
127. Cody Carlson, "This Week in History: Lindbergh Gives Infamous 'Who Are
the War Agitators?' Speech," *Deseret News*, September 12, 2013, https://www
.deseret.com/2013/9/12/20525433/this-week-in-history-lindbergh-gives-infamous
-who-are-the-war-agitators-speech/.
128. Charles Lindbergh, in Carlson, "This Week in History."
129. Carlson, "This Week in History."
130. Charles Lindbergh, "Des Moines Speech—America First Committee," Charles
lindbergh.com, September 11, 1941, http://www.charleslindbergh.com/american
first/speech.asp.

131. Daniel A. Gross, "The U.S. Government Turned Away Thousands of Jewish Refugees, Fearing That They Were Nazi Spies," *Smithsonian Magazine*, November 18, 2015, https://www.smithsonianmag.com/history/us-government-turned-away-thousands-jewish-refugees-fearing-they-were-nazi-spies-180957324/.

132. Gross, "U.S. Government Turned Away Thousands."

133. Gross, "U.S. Government Turned Away Thousands."

134. Erin Blakemore, "A Ship of Jewish Refugees Was Refused US Landing in 1939. This Was Their Fate," History, updated April 15, 2025, https://www.history.com/news/wwii-jewish-refugee-ship-st-louis-1939.

135. Hannah Arendt, in Facing History and Ourselves, "America and the Holocaust," updated February 26, 2021, https://www.facinghistory.org/resource-library/america-holocaust.

136. Editors of Encyclopaedia Britannica, "Spanish Civil War," Britannica, updated April 3, 2025, https://www.britannica.com/event/Spanish-Civil-War.

137. Editors of Encyclopaedia Britannica, "Spanish Civil War."

138. Historyguy71, "The Spanish Civil War: A Comprehensive Overview," History Bite, February 27, 2024, http://www.historybite.com/european-history/the-spanish-civil-war-a-comprehensive-overview/.

139. Editors of Encyclopaedia Britannica, "Causes and Effects of the Spanish-American War," Britannica, accessed April 26, 2025, https://www.britannica.com/summary/Causes-and-Effects-of-the-Spanish-American-War.

140. Jeremy Black, *A Century of Conflict: War, 1914–2014* (Oxford University Press, 2015), 64.

141. Antony Beevor, *The Battle for Spain: The Spanish Civil War 1938–1939* (Penguin Books, 2006), 82.

142. Beevor, *Battle for Spain.*

143. Beevor, *Battle for Spain,* 82–83.

144. Beevor, *Battle for Spain,* 83.

145. Beevor, *Battle for Spain.*

146. Beevor, *Battle for Spain.*

147. R. J. Overy, "Hitler and Air Strategy," *Journal of Contemporary History* 15, no. 3 (July 1980): 409, www.jstor.org/stable/260411.

148. Beevor, *Battle for Spain,* 228.

149. Beevor, *Battle for Spain,* 232.

150. George Steer, "The Tragedy of Guernica: Town Destroyed in Air Attack: Eye-Witness's Account," *Sunday Times*, April 27, 1937, https://www.thetimes.com/article/bombing-of-guernica-original-times-report-from-1937-5j7x3z2k5bv.

151. Beevor, *Battle for Spain,* 233.

152. César Vidal, "La Destrucción de Guernica," trans. Peter Miller, Buber's Basque Page, accessed April 27, 2025, http://www.buber.net/Basque/History/guernica-ix.html.

153. Wolfram von Richthofen, in Vidal, "La Destrucción de Guernica."

154. von Richthofen, in Jörg Diehl, "Practicing Blitzkrieg in Basque Country," Spiegel International, April 26, 2007, http://www.spiegel.de/international/europe/0,1518,479675,00.html.

155. James S. Corum, *The Luftwaffe: Creating the Operational Air War, 1918–1940* (Univeristy Press of Kansas, 1997), 240.

156. James S. Corum, *Wolfram von Richthofen: Master of the German Air War* (Univeristy Press of Kansas, 2008), 134.

157. Corum, *Wolfram von Richthofen*, 200.
158. Corum, *Luftwaffe*, 199–200.

CHAPTER 5

1. Jeremy Black, *A Century of Conflict: War, 1914–2014* (Oxford University Press, 2015), 86–87.
2. China: The Dominated and the Dominating: A Summary of Modern Events and Biographies, "Battle of Shanghai," accessed September 11, 2024, https://china history.co.uk/battleofshanghai.html; Siegphyl, "1937 Battle of Shanghai, Japan's Brutal Attack on China," War History Online, December 15, 2013, https://www.warhistoryonline.com/war-articles/1937-battle-shanghai-japans-brutal-attack-china.html.
3. Eric Niderost, "The Fall of Shanghai: Prelude to the Rape of Nanking and WWII," Warfare History Network, November 2003, https://warfarehistorynetwork.com/article/the-fall-of-shanghai-prelude-to-the-rape-of-nanking-wwii/.
4. Chan Cheng Lin, "Nanjing Massacre and Sook Ching Massacre: Shaping of Chinese Popular Memories in China and Singapore, 1945–2015" (master's thesis, National University of Singapore, December 18, 2015), https://scholarbank.nus.edu.sg/entities/publication/c1d85a55-c675-4093-a31b-a24ecc6d260c.
5. Niall Ferguson, *The War of the World: Twentieth-Century Conflict and the Descent of the West* (Penguin Press, 2006), 476.
6. Iris Chang, *The Rape of Nanking: The Forgotten Holocaust of World War II* (Penguin Books, 1998), 87–88.
7. Walter Zapotoczny, "The Rape of Nanking: Reasons and Recrimination," 2008, https://www.wzaponline.com/yahoo_site_admin/assets/docs/TheRapeofNanking.292125034.pdf, 2.
8. Ferguson, *War of the World*.
9. Toshioki Mukai, in Editorial Team, "The Japanese 'Kill 100 People with Sword' Contest in 1937," *ARGunners Magazine*, August 29, 2015, https://www.argunners.com/the-japanese-kill-100-people-with-sword-contest-in-1937/.
10. "Japanese 'Kill 100 People.'"
11. "Japanese 'Kill 100 People.'"
12. Zapotoczny, "Rape of Nanking."
13. Chang, *Rape of Nanking*, 94–95.
14. Chang, *Rape of Nanking*; Lin, "Nanjing Massacre," 36.
15. Ferguson, *War of the World*, 477.
16. Jean-Louis Margolin, "Japanese Crimes in Nanjing, 1937–38: A Reappraisal," *China Perspectives* 63 (January–February 2006): 3, https://journals.openedition.org/chinaperspectives/571.
17. Margolin, "Japanese Crimes in Nanjing."
18. Lin, "Nanjing Massacre," 35.
19. Lin, "Nanjing Massacre," 35–36.
20. These sites were not discovered until 1962, and excavations occurred from 1963 to 1966. Thirty-five such sites were discovered in and around Singapore. See Lin, "Nanjing Massacre," 36.
21. Lin, "Nanjing Massacre," 44.
22. Lin, "Nanjing Massacre," 44–45.
23. Lin, "Nanjing Massacre," 36.

24. Lin, "Nanjing Massacre."

25. National Library Board (NLB), "Alexandra Hospital Massacre," Singapore Infopedia, April 3, 2014, https://www.nlb.gov.sg/main/article-detail?cmsuuid =7d4fd9a0-7bd0-4533-b0ea-3aa559673b0e.

26. Stuart Lloyd, "Singapore Alexandra Hospital Massacres 1942," Historic UK, February 13, 2022, https://www.historic-uk.com/HistoryUK/HistoryofBritain/ Singapore-Alexandra-Hospital-Massacres-1942/.

27. NLB, "Alexandra Hospital Massacre."

28. NLB, "Alexandra Hospital Massacre."

29. For an in-depth account of one man's experience during this horror, see BBC Open Centre, Hull, "The Alexandra Hospital Massacre," BBC, January 14, 2006, https://www.bbc.co.uk/history/ww2peopleswar/stories/60/a8515460.shtml.

30. Marc Pimentel, "Migration of Doom: Bataan Death March," U.S.S. Salt Lake City CA25, accessed December 20, 2023, https://ussslcca25.com/bataan.htm.

31. Pimentel, "Migration of Doom."

32. Pimentel, "Migration of Doom."

33. Pimentel, "Migration of Doom."

34. Pimentel, "Migration of Doom."

35. History.com Editors, "Bataan Death March," History, updated February 27, 2025, https://www.history.com/topics/world-war-ii/bataan-death-march.

36. James M. Scott, "Battlefield as Crime Scene: The Japanese Massacre in Manila," HistoryNet, January 12, 2019, https://www.historynet.com/worldwar2-japanese -massacre-in-manila/.

37. Scott, "Battlefield as Crime Scene."

38. Henry Keyes, in Scott, "Battlefield as Crime Scene."

39. Scott, "Battlefield as Crime Scene."

40. Scott, "Battlefield as Crime Scene."

41. Scott, "Battlefield as Crime Scene."

42. Scott, "Battlefield as Crime Scene."

43. Scott, "Battlefield as Crime Scene."

44. Scott, "Battlefield as Crime Scene."

45. Scott, "Battlefield as Crime Scene."

46. Japanese Eye Witness, "Execution of Allied Intelligence Officer by the Japanese," Takeo Tanimizu, accessed April 26, 2025, https://rjgeib.com/heroes/tanimizu/japa nese-execution.html.

47. National Museum of the United States Air Force, "The Eight Who Were Captured," accessed October 11, 2024, https://www.nationalmuseum.af.mil/Visit/ Museum-Exhibits/Fact-Sheets/Display/Article/196770the-eight-who-were-captured/.

48. Chang, *Rape of Nanking*, 216.

49. Chang, *Rape of Nanking*, 219.

50. James Bowen, "War Crimes Committed by the Imperial Japanese Navy," Pacific War Historical Society, accessed September 23, 2024, https://www.pacificwar.org/ JapWarCrimes/TenWarCrimes/WarCrimes_Jap_Navy.html.

51. Bowen, "War Crimes."

52. Bowen, "War Crimes."

53. Bowen, "War Crimes."

54. Bowen, "War Crimes."

55. Florian Schmaltz, "Chemical Weapons Research on Soldiers and Concentration Camp Inmates in Nazi Germany," in *One Hundred Years of Chemical Warfare:*

Research, Deployment, Consequences, ed. Breitslav Friedrich, Dieter Hoffmann, Jürgen Renn, Florian Schmaltz, and Martin Worl, 229–58 (Springer International Publishing, 2017), https://doi.org/10.1007/978-3-319-51664-6_13, 235–240.

56. Michael Sturma, "Japanese Treatment of Allied Prisoners During the Second World War: Evaluating the Death Toll," *Journal of Contemporary History* 55, no. 3 (July 2020): 525, https://www.jstor.org/stable/27067639.

57. Sturma, "Japanese Treatment of Allied Prisoners," 527.

58. Sturma, "Japanese Treatment of Allied Prisoners."

59. Sturma, "Japanese Treatment of Allied Prisoners," 529.

60. Sturma, "Japanese Treatment of Allied Prisoners," 532.

61. Sturma, "Japanese Treatment of Allied Prisoners," 530, 533.

62. V. Barras and G. Greub, "History of Biological Warfare and Bioterrorism," *Clinical Microbiology and Infection* 20, no. 6 (June 2014): 497–502, https://doi.org/10.1111/1469-0691.12706.

63. Barras and Greub, "History of Biological Warfare."

64. Barras and Greub, "History of Biological Warfare."

65. Lee A. Gladwin, "American POWs on Japanese Ships Take a Voyage into Hell," *Prologue Magazine* 35, no. 4 (Winter 2003), https://www.archives.gov/publications/prologue/2003/winter/hell-ships.

66. Sally Macdonald, "He Survived—1,800 Fellow Prisoners Aboard Japanese 'Hell Ship' Died 50 Years Ago Today," *Seattle Times*, October 24, 1994, https://archive.seattletimes.com/archive/?date=19941024&slug=1937653.

67. Gladwin, "American POWs on Japanese Ships."

68. Anthony Gaughan, "Execute Against Japan," Faculty Lounge, December 7, 2018, https://www.thefacultylounge.org/2018/12/execute-against-japan.html.

69. Asian Women's Fund, "Who Were the Comfort Women? The Establishment of Comfort Stations," accessed September 25, 2024, https://www.awf.or.jp/e1/facts-01.html.

70. Memorandum from Naosaburo Okabe, in Asian Women's Fund, "Who Were the Comfort Women? Observations and Directions in the Military," accessed September 25, 2024, https://www.awf.or.jp/e1/facts-03.html.

71. Asian Women's Fund, "Who Were the Comfort Women? Observations."

72. Ferguson, *War of the World*, 497–98.

73. Yuki Tanaka, *Hidden Horrors: Japanese War Crimes in World War II* (Westview Press, 1996), 90–91.

74. Tanaka, *Hidden Horrors*, 91–92.

75. Kazuko Watanabe, "Trafficking in Women's Bodies, Then and Now: The Issue of Military 'Comfort Women,'" *Peace and Change* 20, no. 4 (October 1995): 503, https://doi.org/10.1111/j.1468-0130.1995.tb00249.x.

76. Tomomi Yamaguchi, "Japan's Right-Wing Women and the 'Comfort Women' Issue," *Georgetown Journal of Asian Affairs* 6 (2020): 50, 52, https://repository.digital.georgetown.edu/handle/10822/1059392.

77. Yamaguchi, "Japan's Right-Wing Women," 53.

78. Tanaka, *Hidden Horrors*, 103.

79. Tanaka, *Hidden Horrors*.

80. Tanaka, *Hidden Horrors*.

81. Nicola S. Hines, "Unit 731 Justice Long Overdue," *State Bar of Texas International Law Section International Newsletter* 1, no. 1 (Fall 2018): 8–9, https://ilstexas

.org/wp-content/uploads/2018/09/ILS-Quarterly-3Q18-AE18-F2A_with-links.pdf #page=6.
82. Derek Pua, Danielle Dybbro, and Alistair Rogers, *Unit 731: The Forgotten Asian Auschwitz*, 2nd ed. (Pacific Atrocities Education, 2018), 43.
83. Pua, Dybbro, and Rogers, *Unit 731*, 46–49.
84. Pua, Dybbro, and Rogers, *Unit 731*, 50–51.
85. Tanaka, *Hidden Horrors*.
86. Tanaka, *Hidden Horrors*.
87. Tsuchiya Takashi, "The Imperial Japanese Medical Atrocities and Its Enduring Legacy in Japanese Research Ethics," *International Congress of History of Science* (August 29, 2005), https://www.researchgate.net/publication/228480362_The _Imperial_Japanese_Medical_Atrocities_and_Its_Enduring_Legacy_in_Japanese _Research_Ethics; Tanaka, *Hidden Horrors*, 150–51.
88. Tanaka, *Hidden Horrors*, 151–52.
89. Takashi, "The Imperial Japanese Medical Atrocities."
90. Takashi, "The Imperial Japanese Medical Atrocities."
91. Takashi, "The Imperial Japanese Medical Atrocities."
92. David D. Barrett, "Japan's Hellish Unit 731," *WWII Quarterly: Journal of the Second World War* 10, no. 1 (Fall 2018), https://warfarehistorynetwork.com/article/ japans-hellish-unit-731/.
93. Barrett, "Japan's Hellish Unit 731."
94. Facts and Details, "Plague Bombs and Gruesome Experiments at Unit 731," updated September 2016, https://factsanddetails.com/asian/ca67/sub426/entry-5518 .html.
95. Pua, Dybbro, and Rogers, *Unit 731*, 29–30.
96. Barrett, "Japan's Hellish Unit 731."
97. Pua, Dybbro, and Rogers, *Unit 731*, 20.
98. Pua, Dybbro, and Rogers, *Unit 731*.
99. Pua, Dybbro, and Rogers, *Unit 731*, 60.
100. Barrett, "Japan's Hellish Unit 731."
101. Barrett, "Japan's Hellish Unit 731."
102. Pua, Dybbro, and Rogers, *Unit 731*, 39.
103. Tanaka, *Hidden Horrors*, 112.
104. Tanaka, *Hidden Horrors*, 126.
105. Tanaka, *Hidden Horrors*, 112–14.
106. Tanaka, *Hidden Horrors*, 124.
107. Felix Espinoza Jr., in Tanaka, *Hidden Horrors*, 127.
108. Rolando Esteban, "Cannibalism Among Japanese Soldiers in Bukidnon, Philippines, 1945–47," *Asian Studies: Journal of Critical Perspectives on Asia* 52, no. 1 (2016): 68, https://www.asj.upd.edu.ph/mediabox/archive/ASJ_52_1_2016/Cannibalism_Japanese_Soldiers_Bukidnon_1945_1947_Esteban_Rolando2.pdf. This reference is the source of all of the cannibalism types cited in the article.
109. Esteban, "Cannibalism Among Japanese Soldiers," 59.
110. Esteban, "Cannibalism Among Japanese Soldiers," 70–72.
111. Jeanie M. Welch, "Without a Hangman, Without a Rope: Navy War Crimes Trials After World War II," *International Journal of Naval History* 1, no. 1 (April 2002), https://ijnh.seahistory.org/wp-content/uploads/sites/2/2012/01/pdf_welch .pdf (emphasis added).

112. SOFREP News Team, "The Chichijima Incident: Japanese Soldiers Ate US Pilots That Fell into Their Hands," SOFREP, February 13, 2022, https://sofrep.com/news/the-chichijima-incident-japanese-soldiers-ate-us-pilots-that-fell-into-their-hands/.

113. Sueo Matoba in Kaleena Fraga, "Inside the Chichijima Incident, George H. W. Bush's Harrowing Escape from Cannibal Enemies During World War II," ed. John Kuroski, All That's Interesting, updated November 7, 2023, https://allthatinteresting.com/george-bush-cannibalized-chichijima-incident.

114. Fraga, "Inside the Chichijima Incident."

115. Esteban, "Cannibalism Among Japanese Soldiers," 76.

116. Esteban, "Cannibalism Among Japanese Soldiers," 78.

117. Esteban, "Cannibalism Among Japanese Soldiers," 84–88.

118. Esteban, "Cannibalism Among Japanese Soldiers," 88–89.

119. Robert C. Mikesh, *Japan's World War II Balloon Bomb Attacks on North America*, Smithsonian Annals of Flight, no. 9 (Smithsonian Institution Press, 1973), https://repository.si.edu/bitstream/handle/10088/18679/SAoF-0009-Lo_res.pdf?sequence=3, 7.

120. Mikesh, *Japan's Balloon Bomb Attacks*, 17.

121. Mikesh, *Japan's Balloon Bomb Attacks*, 25.

122. Mikesh, *Japan's Balloon Bomb Attacks*.

123. Christopher Klein, "When Japan Launched Killer Balloons in World War II," History, updated January 27, 2025, https://www.history.com/news/japans-killer-wwii-balloons.

124. Mikesh, *Japan's Balloon Bomb Attacks*, 27.

125. Mikesh, *Japan's Balloon Bomb Attacks*; Klein, "Japan Launched Killer Balloons."

126. Mikesh, *Japan's Balloon Bomb Attacks*, 29.

127. Mikesh, *Japan's Balloon Bomb Attacks*.

128. Mikesh, *Japan's Balloon Bomb Attacks*, 35.

129. Editors of Encyclopaedia Britannica, "Okinawa," Britannica, updated April 25, 2025, https://www.britannica.com/place/Okinawa-prefecture-Japan.

130. George Feifer, "The Rape of Okinawa," *World Policy Journal* 17, no. 3 (2000): 34, https://doi.org/10.1215/07402775-2000-4009.

131. For a more in-depth look at the debate over Ryukyuan pacifism, see Gregory Smits, "Examining the Myth of Ryukyuan Pacifism," *Asia-Pacific Journal: Japan Focus* 8, no. 37 (September 13, 2010): 1–20, https://apjjf.org/gregory-smits/3409/article.

132. Evan Muxen, "Okinawa: The Last Battle," U.S. Army, June 22, 2022, https://www.army.mil/article/257747/okinawa_the_last_battle.

133. J. Choho Zukeran, in Justin McCurry, "Told to Commit Suicide, Survivors Now Face Elimination from History," *Guardian*, July 6, 2007, https://www.theguardian.com/world/2007/jul/06/japan.schoolsworldwide.

134. Yuki Kitazawa and Matthew Allen, "A Story That Won't Fade Away: Compulsory Mass Suicide in the Battle of Okinawa," *Asia-Pacific Journal: Japan Focus* 5, no. 7 (July 12, 2007): 3, https://apjjf.org/wp-content/uploads/2023/11/article-2049.pdf.

135. Steve Rabson, "Case Dismissed: Osaka Court Upholds Novelist Oe Kenzaburo for Writing That Japanese Military Ordered 'Group Suicides' in Battle of Okinawa," *Asia-Pacific Journal: Japan Focus* 6, no. 4 (April 1, 2008), https://apjjf.org/wp-content/uploads/2023/11/article-2171.pdf.

136. E. B. Sledge, *With the Old Breed: At Peleliu and Okinawa* (Presidio Press, 2007), 280.

137. Herbert P. Bix, "Remembering the Konoe Memorial: The Battle of Okinawa and Its Aftermath," *Asia-Pacific Journal: Japan Focus* 13, no. 8 (February 23, 2015): 3, https://apjjf.org/2015/13/8/Herbert-P.-Bix/4821.html.

138. Rabson, "Case Dismissed."

139. Directives for the Treatment of Political Commissars ("Commissar Order") (June 6, 1941), German History in Documents and Images, accessed September 23, 2024, https://germanhistorydocs.org/en/nazi-germany-1933-1945/directives-for-the-treatment-of-political-commissars-quot-commissar-order-quot-june-6-1941 (emphasis in original).

140. Decree for the Conduct of Courts-Martial in the District "Barbarossa" and for Special Measures of the Troop, 13 May 1941, in Office of United States Chief of Counsel for Prosecution of Axis Criminality, *Nazi Conspiracy and Aggression*, vol. 3 (United States Government Printing Office, 1946), 637–39.

141. Guidelines for the Behavior of Troops in Russia, in Manfred Messerschmidt and Anne Halley, "The Soldier in the War to Conquer Eastern Europe," *Massachusetts Review* 36, no. 3 (1995): 414, https://www.jstor.org/stable/25090655.

142. Omer Bartov, "The Myths of the Wehrmacht," *History Today* 42, no. 4 (April 1992): 32, https://www.historytoday.com/archive/myths-wehrmacht (spelling in the original).

143. Waitman W. Beorn, "A Calculus of Complicity: The *Wehrmacht*, the Anti-Partisan War, and the Final Solution in White Russia, 1941–42," *Central European History* 44, no. 2 (June 2011): 312, https://doi.org/10.1017/S0008938911000057.

144. Messerschmidt and Halley, "Soldier in the War," 314.

145. Truman Anderson, "Incident at Baranivka," *Journal of Modern History* 71, no. 3 (September 1999): 585–86, http://www.jstor.org/stable/10.1086/235290.

146. Ruckw. Heeresgebiet Sud KTB (Ia), entry of November 9, 1941, USNAT-501/4, in Anderson, "Incident at Baranivka," 603.

147. Anderson, "Incident at Baranivka," 607–8.

148. Anderson, "Incident at Baranivka," 609.

149. Anderson, "Incident at Baranivka," 611.

150. Fred Fallnbigl, in Bartov, "Myths of the Wehrmacht," 35.

151. In Bartov, "Myths of the Wehrmacht," 36.

152. Peter Fritzsche, *Life and Death in the Third Reich* (Harvard University Press, 2008), 197.

153. Stéphanie Trouillard, "The First Major Massacre in the 'Holocaust by Bullets': Babi Yar, 80 Years On," France24, September 29, 2021, https://www.france24.com/en/europe/20210929-the-first-major-massacre-in-the-holocaust-by-bullets-babi-yar-80-years-on.

154. Christopher R. Browning, *Ordinary Men: Reserve Police Battalion 101 and the Final Solution in Poland* (Harper Perennial, 1998), 18.

155. Browning, *Ordinary Men*, 135.

156. Fritzsche, *Life and Death*, 200.

157. Wilhelm Keitel, in Meilan Solly, "Remembering the Khatyn Massacre," *Smithsonian Magazine*, March 22, 2021, https://www.smithsonianmag.com/history/how-1943-khatyn-massacre-became-symbol-nazi-atrocities-eastern-front-180977280/.

158. Solly, "Remembering the Khatyn Massacre."

159. Solly, "Remembering the Khatyn Massacre"; Oleg Koropov, "On March 22, 1943, the Nazi Punitive Detachment Destroyed the Belarusian Village of Khatyn," Belarusian Institute for Strategic Research, March 21, 2025, https://bisr.gov.by/en/mneniya/march-22-1943-nazi-punitive-detachment-destroyed-belarusian-village-khatyn.
160. Solly, "Remembering the Khatyn Massacre."
161. Heinrich Himmler, in Edward B. Westermann, "Partners in Genocide: The German Police and the Wehrmacht in the Soviet Union," Journal of Strategic Studies 31, no. 5 (October 2008): 775, https://doi.org/10.1080/01402390802197977.
162. Westermann, "Partners in Genocide," 776.
163. Browning, Ordinary Men, 225–26.
164. Robert W. Kestling, "Blacks Under the Swastika: A Research Note," Journal of Negro History 83, no. 1 (Winter 1998): 90, http://www.jstor.org/stable/2668561.
165. Kestling, "Blacks Under the Swastika," 85.
166. Adolf Hitler, Mein Kampf, trans. Ralph Manheim, vol. 1 (Mariner Books, 1999), 286.
167. Hitler, Mein Kampf, 624.
168. Kestling, "Blacks Under the Swastika," 92–93.
169. Michael C. Mbabuike and Anna Marie Evans, "Other Victims of the Holocaust," Dialectical Anthropology 25, no. 1 (March 2000): 15, https://doi.org/10.1023/A:1007135521295.
170. Kestling, "Blacks Under the Swastika," 95.
171. Kestling, "Blacks Under the Swastika," 31–33. Of the numerous examples cited as war crimes, all but one were dismissed for various reasons. In the one case that was not dismissed, no additional information can be found regarding the prosecution and outcome of the case.
172. Kestling, "Blacks Under the Swastika," 84.
173. Robert Gellately, Backing Hitler: Consent and Coercion in Nazi Germany (Oxford University Press, 2001), 134–35.
174. Gordon Wright, The Ordeal of Total War 1939–1945 (Harper Torchbooks, 1968), 263–64.
175. R. J. Rummel, "20,946,000 Victims: Nazi Germany, 1933 to 1945," in Democide: Nazi Genocide and Mass Murder (Transaction, 1992), https://www.hawaii.edu/powerkills/NAZIS.CHAP1.HTM.
176. Overcoming the Past: The History and Memory of Nazi Germany, "Ideological Indoctrination," accessed September 25, 2024, https://nazigermany.lmu.build/exhibits/show/castro/ideological-indoctrination.
177. Sarah Farmer, Martyred Village: Commemorating the 1944 Massacre at Oradour-Sur-Glane (University of California Press, 1999), 20–25.
178. Michael Reynolds, "Massacre at Malmedy During the Battle of the Bulge," World War II 17, no. 6 (February 2003): 47, https://www.historynet.com/massacre-at-malmedy-during-the-battle-of-the-bulge/.
179. Jewish Virtual Library, "The SS (Schutzstaffel): The Waffen-SS," accessed September 25, 2024, https://www.jewishvirtuallibrary.org/waffen-ss.
180. Schmaltz, "Chemical Weapons Research," 238.
181. Kenneth Mellanby, "Medical Experiments on Human Beings in Concentration Camps in Nazi Germany," British Medical Journal (January 25, 1947): 148–49, https://www.ncbi.nlm.nih.gov/pmc/articles/PMC2052883/pdf/brmedj03765-0026.pdf.

182. Paul Weindling, Anna von Villiez, Aleksandra Loewenau, and Nichola Farron, "The Victims of Unethical Human Experiments and Coerced Research under National Socialism," *Endeavour* 40, no. 1 (March 2016): 3, tables 1 and 2, https://doi.org/10.1016/j.endeavour.2015.10.005.

183. Holocaust Encyclopedia, "The Doctors Trial: The Medical Case of the Subsequent Nuremberg Proceedings," United States Holocaust Museum, accessed September 27, 2024, https://encyclopedia.ushmm.org/content/en/article/the-doctors-trial-the-medical-case-of-the-subsequent-nuremberg-proceedings.

184. Wiener Holocaust Library, "Science and Suffering: Victims and Perpetrators of Nazi Human Experimentation," accessed April 27, 2025, https://wienerholocaustlibrary.org/exhibition/science-and-suffering-victims-and-perpetrators-of-nazi-human-experimentation/.

185. Ludwig Woltmann, *Politische Anthropologue*, ed. Otto Reche (Leipzig, 1936 [1900]), 16–17, 267, quoted in Richard J. Evans, *The Coming of the Third Reich* (Penguin Books, 2003), 34.

186. Evans, *Coming of Third Reich*, 34–35.

187. H. R. Trevor-Roper, *The Last Days of Hilter* (Macmillan, 1947), 4–5. See also Gerhard L. Weinberg, *A World at Arms: A Global History of World War II*, 2nd ed. (Cambridge University Press, 2008), 21, https://doi.org/10.1017/CBO9780511818639.

188. Trevor-Roper, *Last Days of Hilter*, 5–6.

189. Fritzsche, *Life and Death*, 221; Wright, *Ordeal of Total War*, 264.

190. Adolf Hitler, in Richard Breitman, "Hitler and Genghis Khan," *Journal of Contemporary History* 25, nos. 2/3 (May–June 1990): 337, http://www.jstor.org/stable/260736.

191. Breitman, "Hitler and Genghis Khan," 343.

192. Richard Rhodes, *Masters of Death: The SS-Einsatzgruppen and the Invention of the Holocaust* (Vintage Books, 2002), 99; Mark Roseman, *The Wannsee Conference and the Final Solution: A Reconsideration* (Picador, 2002), 3.

193. Erna Petri, in Gregory S. Gordon, *Atrocity Speech Law: Foundation, Fragmentation, Fruition* (Oxford University Press, 2017), 29–30.

194. Zeki Kiram, in Klaus-Michael Mallmann and Martin Cüppers, *Nazi Palestine: The Plans for the Extermination of the Jews in Palestine*, trans. Krista Smith (Enigma Books, 2010), 30.

195. Mallmann and Cüppers, *Nazi Palestine*, 42.

196. Mallmann and Cüppers, *Nazi Palestine*, 116.

197. Mallmann and Cüppers, *Nazi Palestine*, 130.

198. Voice of Free Arabism, in Jeffrey Herf, "Hate Radio: The Long, Toxic Afterlife of Nazi Propaganda in the Arab World," Think-Israel, November 22, 2009, http://www.think-israel.org/herf.nazipropagandainarabworld.html.

199. Gellately, *Backing Hitler*, 107.

200. Gellately, *Backing Hitler*, 108–10.

201. Gilad Margalit, "The Uniqueness of the Nazi Persecution of the Gypsies," *Romani Studies* 10, no. 2 (December 2000): 192, https://doi.org/10.3828/rs.2000.6.

202. Gellately, *Backing Hitler*, 110.

203. Margalit, "Uniqueness of Nazi Persecution," 199.

204. Gellately, *Backing Hitler*, 110–11.

205. Karina V. Korostelina, "Devastating Civilians at Home: The Plight of Crimean Tatars and Californians of Asian Descent During World War II," in *Civilians and*

Modern War: Armed Conflict and the Ideology of Violence, ed. Daniel Rothbart, Karina V. Korostelina, and Mohammed Cerkaoui, 51–71 (Routledge, 2012), 52–53.

206. Caleb Foote, *Outcasts! The Story of America's Treatment of Her Japanese-American Minority* (Fellowship of Reconciliation, 1943), https://archive.org/details/outcasts storyofa00foot/, 2.

207. Korostelina, "Devastating Civilians at Home," 54–55.

208. Michael Ray, "Executive Order 9066," Britannica, updated March 6, 2025, https://www.britannica.com/topic/Executive-Order-9066.

209. Korostelina, "Devastating Civilians at Home," 54.

210. *Los Angeles Times* in Korostelina, "Devastating Civilians at Home," 55.

211. Bill Yenne, "Fear Itself: The General Who Panicked the West Coast," History-Net, July 11, 2017, https://www.historynet.com/fear-itself-the-general-panicked -west-coast/.

212. John L. DeWitt, in Korostelina, "Devastating Civilians at Home," 59.

213. Ronald H. Bailey and Editors of Time-Life Books, *The Home Front: U.S.A.* (Time-Life Books, 1977), 27.

214. Bailey and Editors of Time-Life Books, *Home Front*.

215. Executive Order No. 9066: Resulting in Japanese-American Incarceration, 1942, https://www.archives.gov/milestone-documents/executive-order-9066.

216. Korostelina, "Devastating Civilians at Home," 59.

217. Korostelina, "Devastating Civilians at Home."

218. Floyd Schmoe, in Korostelina, "Devastating Civilians at Home," 60.

219. Leland Ford, in Korostelina, "Devastating Civilians at Home," 55.

220. J. Edgar Hoover, in Brian Niiya, "J. Edgar Hoover," Densho Encyclopedia, updated May 9, 2024, https://encyclopedia.densho.org/J.%20Edgar%20Hoover.

221. Roger Baldwin, in Korostelina, "Devastating Civilians at Home," 59.

222. Bailey and Editors of Time-Life Books, *Home Front*, 35.

223. Foote, *Outcasts!*

224. George Takei, in Samantha Balaban, "George Takei 'Lost Freedom' Some 80 Years Ago—Now He's Written That Story for Kids," NPR, April 20, 2024, https://www.npr.org/2024/04/20/1245844347/george-takei-my-lost-freedom-picture-book.

225. Takei, in Balaban, "George Takei 'Lost Freedom.'"

226. Billy Mitchell, in Byron King, "Alaska: 'The Most Important Strategic Place in the World,' Part III," Daily Reckoning, February 22, 2008, https://dailyreckoning .com/the-importance-of-alaska-part-ii/.

227. Commission on Wartime Relocation and Internment of Civilians, "War and Evacuation in Alaska," in *Personal Justice Denied*, 317–59 (University of Washington Press, 1997), 317.

228. Commission on Wartime Relocation and Internment of Civilians, "War and Evacuation in Alaska"; Eva Holland, "Agony of the Aleutians: The Forgotten Internment," *Anchorage Daily News*, November 9, 2014, https://www.adn.com/ we-alaskans/article/forgotten-internment/2014/11/09/.

229. Ernest Henry Gruening, in Stephanie Hinnershitz, "The Wartime Internment of Native Alaskans," National WWII Museum, June 30, 2022, https://www.national ww2museum.org/war/articles/wartime-internment-native-alaskans.

230. Simon Bolivar Buckner, in Hinnershitz, "Wartime Internment of Native Alaskans."

231. Erin Blakemore, "The U.S. Forcibly Detained Native Alaskans During World War II," *Smithsonian Magazine*, February 22, 2017, https://www.smithsonianmag.com/ smart-news/us-forcibly-detained-native-alaskans-during-world-war-ii-180962239/.

232. Hinnershitz, "Wartime Internment of Native Alaskans."
233. Hinnershitz, "Wartime Internment of Native Alaskans."
234. Andronik P. Kashevaroff, in Holland, "Agony of the Aleutians."
235. Blakemore, "U.S. Forcibly Detained Native Alaskans."
236. Blakemore, "U.S. Forcibly Detained Native Alaskans"; Holland, "Agony of the Aleutians."
237. Holland, "Agony of the Aleutians."
238. Letter to United States Office of Indian Affairs dated October 1942, in Hinnershitz, "Wartime Internment of Native Alaskans."
239. Letter from Office of Indian Affairs, in Hinnershitz, "Wartime Internment of Native Alaskans."
240. Blakemore, "U.S. Forcibly Detained Native Alaskans."
241. In Holland, "Agony of the Aleutians."
242. Holland, "Agony of the Aleutians."
243. Statement from United States Fish and Wildlife Service, in Holland, "Agony of the Aleutians." See also Blakemore, "U.S. Forcibly Detained Native Alaskans."
244. Holland, "Agony of the Aleutians"; Commission on Wartime Relocation and Internment of Civilians, "War and Evacuation in Alaska," 318.
245. Report from Office of Indian Affairs, in Hinnershitz, "Wartime Internment of Native Alaskans."
246. Holland, "Agony of the Aleutians."
247. Stephen G. Fritz, *Ostkrieg: Hitler's War of Extermination in the East* (University Press of Kentucky, 2011).
248. Antony Beevor, *Berlin: The Downfall 1945* (Penguin Books, 2002), 67.
249. Fritz, *Ostkrieg*, 449–50.
250. Beevor, *Berlin*, 29, 67.
251. Beevor, *Berlin*, 106 (spelling in original).
252. Beevor, *Berlin*.
253. Sturma, "Japanese Treatment of Allied Prisoners," 524.
254. Sturma, "Japanese Treatment of Allied Prisoners."
255. Benjamin B. Fischer, in S. R. O'Konski, "The Katyn Forest Massacre: And Five Betrayals of Poland by Its WWII Allies," World War 2 History Short Stories, July 29, 2019, https://www.ww2history.org/war-in-europe/the-katyn-forest-massacre-and-five-betrayals-of-poland-by-its-wwii-allies/.
256. Gellately, *Backing Hitler*, 255.
257. Editors of Encyclopaedia Britannica, "Tatar," Britannica, updated April 25, 2025, https://www.britannica.com/topic/Tatar.
258. Korostelina, "Devastating Civilians at Home," 61.
259. Korostelina, "Devastating Civilians at Home."
260. *Krasnyi Krym*, in Korostelina, "Devastating Civilians at Home," 62.
261. Korostelina, "Devastating Civilians at Home."
262. Korostelina, "Devastating Civilians at Home," 63.
263. Korostelina, "Devastating Civilians at Home," 64.
264. Korostelina, "Devastating Civilians at Home," 65.
265. Korostelina, "Devastating Civilians at Home," 67.
266. Korostelina, "Devastating Civilians at Home."
267. Robert Harris and Jeremy Paxman, *A Higher Form of Killing: The Secret History of Chemical and Biological Warfare* (Random House Trade Paperbacks, 2002), 66–67.
268. Harris and Paxman, *Higher Form of Killing*, 68–69.

269. Harris and Paxman, *Higher Form of Killing*, 65–66.
270. Edmund P. Russell III, "'Speaking of Annihilation': Mobilizing for War Against Human and Insect Enemies, 1914–1945," *Journal of American History* 82, no. 4 (March 1996): 1511, https://doi.org/10.2307/2945309.
271. Russell, "'Speaking of Annihilation,'" 1521.
272. Ken Stewart, "Zyklon-B," Britannica, updated October 26, 2024, https://www.britannica.com/science/Zyklon-B.
273. Stewart, "Zyklon-B."
274. Harris and Paxman, *Higher Form of Killing*, 83–85.
275. Harris and Paxman, *Higher Form of Killing*, 89–90.
276. Harris and Paxman, *Higher Form of Killing*, 70–76.
277. Harris and Paxman, *Higher Form of Killing*, 102.
278. Harris and Paxman, *Higher Form of Killing*, 103.
279. Harris and Paxman, *Higher Form of Killing*.
280. Harris and Paxman, *Higher Form of Killing*, 104.
281. Harris and Paxman, *Higher Form of Killing*, 106.
282. Harris and Paxman, *Higher Form of Killing*, 107.
283. Hu Long, Zhengyu Liao, Yan Wang, Lina Liao, and Wenli Lai, "Efficacy of Botulinum Toxins on Bruxism: An Evidence-Based Review," *International Dental Journal* 62, no. 1 (February 2012): 1–5, https://doi.org/10.1111/J.1875–595X.2011.00085.X.
284. Harris and Paxman, *Higher Form of Killing*, 95 (emphasis in original).
285. Harris and Paxman, *Higher Form of Killing*.
286. Harris and Paxman, *Higher Form of Killing*, 91–96.
287. Harris and Paxman, *Higher Form of Killing*, 112.
288. Harris and Paxman, *Higher Form of Killing*, 113.
289. Patrick Cockburn, "Churchill's Secret Chemical War," *Independent*, July 26, 2020, https://www.independent.co.uk/independentpremium/long-reads/second-world-war-churchill-secret-mustard-gas-operation-sea-lion-a9632456.html.
290. Harris and Paxman, *Higher Form of Killing*, 123.
291. Harris and Paxman, *Higher Form of Killing*, 121–25.
292. Jennet Conant, "How a WWII Disaster—and Cover-Up—Led to a Cancer Treatment Breakthrough," History, updated January 31, 2025, https://www.history.com/news/wwii-disaster-bari-mustard-gas; Jennet Conant, "How a Chemical Weapons Disaster in WWII Led to a U.S. Cover-Up—and a New Cancer Treatment," *Smithsonian Magazine*, September 2020, https://www.smithsonianmag.com/history/bombing-and-breakthrough-180975505/. Alexander's report, which showed mustard's toxicity on white blood cells, led Colonel Cornelius P. "Dusty" Rhoads to develop Mustargen (mechlorethamine), the first chemotherapeutic drug approved by the FDA. This drug was successfully used to treat non-Hodgkin's lymphoma and opened the door to cancer chemotherapy, as noted in Conant, "Chemical Weapons Disaster," and Conant, "WWII Disaster."
293. Harris and Paxman, *Higher Form of Killing*, 136.
294. Winston Churchill, in Harris and Paxman, *Higher Form of Killing*.
295. Harris and Paxman, *Higher Form of Killing*, 137.
296. Harris and Paxman, *Higher Form of Killing*, 138.
297. Read Harris and Paxman's review of this topic to gain a further understanding of the issues involved and their interactions with each other. See Harris and Paxman, *Higher Form of Killing*, 133–38.

298. Lawrence H. Keeley, *War Before Civilization* (Oxford University Press, 1996), 17.
299. Clay Blair, *Hitler's U-Boat War: The Hunters, 1939–1942* (Random House, 1996), 23.
300. Blair, *Hitler's U-Boat War*, 31.
301. Blair, *Hitler's U-Boat War*, 32.
302. Blair, *Hitler's U-Boat War*, 37.
303. Blair, *Hitler's U-Boat War*, 39.
304. Blair, *Hitler's U-Boat War*.
305. For a more complete review of these rules, see Blair, *Hitler's U-Boat War*, 64–66.
306. Blair, *Hitler's U-Boat War*, 67.
307. Peter Padfield, *War Beneath the Sea: Submarine Conflict During World War II* (John C. Wiley and Sons, 1998), 295.
308. Karl Dönitz, in G. H. Bennett, "The 1942 Laconia Order: The Murder of Ship-wrecked Survivors and the Allied Pursuit of Justice 1945–46," *Law, Crime and History* 1, no. 1 (Spring 2011): 18, https://go.gale.com/ps/i.do?id=GALE%7CA3105 18860&sid=googleScholar&v=2.1&it=r&linkaccess=abs&issn=20459238&p=A ONE&sw=w&userGroupName=anon%7E459a5622&aty=open-web-entry.
309. Padfield, *War Beneath the Sea*, 297–300.
310. Bennett, "1942 Laconia Order," 19.
311. Padfield, *War Beneath the Sea*, 381.
312. Padfield, *War Beneath the Sea*, 382.
313. War Stories with Mark Felton, "U-Boat Massacre—The Case of U-852," posted June 7, 2020, by War Stories with Mark Felton, YouTube, 10 min., 21 sec., https://www.youtube.com/watch?v=Q9LeA7V77E0.
314. War Stories with Mark Felton, "U-Boat Massacre."
315. Bennett, "1942 Laconia Order," 32.
316. Bennett, "1942 Laconia Order," 34.
317. Anthony Miers, in Hidden History, "Commander Anthony Miers: War Criminal or War Hero?" posted August 29, 2024, by Hidden History, YouTube, 11 min., 14 sec., https://www.youtube.com/watch?v=wKTUBQwmnzs.
318. Hidden History, "Commander Anthony Miers"; Uboat.Net, "HMS Torbay (N 79)," accessed September 22, 2024, https://uboat.net/allies/warships/ship/3498 .html.
319. Hidden History, "Commander Anthony Miers."
320. Williamson Murray, "Why Germany's Kriegsmarine Lost the Battle of the At-lantic," HistoryNet, April 28, 2015, https://www.historynet.com/why-germanys -kriegsmarine-lost-the-battle-of-the-atlantic/.
321. Murray, "Why Germany's Kriegsmarine Lost."
322. Jeffrey G. Barlow, "The Navy's Atlantic War Learning Curve," *Naval History* 22, no. 3 (June 2008), https://www.usni.org/magazines/naval-history-magazine/2008/ june/navys-atlantic-war-learning-curve.
323. Barlow, "Navy's Atlantic War."
324. Barlow, "Navy's Atlantic War."
325. Barlow, "Navy's Atlantic War"; Jack Sweetman, "The Battle of the Atlantic," *Naval History* 9, no. 3 (June 1995), https://www.usni.org/magazines/naval-history -magazine/1995/june/battle-atlantic.
326. Blair, *Hitler's U-Boat War*, 695, plate 12.
327. Blair, *Hitler's U-Boat War*, 771.

328. Norm Haskett, "Atlantic, Battle of the (1939–1945)," World War II, accessed April 26, 2025, https://ww2days.com/atlantic-battle-of-1939-1945.html.

329. Gaughan, "Execute Against Japan"; Frederick "Fritz" Roegge, "December 7th, 1941: A Submarine Force Perspective," Sextant, December 7, 2016, https://usn history.navylive.dodlive.mil/Recent/Article-View/Article/2686191/.

330. Gaughan, "Execute Against Japan."

331. Gaughan, "Execute Against Japan."

332. Michael Sturma, "Atrocities, Conscience, and Unrestricted Warfare: US Submarines During the Second World War," War in History 16, no. 4 (November 2009): 447–68, https://doi.org/10.1177/0968344509341686.

333. Sturma, "Atrocities, Conscience, Unrestricted Warfare," 456.

334. Sturma, "Atrocities, Conscience, Unrestricted Warfare," 453.

335. Padfield, War Beneath the Sea, 343.

336. Sturma, "Atrocities, Conscience, Unrestricted Warfare," 448.

337. Sturma, "Atrocities, Conscience, Unrestricted Warfare."

338. Samuel Eliot Morrison, History of United States Naval Operations in World War II, vol 4, Coral Sea, Midway and Submarine Actions May 1942–August 1942 (Little, Brown, 1959), 189; Sturma, "Atrocities, Conscience, Unrestricted Warfare," 449.

339. Morrison, History of United States Naval Operations, 200–201.

340. Morrison, History of Naval Operations, 196–98.

341. David Vergun, "Submarine Warfare Played Major Role in World War II Victory," U.S. Department of Defense, March 16, 2020, https://www.defense.gov/News/Feature-Stories/Story/Article/2114035/.

342. Joint Army-Navy Assessment Committee, "Japanese Naval and Merchant Shipping Losses During WWII by All Causes," Naval History and Heritage Command, February 1947, https://www.history.navy.mil/research/library/online-reading-room/title-list-alphabetically/j/japanese-naval-merchant-shipping-losses-wwii.html#pageiv.

343. Vergun, "Submarine Warfare"; Padfield, War Beneath the Sea, 478–79.

344. Gaughan, "Execute Against Japan."

345. Gaughan, "Execute Against Japan."

346. Padfield, War Beneath the Sea, 385–86.

347. Padfield, War Beneath the Sea, 436.

348. Sturma, "Atrocities, Conscience, Unrestricted Warfare," 466.

349. Dudley Morton, in Sturma, "Atrocities, Conscience, Unrestricted Warfare," 467.

350. Sturma, "Atrocities, Conscience, Unrestricted Warfare," 468.

351. Morrison, History of Naval Operations, 227–28.

352. Robert C. Bornmann and Jan K. Herman, "Operating Under Pressure," Naval History 10, no. 4 (August 1996), https://www.usni.org/magazines/naval-history-magazine/1996/august/operating-under-pressure.

353. Vincent J. Cirillo, "Two Faces of Death: Fatalities from Disease and Combat in America's Principal Wars, 1775 to Present," Perspectives in Biology and Medicine 51, no. 1 (Winter 2008): 129, https://doi.org/10.1353/pbm.2008.0005.

354. Cirillo, "Two Faces of Death," 128.

355. Cirillo, "Two Faces of Death."

356. Cirillo, "Two Faces of Death," 128–29.

357. Cirillo, "Two Faces of Death," 130.

358. Hugh Trenchard, in Tami Davis Biddle, Rhetoric and Reality in Air Warfare (Princeton University Press, 2002), 27.

359. Biddle, Rhetoric and Reality, 33.

360. Trenchard, in Biddle, *Rhetoric and Reality*, 94.
361. Biddle, *Rhetoric and Reality*, 134.
362. Biddle, *Rhetoric and Reality*, 136–37.
363. Biddle, *Rhetoric and Reality*, 140.
364. Stanley Baldwin, in Alexander B. Downes, "Military Culture and Civilian Victimization: The Allied Bombing of Germay in World War II," in *Civilians and Modern War: Armed Conflict and the Ideology of Violence*, ed. Daniel Rothbart, Karina V. Korostelina, and Mohammed D. Cherkaoui, 72–95 (Routledge, 2012), 79.
365. Downes, "Military Culture and Civilian Victimization," 80.
366. Trenchard, in Biddle, *Rhetoric and Reality*, 77.
367. Biddle, *Rhetoric and Reality*, 104.
368. Richard Overy, in Roger Chickering, Stig Förster, and Bernd Greiner, eds., *A World at Total War: Global Conflict and the Politics of Destruction, 1937–1945* (Cambridge University Press, 2010), 278.
369. Beau Grosscup, *Strategic Terror: The Politics and Ethics of Aerial Bombardment* (Zed Books, 2006).
370. Luftwaffe regulation 16: *Luftkriegführung*, in James S. Corum, *The Luftwaffe: Creating the Operational Air War, 1918–1940* (University Press of Kansas, 2008), 143–44.
371. Luftwaffe regulation 16, in Corum, *Luftwaffe*, 144.
372. H. W. Koch, "The Strategic Air Offensive Against Germany: The Early Phase, May-September 1940," *Historical Journal* 34, no. 1 (March 1991): 121, https://doi.org/10.1017/S0018246X00013959.
373. Churchill, in Koch, "Strategic Air Offensive," 119.
374. Grosscup, *Strategic Terror*, 65; R. J. Overy, "Hitler and Air Strategy," *Journal of Contemporary History* 15, no. 3 (July 1980): 405–21, www.jstor.org/stable/260411.
375. Churchill, in Biddle, *Rhetoric and Reality*, 202.
376. Ian Patterson, *Guernica and Total War* (Harvard University Press, 2007), 132, 135.
377. Russell F. Weigley, *The American Way of War: A History of United States Military Strategy and Policy* (Macmillan, 1973), 313.
378. Weigley, *American Way of War*, 317.
379. U.S. Army Air Corps, Air War Plans Division, AWPD-1, August 1941, in *Major Problems in American Military History: Documents and Essays*, ed. John Whiteclay II and G. Kurt Piehler (Houghton Mifflin, 1999), 341 (emphasis in original).
380. US Air Corps Tactical School (ACTS) in Grosscup, *Strategic Terror*, 61.
381. Claire Chennault, in Biddle, *Rhetoric and Reality*, 169.
382. Churchill, in Koch, "Strategic Air Offensive,"134.
383. Jeffrey W. Legro, "Military Culture and Inadvertent Escalation in World War II," *International Security* 18, no. 4 (Spring 1994): 126, https://doi.org/10.2307/2539179.
384. "The Gazette Hall of Fame: Sir Howard Kingsley Wood," *Gazette*, accessed July 1, 2024, https://www.thegazette.co.uk/all-notices/content/224.
385. Keeley, *War Before Civilization*, 108.
386. *The United States Strategic Bombing Survey: Summary Report (European War)* (United States Government Printing Office, September 30, 1945), http://www.anesi.com/ussbs02.htm, 16.
387. Grosscup, *Strategic Terror*, 64.
388. Grosscup, *Strategic Terror*, 64–65.

389. Combined Chiefs of Staff, Memorandum by the Combined Chiefs of Staff, Casablanca, January 21, 1943, https://history.state.gov/historicaldocuments/frus1941 -43/d412.
390. Hitler, in Overy, "Hitler and Air Strategy," 411.
391. Biddle, Rhetoric and Reality, 74–75.
392. Overy, "Hitler and Air Strategy."
393. James S. Corum, Wolfram von Richthofen: Master of the German Air War (University Press of Kansas, 2008), 172–73.
394. William L. Shirer, The Rise and Fall of the Third Reich (Simon and Schuster Paperbacks, 1988), 722.
395. Hew Strachan, "Total War: The Conduct of War, 1939–1945," in A World at Total War: Global Conflict and the Politics of Destruction, ed. Roger Chickering, Stig Förster, and Bernd Greiner, 33–52 (Cambridge University Press, 2010), 40.
396. George E. Blau, ed., trans., The German Campaigns in the Balkans (Spring 1941) (Center of Military History, United States Army, 1986), 49.
397. Vittorio Mussolini, in Patterson, Guernica and Total War, 123.
398. Eric Langenbacher, "The Allies in World War II: The Anglo-American Bombardment of German Cities," in Genocide, War Crimes and the West, ed. Adam Jones, 116–33 (Zed Books, 2004), 118.
399. Weinberg, World at Arms, 616.
400. Langenbacher, "Allies in World War II," 118; United States Strategic Bombing Survey (European War), 3; Weinberg, World at Arms.
401. Tami Davis Biddle, "Dresden 1945: Reality, History, and Memory," Journal of Military History 72, no. 2 (April 2008): 424, https://doi.org/10.1353/jmh.2008.0074.
402. Biddle, "Dresden 1945," 425.
403. Weigley, American Way of War, 355–56.
404. Weigley, American Way of War, 357–58.
405. United States Strategic Bombing Survey (European War), 16.
406. United States Strategic Bombing Survey (European War), 15.
407. Fritzsche, Life and Death, 221.
408. Fritzsche, Life and Death.
409. Tami Davis Biddle, "Air Power Theory: An Analytical Narrative From the First World War to the Present," in U.S. Army War College Guide to National Security Policy and Strategy, 2nd ed., ed. J. Boone Martholomees Jr., 331–60 (United States Army War College, 2008), https://press.armywarcollege.edu/cgi/viewcontent .cgi?article=1077&context=monographs, 342.
410. Biddle, "Air Power Theory," 343.
411. United States Strategic Bombing Survey (European War), 16.
412. Patterson, Guernica and Total War, 153–54.
413. Weinberg, World at Arms, 868; Dan van der Vat, The Pacific Campaign: The U.S.-Japanese Naval War 1941–1945 (Touchstone, 1991), 502.
414. Herman S. Wolk and Richard P. Hallion, "FDR and Truman: Continuity and Context in the A-Bomb Decision," Air Power Journal 9, no. 3 (Fall 1995): 56–62, https://www.airuniversity.af.edu/Portals/10/ASPJ/journals/Volume-09_Issue -1-Se/1995_Vol9_No3.pdf.
415. "Operation Downfall: Planned Invasion of the Islands of Japan in World War II," accessed April 26, 2025, https://i.4pcdn.org/tg/1464699782538.pdf.
416. Gerhard L. Weinberg, "Some Myths of World War II," The Journal of Military History 75, no. 3 (July 2011): 714.

417. John W. Dower, *War Without Mercy: Race and Power in the Pacific War* (Pantheon Books, 1986), 73. See also John Toland, *The Rising Sun: The Decline and Fall of the Japanese Empire, 1936–1945* (Modern Library, 2003), 439n.
418. Ronald H. Spector, *Eagle Against the Sun* (Vintage Books, 1985), 317; Toland, *Rising Sun*, 519; van der Vat, *Pacific Campaign*, 328. Spector and van der Vat use the estimate of 8,000, while Toland cites the much higher value of 22,000.
419. Spector, *Eagle Against the Sun*, 317; van der Vat, *Pacific Campaign*, 328.
420. Thomas R. Searle, "'It Made a Lot of Sense to Kill Skilled Workers': The Fire-bombing of Tokyo in March 1945," *Journal of Military History* 66, no. 1 (January 2002): 116–117, https://doi.org/10.2307/2677346.
421. Searle, "'It Made a Lot of Sense,'" 112.
422. Mitchell, in Searle, "'It Made a Lot of Sense,'" 115.
423. George C. Marshall, in Searle, "'It Made a Lot of Sense,'" 115–16.
424. Searle, "'It Made a Lot of Sense,'" 117.
425. Searle, "'It Made a Lot of Sense,'" 120, 133.
426. *United States Strategic Bombing Survey Summary Report (Pacific War)* (United States Government Printing Office, July 1, 1946), 16–17, 20, http://www.anesi.com/ussbs01.htm; Kenneth P. Werrell, *Blankets of Fire: U.S. Bombers over Japan During World War II* (Smithsonian Institution Press, 1996), 163.
427. Robert A. Pape, "Why Japan Surrendered," *International Security* 18, no. 2 (Fall 1993): 165, https://doi.org/10.2307/2539100. See also Werrell, *Blankets of Fire*, 166–68.
428. Curtis LeMay and McKinlay Kantor, *Mission with LeMay: My Story* (Doubleday, 1965), 367.
429. Pape, "Why Japan Surrendered," 163.
430. *United States Strategic Bombing Survey (Pacific War)*, 21.
431. Searle, "'It Made a Lot of Sense,'" 119, 122.
432. *United States Strategic Bombing Survey (Pacific War)*, 16.
433. Werrell, *Blankets of Fire*, 193; Spector, *Eagle Against the Sun*, 554–55. See also Toland, *Rising Sun*, 762.
434. Corum, *Luftwaffe*, 200.
435. Weinberg, *World at Arms*, 867.
436. See, for example, Howard Zinn, "Just and Unjust War," in *Declarations of Independence: Cross-Examining American Ideology* (HarperCollins, 1990).
437. Paul Fussell and Michael Walzer, "A Defense of the Atomic Bomb and a Dissent," in *Major Problems in American Military History*, ed. John Whiteclay Chambers II and G. Kurt Piehler (Houghton Mifflin, 1999), 370. The question of whether the use of atomic weapons hastened the surrender of Japan is a topic of debate. Some believe that it was not their use but the Soviet invasion of Japanese-held Manchuria and Korea that finally convinced the Japanese of the futility of continued warfare. See Pape, "Why Japan Surrendered," 178–79, 187–88, and Ferguson, *War of the World*, 574.
438. Werrell, *Blankets of Fire*, 209.
439. Biddle, "Dresden 1945," 446.
440. Biddle, "Dresden 1945," 432.
441. Biddle, "Dresden 1945," 428; Langenbacher, "Allies in World War II," 123.
442. Franklin Delano Roosevelt, in Wolk and Hallion, "FDR and Truman," 1; Roosevelt, in Richard Overy, "Allied Bombing and the Destruction of German Cities," in *A World at Total War: Global Conflict and the Politics of Destruction, 1937–1945*,

ed. Roger Chickering, Stig Förster, and Bernd Greiner, 277–96 (Cambridge University Press, 2010), 286.

443. Wolk and Hallion, "FDR and Truman," 2.

444. Arthur Harris, in Overy, "Allied Bombing," 291 (emphasis in original).

445. Corum, *Luftwaffe*, 199.

446. Searle, "'It Made a Lot of Sense,'" 120.

447. LeMay and Kantor, *Mission with LeMay*, 384.

448. Koch, "Strategic Air Offensive," 129.

449. Corum, *Wolfram von Richthofen*, 173.

450. In one such example, the case of Hamburg, the *United States Strategic Bombing Survey (European War)*, 3, gives a death toll of 60,000 to 100,000, while Weinberg, *World at Arms*, 616, gives a figure of 40,000.

451. Koch, "Strategic Air Offensive," 129; Strachan, "Total War," 40.

452. Koch, "Strategic Air Offensive," 128.

453. Overy, "Allied Bombing," 294.

454. Legro, "Military Culture Inadvertent Escalation," 126–28.

455. Legro, "Military Culture Inadvertent Escalation," 129.

456. Koch, "Strategic Air Offensive," 131 (emphasis in original).

457. Gar Alperovitz, Robert L. Messer, and Barton J. Bernstein, "Marshall, Truman, and the Decision to Drop the Bomb," *International Security* 16, no. 3 (Winter 1991–1992): 208–9, 212, http://www.jstor.org/stable/2539092.

458. Chang, *Rape of Nanking*, 101, 166.

459. LeMay and Kantor, *Mission with LeMay*, 380.

460. William W. Ralph, "Improvised Destruction: Arnold, LeMay, and the Firebombing of Japan," *War in History* 13, no. 4 (November 2006): 514, https://doi.org/10.1177/0968344506069971.

461. Ralph, "Improvised Destruction," 517.

462. LeMay and Kantor, *Mission with LeMay*, 312–13.

463. Thomas M. Coffey, *Iron Eagle: The Turbulent Life of General Curtis LeMay* (Avon Books, 1988), 160. See also LeMay and Kantor, *Mission with LeMay*, 12.

464. Searle, "'It Made a Lot of Sense,'" 122–23n.

465. Gordon, *Atrocity Speech Law*, 30.

466. Gordon, *Atrocity Speech Law*.

467. Gordon, *Atrocity Speech Law*, 30–31.

468. Sue Anne Riley, "Loose Lips Sink Ships: American Propaganda in WWII," *Sea Classics* 45, no. 10 (October 2012): 51.

469. Riley, "Loose Lips Sink Ships," 52.

470. Riley, "Loose Lips Sink Ships," 53.

471. Riley, "Loose Lips Sink Ships."

472. Riley, "Loose Lips Sink Ships."

473. Greg Zyla, "What Years Did Car Manufacturers Not Build Cars During the World War II Era?" Auto Round-Up Publications, September 23, 2019, https://www.autoroundup.com/vehicle/what-years-did-car-manufacturers-not-build-cars-during-the-world-war-ii-era-article-1271.aspx.

474. Barak Kushner, *The Thought War: Japanese Imperial Propaganda* (University of Hawai'i Press, 2006), 157.

475. Kushner, *Thought War*, 184.

476. Kushner, *Thought War*, 6.

477. Kushner, *Thought War*, 37.

478. Kushner, *Thought War*, 64.
479. Gordon, *Atrocity Speech Law*, 30.
480. Joel V. Berreman, "Assumptions About America in Japanese War Propaganda to the United States," *American Journal of Sociology* 54, no. 2 (September 1948): 111, http://www.jstor.org/stable/2771359.
481. Berreman, "Assumptions About America," 112.
482. Berreman, "Assumptions About America," 114.
483. Berreman, "Assumptions About America."
484. Randall Bytwerk, ed., trans., *Landmark Speeches of National Socialism* (Texas A&M University Press, 2008), 1.
485. Bytwerk, *Landmark Speeches*, 112.
486. Bytwerk, *Landmark Speeches*, 112–13.

CHAPTER 6

1. Sun Tzu, *The Art of War*, trans. Samuel B. Griffith (Oxford University, 1971), 77.
2. *Merriam-Webster*, "Asymmetric Warfare," accessed September 3, 2024, https://www.merriam-webster.com/dictionary/asymmetric--warfare.
3. Center for International Relations and International Security, "Asymmetric Warfare," accessed October 19, 2024, https://www.ciris.info/learningcenter/asymmetric-warfare/.
4. Editorial Team, "Understanding Mutually Assured Destruction and Its Implications," Total Military Insight, July 20, 2024, https://totalmilitaryinsight.com/mutually-assured-destruction/.
5. Curtis E. LeMay, "The Operational Side of Air Offense," National Security Archive, May 21, 1957, https://nsarchive.gwu.edu/document/20291-national-security-archive-doc-06-operational.
6. LeMay, "Operational Side of Air Offense."
7. John Simkin, "Cuban Missile Crisis," Spartacus Educational, updated August 2020, https://spartacus-educational.com/COLDcubanmissile.htm.
8. Mohammed Harbi, "Massacre in Algeria," *Le Monde diplomatique*, May 2005, https://mondediplo.com/2005/05/14algeria.
9. Jean-Pierre Peyroulou, "Setif and Guelma (May 1945)," SciencesPo, March 26, 2008, https://www.sciencespo.fr/mass-violence-war-massacre-resistance/en/document/setif-and-guelma-may-1945.html.
10. Peyroulou, "Setif and Guelma."
11. Harbi, "Massacre in Algeria."
12. Daniel Zar, "Sétif and Guelma Massacre, French Slaughter Algerians," History and Headlines, May 8, 2020, https://www.historyandheadlines.com/setif-and-guelma-massacre-french-slaughter-algerians/.
13. Peyroulou, "Setif and Guelma."
14. Amy J. Johnson, "Algeria: War of Independence, 1954–1962," in *Encyclopedia of African History*, vol. 1: A–G, ed. Kevin Shillington, 101–3 (Fitzroy Dearborn, 2005).
15. Johnson, "Algeria."
16. For some of these reasons, see Djamila Ould Khettab, "Q&A: What Really Happened to Algeria's Harkis," *Al Jazeera*, August 22, 2015, https://www.aljazeera.com/news/2015/8/22/qa-what-really-happened-to-algerias-harkis.

17. World Peace Foundation, "Algeria: War of Independence," Mass Atrocity Endings, August 7, 2015, https://sites.tufts.edu/atrocityendings/2015/08/07/algeria-war-of-independence/.
18. John Hayward, "Algerian President Tebboune Declares Victory with 95% of Vote in Election 'Farce,'" *Breitbart News*, September 9, 2024, https://www.breitbart.com/africa/2024/09/09/algerian-president-tebboune-declares-victory-with-95-of-vote-in-election-farce/.
19. Hayward, "Algerian President Tebboune."
20. Susanne Beyer, "German Woman Writes Ground-Breaking Account of WW2 Rape," *Der Spiegel International*, February 26, 2010, https://www.spiegel.de/international/germany/harrowing-memoir-german-woman-writes-ground-breaking-account-of-ww2-rape-a-680354.html.
21. Beyer, "German Woman Writes."
22. Beyer, "German Woman Writes."
23. Frank Biess, "Fears of Retribution in Post-War Germany," National WWII Museum New Orleans, September 20, 2021, https://www.nationalww2museum.org/war/articles/fears-of-retribution-in-post-war-germany.
24. Biess, "Fears of Retribution."
25. Directive to Commander-in-Chief of United States Forces of Occupation Regarding the Military Government of Germany, April 1945, https://usa.usembassy.de/etexts/ga3-450426.pdf.
26. Biess, "Fears of Retribution."
27. Biess, "Fears of Retribution."
28. Eve Conant, "Behind the Headlines: Who Are the Crimean Tatars?" *National Geographic*, March 15, 2014, https://www.nationalgeographic.com/history/article/140314-crimea-tatars-referendum-russia-muslim-ethnic-history-culture.
29. Krym Media, "Only 3.3% of Crimeans Mention Ukrainian as Their Native Language," March 19, 2015, http://en.krymedia.ru/nationality/3373760-Only-33-of-Crimeans-Mention-Ukrainian-as-Their-Native-Language; Conant, "Behind the Headlines."
30. Samuel Osborne, "UN Accuses Russia of Multiple Human Rights Abuses in Crimea," *Independent*, November 16, 2016, https://www.independent.co.uk/news/world/europe/russia-ukraine-crimea-putin-human-rights-abuses-un-accusations-claims-a7421406.html.
31. Osborne, "UN Accuses Russia."
32. Osborne, "UN Accuses Russia."
33. Human Rights Watch, "Crimea: Persecution of Crimean Tatars Intensifies," November 14, 2017, https://www.hrw.org/news/2017/11/14/crimea-persecution-crimean-tatars-intensifies.
34. Ukrinform, "UN Documents Torture and Arrests of Crimean Tatars by Russia," December 12, 2017, https://www.ukrinform.net/rubric-society/2362880-un-documents-torture-and-arrests-of-crimean-tatars-by-russia.html.
35. Jeremy Black, *A Century of Conflict: War, 1914–2014* (Oxford University Press, 2015), 123.
36. Black, *Century of Conflict*, 129.
37. Christopher L. Corley, "Acts of Atrocity: Effects on Public Opinion Support During War or Conflict" (master's thesis, Naval Postgraduate School, December 2007), https://web.archive.org/web/20220507164026/https://apps.dtic.mil/dtic/tr/fulltext/u2/a475745.pdf, 15.

38. Corley, "Acts of Atrocity."
39. Mark Oliver, "The My Lai Massacre: 33 Disturbing Photos of the War Crime the U.S. Got Away With," ed. John Kuroski, All That's Interesting, updated May 20, 2021, https://allthatsinteresting.com/my-lai-massacre-photos#1.
40. Oran Henderson, in Oliver, "My Lai Massacre."
41. Ernest Medina, in Corley, "Acts of Atrocity," 18.
42. "Trials: My Lai: A Question of Orders," Time, January 25, 1971, https://time.com/archive/6838471/trials-my-lai-a-question-of-orders/.
43. Oliver, "My Lai Massacre."
44. History.com Editors, "My Lai Massacre," History, updated April 1, 2025, https://www.history.com/topics/vietnam-war/my-lai-massacre-1.
45. "Trials: My Lai."
46. History.com Editors, "My Lai Massacre."
47. Robert Harris and Jeremy Paxman, A Higher Form of Killing: The Secret History of Chemical and Biological Warfare (Random House, Trade Paperbacks, 2002), 194.
48. Harris and Paxman, Higher Form of Killing.
49. History.com Editors, "Agent Orange," History, updated February 27, 2025, https://www.history.com/topics/vietnam-war/agent-orange-1.
50. Blake Stilwell, "Why the US Used Agent Orange in Vietnam and What Makes It So Deadly," Military.com, August 1, 2022, https://www.military.com/history/why-us-used-agent-orange-vietnam-and-what-makes-it-so-deadly.html.
51. History.com Editors, "Agent Orange."
52. Harris and Paxman, Higher Form of Killing, 195.
53. Stilwell, "US Used Agent Orange."
54. Editors of Encyclopaedia Britannica, "Agent Orange," Britannica, updated March 25, 2025, https://www.britannica.com/science/Agent-Orange; Harris and Paxman, Higher Form of Killing, 195.
55. International Humanitarian Law Databases, "Practice Relating to Rule 76 Herbicides," accessed September 12, 2024, https://ihl-databases.icrc.org/en/customary-ihl/v2/rule76.
56. International Humanitarian Law Databases, "Practice Relating to Rule 76."
57. Harris and Paxman, Higher Form of Killing, 198.
58. Harris and Paxman, Higher Form of Killing, 199.
59. My Jewish Learning, "Modern Israeli History: A Timeline," accessed August 23, 2024, https://www.myjewishlearning.com/article/modern-israeli-history-a-timeline/; History Skills, "How the Modern State of Israel Was Created in 1948," accessed August 23, 2024, https://www.historyskills.com/classroom/modern-history/formation-of-modern-israel-reading/.
60. Shmuel Ettinger, "The Balfour Declaration of 1917," My Jewish Learning, accessed August 23, 2024, https://www.myjewishlearning.com/article/the-balfour-declaration/; History Skills, "Modern State of Israel."
61. History Skills, "Modern State of Israel."
62. Jewish Telegraphic Agency, "Fourth Day of Palestine Warfare Finds Jewry in Grave Danger Throughout Country," Jewish Daily Bulletin, August 27, 1929, http://pdfs.jta.org/1929/1929-08-27_1451.pdf.
63. My Jewish Learning, "Modern Israeli History."
64. David B. Green, "This Day in Jewish History: 1940: Italy Bombs Tel Aviv During WWII," Haaretz, September 9, 2013, https://www.haaretz.com/jewish/2013-09-09/ty-article/1940-italy-bombs-tel-aviv-in-wwii/0000017f-f473-d044-adff-f7fb19a10000.

65. Walter Schellenberg, in Klaus-Michael Mallmann and Martin Cüppers, *Nazi Palestine: The Plans for the Extermination of the Jews in Palestine*, trans. Krista Smith (Enigma Books, 2010), 133–34.
66. Maurice Ostroff, "The Siege of Jerusalem," World Machal, accessed August 23, 2024, https://www.machal.org.il/about-machal/the-siege-of-jerusalem/.
67. Benny Morris and Benjamin Z. Kedar, "Cast Thy Bread: Israeli Biological Warfare During the 1948 War," *Middle Eastern Studies* 59, no. 5 (2023): 753, https://doi.org/10.1080/00263206.2022.2122448.
68. Morris and Kedar, "Cast Thy Bread."
69. Morris and Kedar, "Cast Thy Bread," 758.
70. Morris and Kedar, "Cast Thy Bread," 759.
71. Morris and Kedar, "Cast Thy Bread," 762.
72. American Jewish Committee, "Timeline: Key Events in the Israel-Arab and Israeli-Palestinian Conflict," accessed April 23, 2025, https://www.ajc.org/Israel ConflictTimeline.
73. Joshua Caplan, "Live Updates: Hamas Launches Unprecedented Terror Attack on Israel," *Breitbart News*, October 7, 2023, https://www.breitbart.com/mid dle-east/2023/10/07/live-updates-hamas-launches-unprecedented-terror-attack-on -israel/.
74. Rich Tenorio, "Documentary on October 7 Supernova Festival Massacre Makes US Debut," *Times of Israel*, February 23, 2024, https://www.timesofisrael.com/ documentary-on-october-7-supernova-festival-massacre-makes-us-debut/.
75. Rachael Bunyan and Elena Salvoni, "Hamas Vows to Repeat October 7 Attacks and Bring About the 'Annihilation' of Israel While Cynically Saying It 'Did Not Want to Hurt Civilians' as It Slaughtered 1,400 People 'but There Were Complications on the Ground,'" *Daily Mail*, November 1, 2023, https://www.dailymail .co.uk/news/article-12697293/Hamas-leader-dismisses-Gaza-civilian-deaths-neces sary-price-blood-boasts-terror-group-demonstrated-Israel-beatable-changing-Mid dle-East.html.
76. Joel B. Pollak, "Hamas Leaders to NY Times: No Interest in Helping Palestinians in Gaza, Want 'Permanent' War Against Israel," *Breitbart News*, November 8, 2023, https://www.breitbart.com/middle-east/2023/11/08/hamas-leaders-to-ny -times-no-interest-in-helping-palestinians-gaza-want-permanent-war-against-israel/.
77. Joel B. Pollak, "Israel Discovers Weapons Cache in Maternity Ward of Gaza's Shifa Hospital," *Breitbart News*, March 31, 2024, https://www.breitbart.com/ middle-east/2024/03/31/israel-maternity-ward-shifa-hospital/.
78. Joel B. Pollak, "IDF: Hamas Is Destroying Shifa Hospital, Firing from Maternity Ward," *Breitbart News*, March 24, 2024, https://www.breitbart.com/poli tics/2024/03/24/idf-hamas-is-destroying-shifa-hospital-firing-from-maternity-ward/.
79. Joel B. Pollak, "IDF Kills 15 Terrorists in 'War Room' at UNRWA School," *Breitbart News*, May 14, 2024, https://www.breitbart.com/middle-east/2024/05/14/ idf-kills-15-terrorists-in-war-room-at-unrwa-school/.
80. Joel B. Pollak, "Israel Reveals Hamas Intel Tunnel Underneath UNRWA HQ in Gaza," *Breitbart News*, February 10, 2024, https://www.breitbart.com/ middle-east/2024/02/10/israel-reveals-hamas-intel-tunnel-underneath-unrwa-hq -in-gaza/.
81. Joel B. Pollak, "Israel: UNRWA Is Using Hamas to 'Protect' Aid Trucks," *Breitbart News*, March 6, 2024, https://www.breitbart.com/middle-east/2024/03/06/ israel-unrwa-is-using-hamas-to-protect-aid-trucks/; Ian Hanchett, "Israeli

Spox: Hamas Terrorists Aren't Starving Because They Steal Aid, Pressuring Us Incentivizes Their Evil," *Breitbart News*, April 6, 2024, https://www.breit bart.com/clips/2024/04/06/israeli-spox-hamas-terrorists-arent-starving-because-they -steal-aid-pressuring-us-incentivizes-their-evil/; Joel B. Pollak, "Palestinian Ruling Party Admits: Hamas Steals Aid, Kills Aid Workers in Gaza," *Breitbart News*, April 21, 2024, https://www.breitbart.com/politics/2024/04/21/palestinian -ruling-party-admits-hamas-steals-aid-kills-aid-workers-in-gaza/.

82. Joel B. Pollak, "IDF Discovers Hamas, Islamic Jihad Command Center at UNRWA HQ," *Breitbart News*, July 12, 2024, https://www.breitbart.com/mid dle-east/2024/07/12/idf-discovers-hamas-islamic-jihad-command-center-at-unrwa -hq/.

83. Joel B. Pollak, "UN Admits: 9 UNRWA Employees May Have Participated in October 7 Terror," *Breitbart News*, August 5, 2024, https://www.breitbart.com/mid dle-east/2024/08/05/un-admits-9-unrwa-employees-may-have-participated-in -october-7-terror/.

84. Greg Wehner, "Arabic Copy of Hitler's 'Mein Kampf' Found in Children's Room Used by Hamas: Israeli Officials," Fox News, November 12, 2023, https://www .foxnews.com/world/arabic-copy-hitlers-mein-kampf-found-childrens-room-used -hamas-israeli-officials; Jorge Fitz-Gibbon, "Arabic Copy of Adolf Hitler's 'Mein Kampf' Found Inside Child's Room in Gaza," *New York Post*, November 12, 2023, https://nypost.com/2023/11/12/news/arabic-copy-of-adolf-hitlers-mein-kampf -found-inside-childs-room-in-gaza/; "Arabic Annotated Copy of 'Mein Kampf' Found among Possessions of Terrorist in Gaza Home," *Times of Israel*, November 12, 2023, https://www.timesofisrael.com/liveblog_entry/arabic-annotated-copy-of -mein-kampf-found-among-possessions-of-terrorist-in-gaza-home/.

85. Steve Israel, "The Roots of Hamas' Terror Attack Can Be Found in Gaza's Schools," Forward, October 25, 2023, https://forward.com/opinion/566841/ hamas-schools-indoctrination-antisemitic-textbooks-gaza/; Middle East Media Research Institute, "Hamas' Indoctrination of Children to Jihad, Martyrdom, Hatred of Jews," November 3, 2023, https://www.memri.org/reports/hamas -indoctrination-children-jihad-martyrdom-hatred-jews.

86. Anna Schecter, "'Top Secret' Hamas Documents Show That Terrorists Inten- tionally Targeted Elementary Schools and a Youth Center," NBC News, October 13, 2023, https://www.nbcnews.com/news/investigations/top-secret-hamas-docu ments-show-terrorists-intentionally-targeted-elem-rcna120310.

87. Joel B. Pollak, "IDF: Iran Has Fired Missiles at Israel; All Residents Ordered to Bomb Shelters," *Breitbart News*, October 1, 2024, https://www.breitbart.com/ middle-east/2024/10/01/idf-iran-has-fired-missiles-at-israel-all-residents-ordered -to-bomb-shelters/.

88. Joel B. Pollak, "Official Confirmation: Hamas Leader Yahya Sinwar Is Dead; Killed Trying to Flee Gaza," *Breitbart News*, October 17, 2024, https://www.bre itbart.com/middle-east/2024/10/17/official-confirmation-hamas-leader-yahya-sin war-is-dead-killed-trying-to-flee-gaza/.

89. John Hayward, "U.N. Teacher's Passport Found on Hamas Terrorist Master- mind Yahya Sinwar," *Breitbart News*, October 17, 2024, https://www.breitbart .com/national-security/2024/10/17/u-n-teachers-passport-found-hamas-terrorist -mastermind-yahya-sinwar/; Joel B. Pollak, "Report: Yahya Sinwar Bodyguards Included United Nations Employee; UPDATE: Fake Passport?" *Breitbart News*,

October 17, 2024, https://www.breitbart.com/national-security/2024/10/17/report-yahya-sinwar-bodyguards-included-united-nations-employee/.

90. Pollak, "Report: Yahya Sinwar Bodyguards"; Joel B. Pollak, "Sinwar's Likely Death Comes Days After Biden, Harris Threatened Israel with Arms Embargo," *Breitbart News*, October 17, 2024, https://www.breitbart.com/middle-east/2024/10/17/sinwars-likely-death-comes-days-after-biden-harris-threatened-israel-with-arms-embargo/.

91. Joel B. Pollak, "Biden Tries to Take Credit for Yahya Sinwar's Killing, but Opposed Israel's Operations; Harris Warned of 'Consequences,'" *Breitbart News*, October 17, 2024, https://www.breitbart.com/middle-east/2024/10/17/biden-tries-to-take-credit-for-yahya-sinwars-killing-but-opposed-israels-operations-harris-warned-consequences/.

92. United Nations Office for the Coordination of Humanitarian Affairs (OCHA), "Humanitarian Situation Update #221: Gaza Strip," September 23, 2024, https://www.ochaopt.org/content/humanitarian-situation-update-221-gaza-strip.

93. Editors of Encyclopaedia Britannica, "Nationalist Collapse and the Establishment of the People's Republic of China (1949)," Britannica, updated February 11, 2025, https://www.britannica.com/event/Chinese-Civil-War/Nationalist-collapse-and-the-establishment-of-the-Peoples-Republic-of-China-1949.

94. denton.2@osu.edu, "Who Killed More: Hitler, Stalin, or Mao," Modern Chinese Literature and Culture Resource Center, Ohio State University, February 8, 2018, https://u.osu.edu/mclc/2018/02/08/who-killed-more-hitler-stalin-or-mao/.

95. Kaleena Fraga, "Who Killed the Most People in History? It's Not as Straightforward as You Think," ed. Adam Farley, All That's Interesitng, July 29, 2022, https://allthatsinteresting.com/who-killed-the-most-people-in-history.

96. "From the Archives, 1950: China Invades Tibet," *Sydney Morning Herald*, October 30, 1950, https://www.smh.com.au/world/asia/from-the-archives-1950-china-invades-tibet-20201014-p56560.html.

97. Facts and Details, "Chinese Invasion of Tibet in 1950 and Its Aftermath," updated September 2022, https://factsanddetails.com/china/cat6/sub32/entry-8413.html.

98. Facts and Details, "Chinese Invasion of Tibet."

99. Human Rights Watch, "Tibet: Mass Relocations of Tibetans Not Voluntary," May 21, 2024, https://www.hrw.org/news/2024/05/22/tibet-mass-relocations-tibetans-not-voluntary.

100. Human Rights Watch, "Tibet: Mass Relocations."

101. History.com Editors, "Tibetans Revolt Against Chinese Occupation," History, updated March 2, 2025, https://www.history.com/this-day-in-history/rebellion-in-tibet.

102. Human Rights Watch, "China and Tibet," accessed October 5, 2024, https://www.hrw.org/asia/china-and-tibet.

103. Editors of Encyclopaedia Britannica, "Tiananmen Square Incident," Britannica, updated April 20, 2025, https://www.britannica.com/event/Tiananmen-Square-incident.

104. Editors of Encyclopaedia Britannica, "Tiananmen Square Incident."

105. Editors of Encyclopaedia Britannica, "Tiananmen Square Incident"; History.com Editors, "Tiananmen Square Protests," History, updated March 6, 2025, https://www.history.com/topics/asian-history/tiananmen-square; Frances Martel, "Tim Walz Dodges Question on Tiananmen Square Lie: 'I'm a Knucklehead,'" *Breitbart*

News, October 1, 2024, https://www.breitbart.com/2024-election/2024/10/01/tim-walz-dodges-question-tiananmen-square-lie-im-knucklehead/.

106. Editors of Encyclopaedia Britannica, "Tiananmen Square Incident"; John Hayward, "Tiananmen Square Massacre Families Tell Xi Jinping: 'We Will Never Forget,'" *Breitbart News*, June 3, 2024, https://www.breitbart.com/asia/2024/06/03/tiananmen-square-massacre-families-tell-xi-jinping-we-will-never-forget/.

107. "Weekly Forum Editorial," *Christian Times*, June 2, 2024, https://christiantimes.org.hk/Common/Reader/Version/Show.jsp?Pid=2&Version=1918&Charset=big5_hkscs.

108. Hayward, "Tiananmen Square Massacre Families"; John Hayward, "Taiwan Holds Last Remaining Tiananmen Remembrance in Chinese-Speaking World," *Breitbart News*, June 4, 2024, https://www.breitbart.com/asia/2024/06/04/taiwan-holds-last-remaining-tiananmen-remembrance-chinese-speaking-world/.

109. BBC News, "Who Are the Uyghurs and Why Is China Being Accused of Genocide?" May 24, 2022, https://www.bbc.com/news/world-asia-china-22278037.

110. BBC News, "Who Are the Uyghurs."

111. BBC News, "Who Are the Uyghurs."

112. Human Rights Watch, "'Break Their Lineage, Break Their Roots,'" April 19, 2021, https://www.hrw.org/report/2021/04/19/break-their-lineage-break-their-roots/chinas-crimes-against-humanity-targeting.

113. Human Rights Watch, "'Break Their Lineage.'"

114. Black, *Century of Conflict*, 109.

115. Max Fisher, "Americans Have Forgotten What We Did to North Korea," *Vox*, August 3, 2015, https://www.vox.com/2015/8/3/9089913/north-korea-us-war-crime.

116. Fisher, "Americans Have Forgotten."

117. Fisher, "Americans Have Forgotten."

118. Fisher, "Americans Have Forgotten."

119. Diplomatic cable sent by North Korea's foreign minister to the United Nations, January 1951, in Fisher, "Americans Have Forgotten."

120. Mike Bassett, in Tim Shorrock, "Can the United States Own Up to Its War Crimes During the Korean War?" *Nation*, March 30, 2015, https://www.thenation.com/article/archive/can-united-states-own-its-war-crimes-during-korean-war/.

121. Fisher, "Americans Have Forgotten"; Guenter Lewy, *America in Vietnam* (Oxford University Press, 1980), https://archive.org/details/americainvietnam00lewy/page/450/mode/2up, 450.

122. Lewy, *America in Vietnam*.

123. Black, *Century of Conflict*, 113.

124. United States Institute of Peace, "Truth Commission: South Korea 2005," April 18, 2012, https://www.usip.org/publications/2012/04/truth-commission-south-korea-2005.

125. "Korea Bloodbath Probe Ends; US Escapes Much Blame," *San Diego Union-Tribune*, updated August 31, 2016, https://www.sandiegouniontribune.com/2010/07/10/korea-bloodbath-probe-ends-us-escapes-much-blame/.

126. Jeremy Kuzmarov, "The Korean War: Barbarism Unleashed," United States Foreign Policy: History and Resource Guide, 2016, https://peacehistory-usfp.org/korean-war/.

127. Kuzmarov, "Korean War."

128. Kuzmarov, "Korean War."

129. Kuzmarov, "Korean War"; Elisa Joy Holland, "Massacre at Nogun-Ri," Asia Society, accessed October 5, 2024, https://asiasociety.org/education/massacre-nogun-ri.
130. Fisher, "Americans Have Forgotten," 2.
131. Fisher, "Americans Have Forgotten," 4–13.
132. Yoonjung Seo, Andrew Raine, and Gawon Bae, "Torture, Forced Abortions and Insects for Food: Life Inside North Korean Jails, Says This NGO," CNN World, updated March 24, 2023, https://www.cnn.com/2023/03/23/asia/north-korea-torture-prison-report-intl-hnk-dst/index.html.
133. Sumit Ganguly, "Pakistan's Forgotten Genocide—A Review Essay," *International Security* 39, no. 2 (Fall 2014): 169–70, https://doi.org/10.1162/ISEC_a_00175.
134. Haroon Habib, "In East Pakistan in 1971: A 'Forgotten' Genocide," *Frontline*, January 30, 2022, https://frontline.thehindu.com/world-affairs/in-east-pakistan-in-1971-a-forgotten-genocide-bangladesh-liberation-war/article38307183.ece.
135. Ganguly, "Pakistan's Forgotten Genocide," 170.
136. Ganguly, "Pakistan's Forgotten Genocide." Ganguly covers exensively the US role in the East Pakistani genocide.
137. Ganguly, "Pakistan's Forgotten Genocide"; Habib, "East Pakistan in 1971"; Jalal Alamgir and Bina D'Costa, "The 1971 Genocide: War Crimes and Political Crimes," *Economic and Political Weekly* 46, no. 13 (March 26–April 1, 2011): 38, https://www.jstor.org/stable/41152283; Asfandiyar Khan, "The Splitting of East-Pakistan from West-Pakistan in 1971: The Role of India," Modern Diplomacy, February 7, 2020, https://moderndiplomacy.eu/2020/02/07/the-splitting-of-east-pakistan-from-west-pakistan-in-1971-the-role-of-india/; Black, *Century of Conflict*, 2.
138. David Owen in Paul Raffaele, "Uganda: The Horror," *Smithsonian Magazine*, February 2005, https://www.smithsonianmag.com/history/uganda-the-horror-85439313/.
139. Raffaele, "Uganda: The Horror."
140. Raffaele, "Uganda: The Horror."
141. Karim A. A. Khan, Mame Mandiaye Niang, Leonie Von Braun, Paolina Massidda, and Sarah Pellet, *The Prosecutor v. Joseph Kony*, International Criminal Court, January 19, 2024, https://www.icc-cpi.int/sites/default/files/CourtRecords/CR2024_00006.PDF, 25–32. The charges laid out go into great detail about the events leading to these charges.
142. Reuters, "Court Charges at Least 42 Ugandan Youths over Anti-Graft Protest," July 24, 2024, https://www.reuters.com/world/africa/court-charges-least-42-ugandan-youths-over-anti-graft-protest-2024-07-24/.
143. John Hayward, "'Thieves and Parasites': Marxist Uganda Arrests Dozens Protesting Corruption," *Breitbart News*, July 25, 2024, https://www.breitbart.com/africa/2024/07/25/thieves-and-parasites-marxist-uganda-arrests-dozens-protesting-corruption/.
144. Hayward, "'Thieves and Parasites.'"
145. Jessica Pearce Rotondi, "How Dith Pran's Remarkable Survival Story Exposed Cambodia's Killing Fields," History, updated March 6, 2025, https://www.history.com/news/dith-pran-killing-fields-cambodia-khmer-rouge.
146. Genos Center, "The Cambodian Genocide: A Look at the Killing Fields," Genos Center, July 21, 2023, https://genoscenter.org/the-cambodian-genocide-a-look-at-the-killing-fields/.

147. History.com Editors, "Khmer Rouge," History, updated April 15, 2025, https://www.history.com/topics/cold-war/the-khmer-rouge.
148. United States Holocaust Memorial Museum, "Cambodia 1975–1979," updated April 2018, https://www.ushmm.org/genocide-prevention/countries/cambodia/cambodia-1975.
149. Teeda Butt Mam, in Jon Streeter and Joe Parker, "The Khmer Rouge: 'To Destroy You Is No Loss,'" 15-Minute History Podcast, March 25, 2024, https://www.15minutehistorypodcast.org/episodes/the-khmer-rouge.
150. For a complete examination of this crisis from which these examples are taken, see Mark Bowden, Guests of the Ayatollah: The Iran Hostage Crisis: The First Battle in America's War with Militant Islam (Atlantic Monthly Press, 2006).
151. Michael Brill, "Part I: 'We Attacked Them with Chemical Weapons and They Attacked Us with Chemical Weapons': Iraqi Records and the History of Iran's Chemical Weapons Program," Sources and Methods (blog), Wilson Center, March 29, 2022, https://www.wilsoncenter.org/blog-post/part-i-we-attacked-them-chemical-weapons-and-they-attacked-us-chemical-weapons-iraqi.
152. Charles Duelfer and United States Central Intelligence Agency, Comprehensive Report of the Special Advisor to the DCI on Iraq's WMD with Addendums, vol. 3, US Government Office, September 30, 2004, https://www.govinfo.gov/content/pkg/GPO-DUELFERREPORT/pdf/GPO-DUELFERREPORT-3.pdf.
153. Richard Stone, "Seeking Answers for Iran's Chemical Weapons Victims—Before Time Runs Out," Science, January 4, 2018, https://www.science.org/content/article/seeking-answers-iran-s-chemical-weapons-victims-time-runs-out.
154. Stone, "Seeking Answers."
155. Stone, "Seeking Answers"; Editors of Encyclopaedia Britannica, "Iran-Iraq War," Britannica, updated April 25, 2025, https://www.britannica.com/event/Iran-Iraq-War.
156. Human Rights Watch, "Iran: 1988 Mass Executions Evident Crimes Against Humanity," June 8, 2022, https://www.hrw.org/news/2022/06/08/iran-1988-mass-executions-evident-crimes-against-humanity; Struan Stevenson, "The Forgotten Mass Execution of Prisoners in Iran in 1988," Diplomat, July 31, 2013, https://thediplomat.com/2013/07/the-forgotten-mass-execution-of-prisoners-in-iran-in-1988/; Amnesty International, "Iran: Blood-Soaked Secrets: Why Iran's 1988 Prison Massacres Are Ongoing Crimes Against Humanity," December 4, 2018, https://www.amnesty.org/en/documents/mde13/9421/2018/en/.
157. Gregory Gordon, Atrocity Speech Law: Foundation, Fragmentation, Fruition (Oxford University Press, 2017), 43.
158. Noel Malcolm, in Gordon, Atrocity Speech Law, 44.
159. R. Jeffrey Smith, "Srebrenica Genocide," Britannica, updated March 28, 2025, https://www.britannica.com/event/Srebrenica-massacre#ref294001.
160. Smith, "Srebrenica Genocide."
161. John R. Lampe, "War Crimes and Trials," Britannica, updated April 22, 2025, https://www.britannica.com/event/Bosnian-War/War-crimes-and-trials.
162. Smith, "Srebrenica Genocide."
163. Lampe, "War Crimes and Trials."
164. Raf Casert and Dusan Stojanovic, "UN Court Overturns Acquittal of Serb Ultranationalist," Associated Press, updated April 11, 2018, https://apnews.com/general-news-2774f56f5628497790fa1bbf249dbcac.
165. Gordon, Atrocity Speech Law, 45.

166. Reuters, "U.N. Judges Overturn Acquittal of Serbian Ultra-Nationalist for Role in Wars," April 11, 2018, https://www.reuters.com/article/world/un-judges-over turn-acquittal-of-serbian-ultra-nationalist-for-role-in-wars-idUSKBN1HI1WT/.
167. Gordon, *Atrocity Speech Law*, 46–47.
168. Gordon, *Atrocity Speech Law*, 47.
169. University of Central Arkansas, Government, Public Service, and International Studies, "42. Rwanda (1962–Present)," accessed July 7, 2024, https://uca.edu/ politicalscience/home/research-projects/dadm-project/sub-saharan-africa-region/ rwanda-1962-present/.
170. University of Central Arkansas Government, Public Service, and International Studies, "42. Rwanda."
171. Alison Des Forges in Gordon, *Atrocity Speech Law*.
172. Mandy Southgate, "Rwandan Genocide: The Hutu Ten Commandments," *A Passion to Understand* (blog), August 13, 2011, https://passiontounderstand.blogspot .com/2011/08/rwandan-genocide-hutu-ten-commandments.html.
173. Gordon, *Atrocity Speech Law*, 49.
174. Gordon, *Atrocity Speech Law*, 50.
175. Gordon, *Atrocity Speech Law*, 51–52.
176. Gordon, *Atrocity Speech Law*, 53.
177. Holocaust Encyclopedia, "The Rwanda Genocide," United States Holocaust Memorial Museum, updated April 5, 2021, https://encyclopedia.ushmm.org/content/ en/article/the-rwanda-genocide.
178. Holocaust Encylcopedia, "Rwanda Genocide."
179. Dallaire/UNAMIR/Kigali, "Request for Protection for Informant," outgoing code cable to Baril/DPKO/UNations New York, January 11, 1994, https://nsarchive2 .gwu.edu/NSAEBB/NSAEBB452/docs/doc03.pdf.
180. Dallaire/UNAMIR/Kigali, "Request for Protection," point 11.
181. Annan, UNations, New York, "Contacts with Informant," outgoing code cable to Booh-Booh/Dallaire, UNAMIR Kigali, January 11, 1994, https://nsarchive2.gwu .edu/NSAEBB/NSAEBB452/docs/doc05.pdf, point 11.
182. Annan, UNations, New York, "Contacts with Informant," point 8 (emphasis added).
183. Collin Woldt, "Never Again: Revisiting the 1994 Rwandan Genocide," *Columbia Political Review*, June 4, 2021, https://www.cpreview.org/articles/2021/6/ never-again-revisiting-the-1994-rwandan-genocide.
184. See Woldt, "Never Again."
185. Woldt, "Never Again."
186. Zofsha Merchant and Joanna Michalopoulos, "Democratic Republic of the Congo," World Without Genocide, updated October 2023, https://worldwithout genocide.org/genocides-and-conflicts/congo.
187. Reuters, "Witnesses Tell of Congo Massacre," CNN, April 8, 2003, https://www .cnn.com/2003/WORLD/africa/04/08/congo.massacre.reut/index.html.
188. Black, *Century of Conflict*, 187.
189. Peter Moszynski, "5.4 Million People Have Died in Democratic Republic of Congo Since 1998 Because of Conflict, Report Says," *British Medical Journal* 336, no. 7638 (February 2, 2008): 235, https://doi.org/10.1136/BMJ.39475.524282.DB.
190. Nada Al-Nashif, "Deputy High Commissioner Updates Council on the Democratic Republic of the Congo," United Nations Human Rights Office of the High Commissioner, April 1, 2025, https://www.ohchr.org/en/statements

-and-speeches/2025/04/deputy-high-commissioner-updates-council-democratic
-republic-congo.

191. John Hayward, "Congo Seeks to Execute Americans for Alleged 'Coup' as Joe Biden Soaks Up Sun," *Breitbart News*, August 28, 2024, https://www.breitbart .com/africa/2024/08/28/congo-seeks-to-execute-americans-for-alleged-coup-as-joe -biden-soaks-up-sun/.

192. Hayward, "Congo Seeks to Execute."

193. Black, *Century of Conflict*, 179.

194. Carl Conetta, *The Wages of War: Iraqi Combatant and Noncombatant Fatalities in the 2003 Conflict* (Project on Defense Alternatives, Commonwealth Institute, October 20, 2003), https://www.comw.org/pda/fulltext/0310rm8.pdf, 3.

195. Conetta, *Wages of War*, 16.

196. Black, *Century of Conflict*, 179–80.

197. Sarah Sanbar, "Twenty Years On, Iraq Bears Scars of US-Led Invasion," Human Rights Watch, March 19, 2023, https://www.hrw.org/news/2023/03/19/ twenty-years-iraq-bears-scars-us-led-invasion.

198. Sanbar, "Twenty Years On"; Amnesty International UK, "Iraq: Twenty Years On, Still No Justice for War Crimes by US-Led Coalition," March 20, 2023, https://www.amnesty.org.uk/press-releases/iraq-twenty-years-still -no-justice-war-crimes-us-led-coalition.

199. James E. Waller, "It Can Happen Here: Assessing the Risk of Genocide in the US," Center for Develoment of International Law, February 24, 2017, https://world withoutgenocide.org/wp-content/uploads/2017/03/Waller-Assessing-the-Risk-of -Genocide.pdf.

200. Marissa Evans, "Darfur Genocide (2003–)," BlackPast, March 25, 2009, https:// www.blackpast.org/global-african-history/darfur-genocide-2003/.

201. Black, *Century of Conflict*, 186.

202. Black, *Century of Conflict*; Evans, "Darfur Genocide."

203. Beverly Ochieng, Wedaeli Chibelushi, and Natasha Booty, "Sudan War: A Simple Guide to What Is Happening," BBC News, March 25, 2025, https://www.bbc.com/ news/world-africa-59035053.

204. John Hayward, "Biden-Harris Admin Finally Notices Sudan Disaster After Doctors Without Borders Bails on Refugee Camp," *Breitbart News*, October 14, 2024, https://www.breitbart.com/africa/2024/10/14/biden-harris-admin-finally-notices -sudan-disaster-after-doctors-without-borders-bails-on-refugee-camp/.

205. Hayward, "Biden-Harris Admin."

206. Ochieng, Chibelushi, and Booty, "Sudan War."

207. Ochieng, Chibelushi, and Booty, "Sudan War"; Hayward, "Biden-Harris Admin."

208. Hayward, "Biden-Harris Admin."

209. Jeffrey Gettleman, "Disputed Vote Plunges Kenya into Bloodshed," *New York Times*, December 31, 2007, https://www.nytimes.com/2007/12/31/world/africa/ 31kenya.html.

210. Human Rights Watch, "'I Just Sit and Wait to Die': Reparations for Survivors of Kenya's 2007–2008 Post-Election Sexual Violence," February 15, 2016, https:// www.hrw.org/report/2016/02/15/i-just-sit-and-wait-die/reparations-survivors-kenyas -2007-2008-post-election.

211. Evelyn Middleton, "Côte d'Ivoire (Ivory Coast)," World Without Genocide, updated October 2020, https://worldwithoutgenocide.org/genocides-and-conflicts/ ivory-coast.

212. Gordon, *Atrocity Speech Law*, 58.
213. Human Rights Watch, "'They Killed Them Like It Was Nothing,'" October 5, 2011, https://www.hrw.org/report/2011/10/05/they-killed-them-it-was-nothing/ need-justice-cote-divoires-post-election-crimes.
214. Middleton, "Côte d'Ivoire."
215. Middleton, "Côte d'Ivoire."
216. Middleton, "Côte d'Ivoire."
217. Middleton, "Côte d'Ivoire."
218. Middleton, "Côte d'Ivoire."
219. United States Holocaust Memorial Museum, "Côte d'Ivoire," accessed July 11, 2024, https://www.ushmm.org/genocide-prevention/countries/cote-divoire.
220. Editors of Encyclopaedia Britannica, "Syrian Civil War," Britannica, updated April 16, 2025, https://www.britannica.com/event/Syrian-Civil-War.
221. Editors of Encyclopaedia Britannica, "Syrian Civil War."
222. CFR.org Editors, "Syria's Civil War: The Descent into Horror," Council on Foreign Relations, December 20, 2024, https://www.cfr.org/article/syrias-civil-war.
223. Editors of Encyclopaedia Britannica, "Syrian Civil War."
224. Syrian Network for Hurman Rights, "Documentation of 72 Torture Methods the Syrian Regime Continues to Practice in Its Detention Centers and Military Hospitals," October 21, 2019, https://snhr.org/blog/2019/10/21/54362/, 8–35.
225. Syrian Network for Hurman Rights, "Documentation of 72 Torture Methods," 7.
226. Syrian Network for Hurman Rights, "Documentation of 72 Torture Methods," 2, 7, 8.
227. Julian Borger and Kareem Shaheen, "Russia Accused of War Crimes in Syria at UN Security Council Session," *Guardian*, September 26, 2016, https://www.theguardian.com/world/2016/sep/25/russia-accused-war-crimes-syria-un-security-council-aleppo.
228. Angela Dewan and Hilary McGann, "Amnesty International Says US-Led Strikes on Raqqa May Amount to War Crimes," CNN, June 5, 2018, https://www.cnn.com/2018/06/05/middleeast/us-led-coalition-raqqa-war-crimes-intl/index.html.
229. Michelle Bachelet, "Presentation of the Report on Civilian Deaths in the Syrian Arab Republic," United Nations Human Rights Office of the High Commssioner, June 30, 2022, https://www.ohchr.org/en/statements/2022/06/presentation-report-civilian-deaths-syrian-arab-republic.
230. United States Department of State, Bureau of Democracy, Human Rights and Labor, *Syria 2013 Human Rights Report* (United States Department of State, 2013), 3.
231. United States Department of State, Bureau of Democracy, Human Rights and Labor, *Syria 2013 Human Rights Report*, 5.
232. Thomas Fuller, "Crisis in Myanmar over Buddhist-Muslim Clash," *New York Times*, June 10, 2012, https://www.nytimes.com/2012/06/11/world/asia/state-of-emergency-declared-in-western-myanmar.html.
233. Gordon, *Atrocity Speech Law*, 58.
234. Gordon, *Atrocity Speech Law*; BBC News, "Why Is There Communal Violence in Myanmar?" July 3, 2014, https://www.bbc.com/news/world-asia-18395788.
235. Lindsay Maizland, "Myanmar's Troubled History: Coups, Military Rule, and Ethnic Conflict," Council on Foreign Relations, updated January 31, 2022, https://www.cfr.org/backgrounder/myanmar-history-coup-military-rule-ethnic-conflict-rohingya.

236. Laura Smith-Spark, Phil Black, and Frederik Pleitgen, "Russia Flexes Military Muscle as Tensions Rise in Ukraine's Crimea," CNN World, updated February 26, 2014, https://edition.cnn.com/2014/02/26/world/europe/ukraine-pol itics; Interfax-Ukraine, "Checkpoints Put at All Entrances to Sevastopol," *Kyiv Post*, February 26, 2014, https://archive.kyivpost.com/article/content/ukrainepol itics//checkpoints-put-at-all-entrances-to-sevastopol-337655.html; Alissa de Carbonnel and Alessandra Prentice, "Armed Men Seize Two Airports in Ukraine's Crimea, Yanukovich Reappears," Reuters, February 28, 2014, https:// www.yahoo.com/news/armed-standoff-pro-russian-region-raises-ukraine-tension -033318395.html.
237. Matt Smith and Alla Eshchenko, "Ukraine Cries 'Robbery' as Russia Annexes Crimea," CNN World, updated March 18, 2014, https://www.cnn.com/2014/03/18/ world/europe/ukraine-crisis/index.html.
238. Matilda Bogner, "Situation in Ukraine," United Nations Human Rights Office of the High Commissioner, March 25, 2022, https://www.ohchr.org/en/ statements/2022/03/situation-ukraine.
239. Bogner, "Situation in Ukraine."
240. In Carlotta Gall and Andrew E. Kramer, "In the Kyiv Suburb of Bucha, 'They Shot Everyone They Saw,'" *New York Times*, April 3, 2022, https://web.archive .org/web/20220412065516/https://www.nytimes.com/2022/04/03/world/europe/ ukraine-russia-war-civilian-deaths.html.
241. Reuters, "Ukraine Probing over 122,000 Suspected War Crimes, Says Prosecutor," February 23, 2024, https://www.reuters.com/world/europe/ ukraine-probing-over-122000-suspected-war-crimes-says-prosecutor-2024-02-23/.
242. Nick Cumming-Bruce, "'Welcome to Hell': U.N. Panel Says Russian War Crimes Are Widespread," *New York Times*, March 15, 2024, https://www.nytimes .com/2024/03/15/world/europe/russia-war-crimes.html.

CHAPTER 7

1. Henry David Thoreau, "The Savage in Man Is Never Quite Eradicated," Fix-Quotes, accessed June 1, 2025, https://fixquotes.com/quotes/the-savage-in-man-is -never-quite-eradicated-35771.htm.
2. Jack Maden, "Mengzi vs. Xunzi on Human Nature: Are We Good or Evil?" Philosophy Break, October 2023, https://philosophybreak.com/articles/mengzi -xunzi-on-human-nature-are-we-good-or-evil/.
3. Mengzi, in Maden, "Mengzi vs. Xunzi."
4. Xunzi, in Maden, "Mengzi vs. Xunzi."
5. Maden, "Mengzi vs. Xunzi."
6. Buddha, in Frederick Meyer, "Is Human Nature Inherently Good or Bad? The Buddhist View of Human Nature," Shambhala, June 9, 2023, https://shambhala .org/community/blog/is-human-nature-inherently-good-or-bad-the-buddhist-view -of-human-nature/.
7. Meyer, "Is Human Nature Inherently Good."
8. Clancy Martin and Alan Strudler, "Are Humans Good or Evil? A Brief Philosophical Debate," *Harper's Magazine*, October 9, 2014, https://harpers.org/2014/10/ are-humans-good-or-evil/.
9. Martin and Strudler, "Are Humans Good or Evil?"
10. Martin and Strudler, "Are Humans Good or Evil?"

11. James E. Waller, "It Can Happen Here: Assessing the Risk of Genocide in the US," Center for Develoment of International Law, February 24, 2017, https://worldwithoutgenocide.org/wp-content/uploads/2017/03/Waller-Assessing-the-Risk-of-Genocide.pdf, 2 (emphasis in original).

12. Waller, "It Can Happen Here," 5–7.

13. Waller, "It Can Happen Here."

14. Christopher Browning, "Dangers I Didn't See Coming: 'Tyranny of the Minority' and an Irrelevant Press," *Vox*, January 18, 2017, https://www.vox.com/the-big-idea/2017/1/18/14303960/tyranny-minority-trump-democracy-gerrymander-media.

15. Waller, "It Can Happen Here," 13.

16. U.S. Const., art. II, § 1. The pertinent portion is clauses 2 and 3, which state,

> Each State shall appoint, in such Manner as the Legislature thereof may direct, a Number of Electors, equal to the whole Number of Senators and Representatives to which the State may be entitled in the Congress: but no Senator or Representative, or Person holding an Office of Trust or Profit under the United States, shall be appointed an Elector.
>
> The Electors shall meet in their respective States, and vote by Ballot for two Persons, of whom one at least shall not be an Inhabitant of the same State with themselves. And they shall make a List of all the Persons voted for, and of the Number of Votes for each; which List they shall sign and certify, and transmit sealed to the Seat of the Government of the United States, directed to the President of the Senate. The President of the Senate shall, in the Presence of the Senate and House of Representatives, open all the Certificates, and the Votes shall then be counted. The Person having the greatest Number of Votes shall be the President, if such Number be a Majority of the whole Number of Electors appointed; and if there be more than one who have such Majority, and have an equal Number of Votes, then the House of Representatives shall immediately chuse by Ballot one of them for President; and if no Person have a Majority, then from the five highest on the List the said House shall in like Manner chuse the President. But in chusing the President, the Votes shall be taken by States, the Representation from each State having one Vote; A quorum for this Purpose shall consist of a Member or Members from two thirds of the States, and a Majority of all the States shall be necessary to a Choice. In every Case, after the Choice of the President, the Person having the greatest Number of Votes of the Electors shall be the Vice President. But if there should remain two or more who have equal Votes, the Senate shall chuse from them by Ballot the Vice President. (spelling in original)

17. Ballotpedia, "Splits Between the Electoral College and Popular Vote," accessed July 30, 2024, https://ballotpedia.org/Splits_between_the_Electoral_College_and_popular_vote.

18. National Archives, "Distribution of Electoral Votes," reviewed November 4, 2024, https://www.archives.gov/electoral-college/allocation.

19. Ballotpedia, "Splits Between Electoral College."

20. Waller, "It Can Happen Here," 13.
21. Waller, "It Can Happen Here," 20–21; IPT News, "Federal Judge Agrees: CAIR Tied to Hamas," Investigative Project on Terrorism, November 22, 2010, https://www.investigativeproject.org/2340/federal-judge-agrees-cair-tied-to-hamas.
22. Waller, "It Can Happen Here," 19–20.
23. Tammy Duckworth, Kirsten Gillibrand, and John Thune, Congressional Record—Senate, no. S5648 (November 29, 2023), https://www.congress.gov/118/crec/2023/11/29/169/196/CREC-2023-11-29-pt1-PgS5648.pdf.
24. Waller, "It Can Happen Here," 24.
25. Deb Riechmann, Mathew Lee, and Jonathan Lemire, "Israel Signs Pacts with 2 Arab States: A 'New' Mideast?" Associated Press, September 15, 2020, https://apnews.com/article/bahrain-israel-united-arab-emirates-middle-east-elections-7544b322a254ebea1693e387d83d9d8b.
26. "Gallup Poll Reveals Obama Has Turned Back Clock on Race Relations," Investor's Business Daily, April 12, 2016, https://www.investors.com/politics/editorials/gallup-poll-reveals-obama-has-turned-back-clock-on-race-relations/.
27. White House, "Fact Sheet: President Biden Announces Bold Plan to Reform the Supreme Court and Ensure No President Is Above the Law," September 29, 2024, https://bidenwhitehouse.archives.gov/briefing-room/statements-releases/2024/07/29/fact-sheet-president-biden-announces-bold-plan-to-reform-the-supreme-court-and-ensure-no-president-is-above-the-law/; Amy Howe, "Biden Proposes Supreme Court Reforms," SCOTUSblog (blog), July 29, 2024, https://www.scotusblog.com/2024/07/biden-proposes-supreme-court-reforms/; Ken Klukowski, "Biden and Harris Propose Abolishing Supreme Court as Independent Branch of Government," Breitbart News, July 29, 2024, https://www.breitbart.com/politics/2024/07/29/biden-harris-propose-abolishing-supreme-court-independent-branch/.
28. Jason Willick, "Biden's Supreme Stunt," Washington Post, July 30, 2024, https://www.washingtonpost.com/opinions/2024/07/30/joe-biden-supreme-court-reform/.
29. Waller, "It Can Happen Here," 2.
30. Rachael Bunyan and Elena Salvoni, "Hamas Vows to Repeat October 7 Attacks and Bring About the 'Annihilation' of Israel While Cynically Saying It 'Did Not Want to Hurt Civilians' as It Slaughtered 1,400 People 'but There Were Complications on the Ground,'" Daily Mail, November 1, 2023. https://www.dailymail.co.uk/news/article-12697293/Hamas-leader-dismisses-Gaza-civilian-deaths-necessary-price-blood-boasts-terror-group-demonstrated-Israel-beatable-changing-Middle-East.html; Joshua Klein, "'Muslim Terrorists Are in Our Midst': Florida Imam Calls for Annihilation of All Jews," Breitbart News, May 13, 2024, https://www.breitbart.com/politics/2024/05/13/muslim-terrorists-are-in-our-midst-florida-imam-calls-for-annihilation-of-all-jews/.
31. Joel B. Pollak, "IDF Discovers Subway Tracks Underneath Northern Gaza," Breitbart News, September 3, 2024, https://www.breitbart.com/middle-east/2024/09/03/idf-discovers-subway-tracks-underneath-northern-gaza/.
32. Joel B. Pollak, "Israel Discovers Weapons Cache in Maternity Ward of Gaza's Shifa Hospital," Breitbart News, March 31, 2024, https://www.breitbart.com/middle-east/2024/03/31/israel-maternity-ward-shifa-hospital/; Joel B. Pollak, "IDF Kills 15 Terrorists in 'War Room' at UNRWA School," Breitbart News, May 14, 2024, https://www.breitbart.com/middle-east/2024/05/14/idf-kills-15-terrorists-in-war-room-at-unrwa-school/; Ian Hanchett, "Israeli Spox: Hamas Terrorists

Aren't Starving Because They Steal Aid, Pressuring Us Incentivizes Their Evil," *Breitbart News*, April 6, 2024, https://www.breitbart.com/clips/2024/04/06/israeli-spox-hamas-terrorists-arent-starving-because-they-steal-aid-pressuring-us-incentivizes-their-evil/.

33. Joel B. Pollak, "Hamas Leaders to NY Times: No Interest in Helping Palestinians in Gaza, Want 'Permanent' War Against Israel," *Breitbart News*, November 8, 2023, https://www.breitbart.com/middle-east/2023/11/08/hamas-leaders-to-ny-times-no-interest-in-helping-palestinians-gaza-want-permanent-war-against-israel/.

34. Joshua Klein, "IDF Confirms: All 3 Rescued Male Israeli Hostages Held by Gaza 'Journalist,'" *Breitbart News*, June 9, 2024, https://www.breitbart.com/politics/2024/06/09/idf-confirmed-3-male-israeli-hostages-held-gaza-journalist-hamas/.

35. Joel B. Pollak, "UN Report Confirms: Hamas Committed Rape Against Israelis, Hostages," *Breitbart News*, March 4, 2024, https://www.breitbart.com/middle-east/2024/03/04/un-report-confirms-hamas-committed-rape-against-israelis-hostages/; Joel B. Pollak, "Hamas Filmed Six Hostages Before Executions; One Clip Released by Family," *Breitbart News*, September 2, 2024, https://www.breitbart.com/middle-east/2024/09/02/hamas-filmed-six-hostages-before-executions-one-clip-released-by-family/; Joel B. Pollak, "Report: Eyewitness Says Hamas Executed Israeli Woman During Gang Rape," *Breitbart News*, November 9, 2023, https://www.breitbart.com/middle-east/2023/11/09/warning-graphic-content-eyewitness-says-hamas-executed-israeli-woman-during-gang-rape/.

36. Wendell Husebo, "Report: Israeli Soldiers Find Dead Babies After Palestinian Hamas Attack," *Breitbart News*, October 10, 2023, https://www.breitbart.com/politics/2023/10/10/report-israeli-soldiers-find-dead-babies-after-palestinian-hamas-attack/.

37. Joel B. Pollak, "UN Admits: 9 UNRWA Employees May Have Participated in October 7 Terror," *Breitbart News*, August 5, 2024. https://www.breitbart.com/middle-east/2024/08/05/un-admits-9-unrwa-employees-may-have-participated-in-october-7-terror/; Joel B. Pollak, "IDF Discovers Hamas, Islamic Jihad Command Center at UNRWA HQ," *Breitbart News*, July 12, 2024, https://www.breitbart.com/middle-east/2024/07/12/idf-discovers-hamas-islamic-jihad-command-center-at-unrwa-hq/.

38. Joel B. Pollak, "CAIR Condemns Israeli Hostage Rescue: 'Horrific Massacre,'" *Breitbart News*, June 8, 2024, https://www.breitbart.com/middle-east/2024/06/08/cair-condemns-israeli-hostage-rescue-horrific-massacre/; Joseph Goebbels, "Joseph Goebbels Quotes," AZ Quotes, accessed September 3, 2024, https://www.azquotes.com/author/5626-Joseph_Goebbels.

39. John Hayward, "W.H.O. Chief Tedros Demands Israel Stop Pursuing Hamas," *Breitbart News*, May 1, 2024, https://www.breitbart.com/middle-east/2024/05/01/w-h-o-chief-tedros-demands-israel-stop-pursuing-hamas/.

40. Joshua Klein, "BBC Mocked for 'Outrageous' Suggestion IDF Should Have Warned Gazans Before Hostage Rescue Mission," *Breitbart News*, June 10, 2024, https://www.breitbart.com/the-media/2024/06/10/bbc-mocked-for-outrageous-suggestion-idf-should-have-warned-gazans-before-hostage-rescue-mission/.

41. Joel B. Pollak, "Biden Summons Israel for Talks; Says Netanyahu's Position on Hamas Is 'Nonsense,'" *Breitbart News*, March 18, 2024, https://www.breitbart.com/middle-east/2024/03/18/biden-summons-israel-for-talks-says-netanyahus-position-on-hamas-is-nonsense/.

42. U.S. Department of Justice, "Justice Department Announces Terrorism Charges Against Senior Leaders of Hamas," press release, September 3, 2024, https://www.justice.gov/opa/pr/justice-department-announces-terrorism-charges-against-senior-leaders-hamas.
43. Saphora Smith and Emily R. Siegel, "'Never Again': Anti-Semitism Surges as Memories of Holocaust Fade," NBC News, updated January 25, 2020, https://www.nbcnews.com/news/world/never-again-anti-semitism-surges-memories-holocaust-fade-n1122081.
44. Stuart Winer, "Fear in Berlin as Star of David Scrawled at Entrances of Buildings Where Jews Reside," Times of Israel, October 15, 2023, https://www.timesofisrael.com/fear-in-berlin-as-star-of-david-scrawled-at-entrances-of-buildings-where-jews-reside/.
45. Ian Riñon, "Another Kristallnacht? Star of David Graffitied on Jewish Homes in Berlin," Headlines and Global News, updated October 15, 2023, https://www.hngn.com/articles/252954/20231015/another-kristallnacht-star-david-graffitied-jewish-homes-berlin.htm.
46. George Washington, "Fifth Annual Message of George Washington," Yale Law School, Lillian Goldman Law Library, December 3, 1793, https://avalon.law.yale.edu/18th_century/washs05.asp.
47. Ronald Reagan Presidential Library and Museum, "Peace Through Strength," accessed April 27, 2025, https://www.reaganlibrary.gov/permanent-exhibits/peace-through-strength.
48. Lawrence H. Keeley, War Before Civilization (Oxford University Press, 1996), 174.
49. René Ostberg, "Pride," Britannica, updated April 14, 2025, https://www.britannica.com/topic/pride-deadly-sin; René Ostberg, "Greed," Britannica, updated February 21, 2025, https://www.britannica.com/topic/greed; René Ostberg, "Lust," Britannica, updated April 11, 2025, https://www.britannica.com/topic/lust-deadly-sin; René Ostberg, "Envy," in Britannica, updated December 3, 2024, https://www.britannica.com/topic/envy; René Ostberg, "Gluttony," Britannica, updated April 22, 2025, https://www.britannica.com/topic/gluttony; René Ostberg, "Wrath," Britannica, updated August 20, 2024, https://www.britannica.com/topic/wrath; René Ostberg, "Sloth," Britannica, updated December 17, 2024, https://www.britannica.com/topic/sloth-behaviour.
50. The United States Strategic Bombing Survey Summary Report (European War) (United States Government Printing Office, September 30, 1945), http://www.anesi.com/ussbs02.htm, 17.
51. Curtis LeMay and McKinlay Kantor, Mission with LeMay: My Story (Doubleday, 1965), 382.
52. George Orwell, in Niall Ferguson, The War of the World: Twentieth-Century Conflict and the Descent of the West (Penguin Press, 2006), 532.
53. Cherokee tale, in Israel W. Charny, "The Nature of Man: Is Man by Nature Good, or Basically Bad?" Psychology Today, March 27, 2018, https://www.psychologytoday.com/us/blog/warrior-life/201803/the-nature-man-is-man-nature-good-or-basically-bad.

BIBLIOGRAPHY

ABC News. "Germany Set to Make Final World War I Reparation Payment." September 29, 2010. https://abcnews.go.com/International/germany-makes -final-reparation-payments-world-war/story?id=11755920.

Adena, Maja, Ruben Enikolopov, Maria Petrova, Veronica Santarosa, and Ekaterina Zhuravshaya. "Radio and the Rise of Nazis in Prewar Germany." *Quarterly Journal of Economics* (2013): 1–134. https://doi.org/10.2139/ssrn.2242446.

Alamgir, Jalal, and Bina D'Costa. "The 1971 Genocide: War Crimes and Political Crimes." *Economic and Political Weekly* 46, no. 13 (March 26–April 1, 2011): 38–41. https://www.jstor.org/stable/41152283.

Al-Nashif, Nada. "Deputy High Commissioner Updates Council on the Democratic Republic of the Congo." United Nations Human Rights Office of the High Commissioner. April 1, 2025. https://www.ohchr.org/en/statements-and-speeches/2025/04/deputy-high-commissioner-updates-council-democratic-republic-congo.

Alperovitz, Gar, Robert L. Messer, and Barton J. Bernstein. "Marshall, Truman, and the Decision to Drop the Bomb." *International Security* 16, no. 3 (Winter 1991–1992): 204–21. http://www.jstor.org/stable/2539092.

American Battlefield Trust. "Revolutionary War Strategy." Accessed April 23, 2025. https://www.battlefields.org/learn/articles/revolutionary-war-strategy.

American Battlefield Trust. "Sand Creek: Sand Creek Massacre, Chivington Massacre." Accessed April 23, 2025. https://www.battlefields.org/learn/civil-war/battles/sand-creek.

American Civil War. "Confederate Prisoner of War Camps." Accessed August 21, 2024. https://www.mycivilwar.com/pow/confederate.html.

American Civil War. "Union Prisoner of War Camps." Accessed August 21, 2024. https://www.mycivilwar.com/pow/union.html.

American Experience. "The Mountain Meadows Massacre." PBS. Accessed April 27, 2024. https://www.pbs.org/wgbh/americanexperience/features/mormons-massacre/.

American Jewish Committee. "Timeline: Key Events in the Israel-Arab and Israeli-Palestinian Conflict." Accessed April 23, 2025. https://www.ajc.org/Israel ConflictTimeline.

Amnesty International. "Iran: Blood-Soaked Secrets: Why Iran's 1988 Prison Massacres Are Ongoing Crimes Against Humanity." December 4, 2018. https://www.amnesty.org/en/documents/mde13/9421/2018/en/.

Amnesty International UK. "Iraq: Twenty Years On, Still No Justice for War Crimes by US-Led Coalition." March 20, 2023. https://www.amnesty.org.uk/press-releases/iraq-twenty-years-still-no-justice-war-crimes-us-led-coalition.

Anderson, Truman. "Incident at Baranivka." *Journal of Modern History* 71, no. 3 (September 1999): 585–623. http://www.jstor.org/stable/10.1086/235290.

Andrei, Cristina, and Decebal Nedu. "The Campaign of Marcus Atilius Regulus in Africa: Military Operations by Sea and by Land (256–255 B.C.)." *Constanta Maritime University Annals* 13 (2010). https://www.academia.edu/43696687/THE_CAMPAIGN_OF_MARCUS_ATILIUS_REGULUS_IN_AFRICA_MILITARY_OPERATIONS_BY_SEA_AND_BY_LAND_256_255_B_C.

Annan, UNations, New York. "Contacts with Informant." Outgoing code cable to Booh-Booh/Dallaire, UNAMIR Kigali. January 11, 1994. https://nsarchive2.gwu.edu/NSAEBB/NSAEBB452/docs/doc05.pdf.

Articles 231–247 and Annexes [Treaty of Versailles]. World War I Document Archives. Accessed April 23, 2025. https://wwi.lib.byu.edu/index.php/Articles_231_-_247_and_Annexes.

Articles of War: An Act for Establishing Rules and Articles for the Government of the Armies of the United States. 9th Cong. Approved April 10, 1806. https://military-justice.ca/wp-content/uploads/2019/01/U.S.-ARTICLES-OF-WAR.pdf.

Asian Women's Fund. "Who Were the Comfort Women? The Establishment of Comfort Stations." Accessed September 25, 2024. https://www.awf.or.jp/e1/facts-01.html.

Asian Women's Fund. "Who Were the Comfort Women? Observations and Directions in the Military." Accessed September 25, 2024. https://www.awf.or.jp/e1/facts-03.html.

Baar, Aaron. "Global Ad Spending on Track to Top $1T for First Time, WARC Says." MarketingDive. August 28, 2023. https://www.marketingdive.com/news/global-ad-spending-2023-2024-1t-trillion-warc/692010/.

Bachelet, Michelle. "Presentation of the Report on Civilian Deaths in the Syrian Arab Republic." United Nations Human Rights Office of the High Commssioner. June 30, 2022. https://www.ohchr.org/en/statements/2022/06/presentation-report-civilian-deaths-syrian-arab-republic.

Bailey, Ronald H., and Editors of Time-Life Books. *The Home Front, U.S.A.* Time-Life Books, 1977.

Balaban, Samantha. "George Takei 'Lost Freedom' Some 80 Years Ago—Now He's Written That Story for Kids." NPR. April 20, 2024. https://www.npr.org/2024/04/20/1245844347/george-takei-my-lost-freedom-picture-book.

Ballotpedia. "Splits Between the Electoral College and Popular Vote." Accessed July 30, 2024. https://ballotpedia.org/Splits_between_the_Electoral_College_and_popular_vote.

Barlow, Jeffrey G. "The Navy's Atlantic War Learning Curve." *Naval History* 22, no. 3 (June 2008). https://www.usni.org/magazines/naval-history-magazine/2008/june/navys-atlantic-war-learning-curve.

Barras, V., and G. Greub. "History of Biological Warfare and Bioterrorism." *Clinical Microbiology and Infection* 20, no. 6 (June 2014): 497–502. https://doi.org/10.1111/1469-0691.12706.

Barrett, David D. "Japan's Hellish Unit 731." *WWII Quarterly: Journal of the Second World War* 10, no. 1 (Fall 2018). https://warfarehistorynetwork.com/article/japans-hellish-unit-731/.

Bartov, Omer. *Mirrors of Destruction: War, Genocide, and Modern Identity.* Oxford University Press, 2000.

Bartov, Omer. "The Myths of the Wehrmacht." *History Today* 42, no. 4 (April 1992): 30–36.https://www.historytoday.com/archive/myths-wehrmacht.

BBC News. "Who Are the Uyghurs and Why Is China Being Accused of Genocide?" May 24, 2022. https://www.bbc.com/news/world-asia-china-22278037.

BBC News. "Why Is There Communal Violence in Myanmar?" July 3, 2014. https://www.bbc.com/news/world-asia-18395788.

BBC Open Centre, Hull. "The Alexandra Hospital Massacre." BBC. January 14, 2006. https://www.bbc.co.uk/history/ww2peopleswar/stories/60/a8515460.shtml.

Beevor, Antony. *The Battle for Spain: The Spanish Civil War 1938–1939.* Penguin Books, 2006.

Beevor, Antony. *Berlin: The Downfall 1945.* Penguin Books, 2002.

Belleau, Jean-Philippe. "Massacres Perpetrated in the 20th Century in Haiti." Sciences Po. April 2, 2008. https://www.sciencespo.fr/mass-violence-war-massacre-resistance/en/document/massacres-perpetrated-20th-century-haiti.html.

Bennett, G. H. "The 1942 Laconia Order: The Murder of Shipwrecked Survivors and the Allied Pursuit of Justice 1945–46." *Law, Crime and History* 1, no. 1 (Spring 2011): 16–34. https://go.gale.com/ps/i.do?id=GALE%7CA310518860&sid=googleScholar&v=2.1&it=r&linkaccess=abs&issn=20459238&p=AONE&sw=w&userGroupName=anon%7E459a5622&aty=open-web-entry.

Beorn, Waitman W. "A Calculus of Complicity: The *Wehrmacht*, the Anti-Partisan War, and the Final Solution in White Russia, 1941–42." *Central European History* 44, no. 2 (June 2011): 308–37. https://doi.org/10.1017/S0008938911000057.

Berreman, Joel V. "Assumptions About America in Japanese War Propaganda to the United States." *American Journal of Sociology* 54, no. 2 (September 1948): 108–17. http://www.jstor.org/stable/2771359.

Beyer, Susanne. "German Woman Writes Ground-Breaking Account of WW2 Rape." *Der Spiegel International*, February 26, 2010. https://www.spiegel.de/international/germany/harrowing-memoir-german-woman-writes-ground-breaking-account-of-ww2-rape-a-680354.html.

Biddle, Tami Davis. "Air Power Theory: An Analytical Narrative from the First World War to the Present." In *U.S. Army War College Guide to National Security Policy and Strategy*, 2nd ed., edited by J. Boone Martholomees Jr., 331–60. United States Army War College, 2008. https://press.armywarcollege.edu/cgi/viewcontent.cgi?article=1077&context=monographs.

Biddle, Tami Davis. "Dresden 1945: Reality, History, and Memory." *Journal of Military History* 72, no. 2 (April 2008): 413–49. https://doi.org/10.1353/jmh.2008.0074.

Biddle, Tami Davis. *Rhetoric and Reality in Air Warfare.* Princeton University Press, 2002.

Biess, Frank. "Fears of Retribution in Post-War Germany." National WWII Museum New Orleans. September 20, 2021. https://www.nationalww2museum.org/war/articles/fears-of-retribution-in-post-war-germany.

Bill of Rights Institute. "A Deep Stain on the American Character: John Marshall and Justice for Native Americans—Handout A: Narrative." Accessed April 23, 2025.

https://billofrightsinstitute.org/activities/a-deep-stain-on-the-american-character -john-marshall-and-justice-for-native-americans-handout-a-narrative.

Bix, Herbert P. "Remembering the Konoe Memorial: The Battle of Okinawa and Its Aftermath." *Asia-Pacific Journal: Japan Focus* 13, no. 8 (February 23, 2015). https:// apjjf.org/2015/13/8/Herbert-P.-Bix/4821.html.

Black, Jeremy. *The Age of Total War, 1860–1945.* Praeger Security International, 2006.

Black, Jeremy. *A Century of Conflict: War, 1914–2014.* Oxford University Press, 2015.

Blair, Clay. *Hitler's U-Boat War: The Hunters, 1939–1942.* Random House, 1996.

Blakemore, Erin. "A Ship of Jewish Refugees Was Refused US Landing in 1939. This Was Their Fate." History. Updated April 15, 2025. https://www.history.com/news/ wwii-jewish-refugee-ship-st-louis-1939.

Blakemore, Erin. "The U.S. Forcibly Detained Native Alaskans During World War II." *Smithsonian Magazine,* February 22, 2017. https://www.smithsonianmag.com/ smart-news/us-forcibly-detained-native-alaskans-during-world-war-ii-180962239/.

Blau, George E., ed., trans. *The German Campaigns in the Balkans (Spring 1941).* Center of Military History, United States Army, 1986.

Bogner, Matilda. "Situation in Ukraine." United Nations Human Rights Office of the High Commissioner. March 25, 2022. https://www.ohchr.org/en/ statements/2022/03/situation-ukraine.

Bonaparte, Napoleon. *Napoleon on the Art of War.* Edited and translated by Jay Luvaas. Touchstone, 2001.

Borger, Julian, and Kareem Shaheen. "Russia Accused of War Crimes in Syria at UN Security Council Session." *Guardian,* September 26, 2016. https://www.theguardian .com/world/2016/sep/25/russia-accused-war-crimes-syria-un-security-council-aleppo.

Bornmann, Robert C., and Jan K. Herman. "Operating Under Pressure." *Naval History* 10, no. 4 (August 1996). https://www.usni.org/magazines/naval-history -magazine/1996/august/operating-under-pressure.

Bowden, Mark. *Guests of the Ayatollah: The Iran Hostage Crisis: The First Battle in America's War with Militant Islam.* Atlantic Monthly Press, 2006.

Bowen, James. "War Crimes Committed by the Imperial Japanese Navy." Pacific War Historical Society. Accessed September 23, 2024. https://www.pacificwar.org/Jap WarCrimes/TenWarCrimes/WarCrimes_Jap_Navy.html.

Breitman, Richard. "Hitler and Genghis Khan." *Journal of Contemporary History* 25, nos. 2/3 (May–June 1990): 337–51. http://www.jstor.org/stable/260736.

Brenckle, Matthew. "British and American Naval Prisoners of War During the War of 1812." USS Constitution Museum. 2005. https://ussconstitutionmuseum.org/ wp-content/uploads/2020/05/British-and-American-Naval-Prisoners-of-War-During -the-War-of-1812.pdf.

Brill, Michael. "Part I: 'We Attacked Them with Chemical Weapons and They Attacked Us with Chemical Weapons': Iraqi Records and the History of Iran's Chemical Weapons Program." *Sources and Methods* (blog). Wilson Center. March 29, 2022. https://www.wilsoncenter.org/blog-post/part-i-we-attacked-them-chemical -weapons-and-they-attacked-us-chemical-weapons-iraqi.

Brinsfield, John W. "The Military Ethics of General William T. Sherman: A Reassessment." *Parameters* 12, no. 1 (July 4, 1982): 36–48. https://doi.org/10.55540/ 0031-1723.1280.

Browning, Christopher R. "Dangers I Didn't See Coming: 'Tyranny of the Minority' and an Irrelevant Press." *Vox,* January 18, 2017. https://www.vox.com/the-big-idea/ 2017/1/18/14303960/tyranny-minority-trump-democracy-gerrymande-media.

Browning, Christopher R. *Ordinary Men: Reserve Police Battalion 101 and the Final Solution in Poland*. Harper Perennial, 1998.

Bryce, J. B. *Report of the Committee on Alleged German Outrages Appointed by His Britannic Majesty's Government and Presided over by the Right Hon. Viscount Bryce.* Macmillan, 1915. https://archive.org/details/reportofcommitte00grea/page/n3/mode/2up.

Buenviaje, Dino E., and James F. Willis. "Laws of Warfare: World War I." In *World at War: Understanding Conflict and Society*. ABC-CLIO, 2024.

Bunyan, Rachael, and Elena Salvoni. "Hamas Vows to Repeat October 7 Attacks and Bring About the 'Annihilation' of Israel While Cynically Saying It 'Did Not Want to Hurt Civilians' as It Slaughtered 1,400 People 'but There Were Complications on the Ground.'" *Daily Mail*, November 1, 2023. https://www.dailymail.co.uk/news/article-12697293/Hamas-leader-dismisses-Gaza-civilian-deaths-necessary-price-blood-boasts-terror-group-demonstrated-Israel-beatable-changing-Middle-East.html.

Burch, Ernest S., Jr. "Traditional Native Warfare in Western Alaska." In *North American Indigenous Warfare and Ritual Violence*, edited by Richard J. Chacon and Rubén G. Mendoza, 11–29. University of Arizona Press, 2013.

Burleigh, Michael. "Euthanasia and the Third Reich." *History Today* 40, no. 2 (1990): 11–16. https://www.historytoday.com/archive/euthanasia-and-third-reich.

Bytwerk, Randall L. "The Argument for Genocide in Nazi Propaganda." *Quarterly Journal of Speech* 91, no. 1 (February 2005): 37–62. http://dx.doi.org/10.1080/00335630500157516.

Bytwerk, Randall L. *Bending Spines: The Propagandas of Nazi Germany and the German Democratic Republic*. Michigan State University Press, 2004.

Bytwerk, Randall, ed., trans. *Landmark Speeches of National Socialism*. Texas A&M University Press, 2008.

BYU Department of History. "The Mountain Meadows Massacre." BYU College of Family, Home, and Social Sciences. 2024. https://history.byu.edu/mountain meadowsmassacre.

Caplan, Joshua. "Live Updates: Hamas Launches Unprecedented Terror Attack on Israel." *Breitbart News*, October 7, 2023. https://www.breitbart.com/middle-east/2023/10/07/live-updates-hamas-launches-unprecedented-terror-attack-on-israel/.

Carbonnel, Alissa de, and Alessandra Prentice. "Armed Men Seize Two Airports in Ukraine's Crimea, Yanukovich Reappears." Reuters. February 28, 2014. https://www.yahoo.com/news/armed-standoff-pro-russian-region-raises-ukraine-tension-033318395.html.

Carlson, Cody. "This Week in History: Lindbergh Gives Infamous 'Who Are the War Agitators?' Speech." *Deseret News*, September 12, 2013. https://www.deseret.com/2013/9/12/20525433/this-week-in-history-lindbergh-gives-infamous-who-are-the-war-agitators-speech/.

Carrigan, William D., and Clive Webb. "The Lynching of Persons of Mexican Origin or Descent in the United States, 1848 to 1928." *Journal of Social History* 37, no. 2 (Winter 2003): 411–38. https://www.jstor.org/stable/3790404.

Cartwright, Mark. "Greek Fire." World History Encyclopedia. November 14, 2017. https://www.worldhistory.org/Greek_Fire/.

Casert, Raf, and Dusan Stojanovic. "UN Court Overturns Acquittal of Serb Ultranationalist." Associated Press. Updated April 11, 2018. https://apnews.com/general-news-2774f56f5628497790fa1bbf249dbcac.

Center for International Relations and International Security. "Asymmetric Warfare." Accessed October 19, 2024. https://www.ciris.info/learningcenter/asymmetric-warfare/.

Centers for Disease Control and Prevention (CDC). "Mustard Gas." September 6, 2024. https://www.cdc.gov/chemical-emergencies/chemical-fact-sheets/mustard-gas.html.

Centers for Disease Control and Prevention (CDC). "Phosgene." September 6, 2024. https://emergency.cdc.gov/agent/phosgene/basics/facts.asp.

CFR.org Editors. "Syria's Civil War: The Descent into Horror." Council on Foreign Relations. December 20, 2024. https://www.cfr.org/article/syrias-civil-war.

Chacon, Richard J., and Rubén G. Mendoza, eds. *North American Indigenous Warfare and Ritual Violence*. University of Arizona Press, 2013.

Chambers, John Whiteclay, II, and G. Kurt Piehler, eds. *Major Problems in American Military History*. Houghton Mifflin, 1999.

Chang, Iris. *The Rape of Nanking: The Forgotten Holocaust of World War II*. Penguin Books, 1998.

Charny, Israel W. "The Nature of Man: Is Man by Nature Good, or Basically Bad?" *Psychology Today*, March 27, 2018. https://www.psychologytoday.com/us/blog/warrior-life/201803/the-nature-man-is-man-nature-good-or-basically-bad.

Charters, Erica. "Military Medicine and the Ethics of War: British Colonial Warfare During the Seven Years War (1756–63)." *Canadian Bulletin of Medical History* 27, no. 2 (Fall 2010): 273–98. https://doi.org/10.3138/cbmh.27.2.273.

Chase, Paul A., and James Wood II. "Prisoners of War (POWs) During the American Revolution." Sons of the American Revolution. Accessed August 21, 2024. https://www.sar.org/wp-content/uploads/2020/06/Prisoner-of-War-Briefing-by-Paul-Chase.pdf.

Chickering, Roger, Stig Förster, and Bernd Greiner, eds. *A World at Total War: Global Conflict and the Politics of Destruction, 1937–1945*. Cambridge University Press, 2010.

China: The Dominated and the Dominating: A Summary of Modern Events and Biographies. "Battle of Shanghai." Accessed September 11, 2024. https://chinahistory.co.uk/battleofshanghai.html.

Christian Times. "Weekly Forum Editorial." June 2, 2024. https://christiantimes.org.hk/Common/Reader/Version/Show.jsp?Pid=2&Version=1918&Charset=big5_hkscs.

Cirillo, Vincent J. "Two Faces of Death: Fatalities from Disease and Combat in America's Principal Wars, 1775 to Present." *Perspectives in Biology and Medicine* 51, no. 1 (Winter 2008): 121–33. https://doi.org/10.1353/pbm.2008.0005.

Cockburn, Patrick. "Churchill's Secret Chemical War." *Independent*, July 26, 2020. https://www.independent.co.uk/independentpremium/long-reads/second-world-war-churchill-secret-mustard-gas-operation-sea-lion-a9632456.html.

Coffey, Thomas M. *Iron Eagle: The Turbulent Life of General Curtis LeMay*. Avon Books, 1988.

Cohen, Jennie. "6 Things You May Not Know About the Spanish American War." History. Updated February 18, 2025. https://www.history.com/news/6-things-you-may-not-know-about-the-spanish-american-war.

Combined Chiefs of Staff. Memorandum by the Combined Chiefs of Staff. Casablanca. January 21, 1943. https://history.state.gov/historicaldocuments/frus1941-43/d412.

Commission on Wartime Relocation and Internment of Civilians. "War and Evacuation in Alaska." In *Personal Justice Denied*, 317–59. University of Washington Press, 1997.

Conant, Eve. "Behind the Headlines: Who Are the Crimean Tatars?" *National Geographic*, March 15, 2014. https://www.nationalgeographic.com/history/article/140314-crimea-tatars-referendum-russia-muslim-ethnic-history-culture.

Conant, Jennet. "How a Chemical Weapons Disaster in WWII Led to a U.S. Cover-Up—and a New Cancer Treatment." *Smithsonian Magazine*, September 2020. https://www.smithsonianmag.com/history/bombing-and-breakthrough-180975505/.

Conant, Jennet. "How a WWII Disaster—and Cover-Up—Led to a Cancer Treatment Breakthrough." History. Updated January 31, 2025. https://www.history.com/news/wwii-disaster-bari-mustard-gas.

Conetta, Carl. *The Wages of War: Iraqi Combatant and Noncombatant Fatalities in the 2003 Conflict*. Project on Defense Alternatives, Commonwealth Institute, October 20, 2003. https://www.comw.org/pda/fulltext/0310rm8.pdf.

Convention Relative to the Treatment of Prisoners of War. International Humanitarian Law Databases. July 27, 1929. https://ihl-databases.icrc.org/assets/treaties/305-IHL-GC-1929-2-EN.pdf.

Corley, Christopher L. "Acts of Atrocity: Effects on Public Opinion Support During War or Conflict." Master's thesis, Naval Postgraduate School, December 2007. https://web.archive.org/web/20220507164026/https://apps.dtic.mil/dtic/tr/fulltext/u2/a475745.pdf.

Correspondence with the German Government Respecting the Death by Burning of J. P. Genower, Able Seaman, When Prisoner of War at Brandenburg Camp. Parliamentary Paper, Misc. no. 6. London, 1918. https://dn790003.ca.archive.org/0/items/correspondenceger00grea/correspondenceger00grea.pdf.

Corum, James S. *The Luftwaffe: Creating the Operational Air War, 1918–1940*. Univeristy Press of Kansas, 1997.

Corum, James S. *Wolfram von Richthofen: Master of the German Air War*. University Press of Kansas, 2008.

Coulter, C. R. "Our Policy in the Philippines." *San Francisco Call*, August 1, 1899. https://cdnc.ucr.edu/?a=d&d=SFC18990801.2.111.

Cretu, Doina Anca. "Health, Disease, Mortality; Demographic Effects." International Encyclopedia of the First World War. Updated November 17, 2020. https://encyclopedia.1914-1918-online.net/article/health-disease-mortality-demographic-effects/.

Cumming-Bruce, Nick. "'Welcome to Hell': U.N. Panel Says Russian War Crimes Are Widespread." *New York Times*, March 15, 2024. https://www.nytimes.com/2024/03/15/world/europe/russia-war-crimes.html.

Dallaire/UNAMIR/Kigali. "Request for Protection for Informant." Outgoing code cable to Baril/DPKO/UNations New York. January 11, 1994. https://nsarchive2.gwu.edu/NSAEBB/NSAEBB452/docs/doc03.pdf.

Davis, Robert Scott. "Andersonville Prison." New Georgia Encyclopedia. Updated July 15, 2020. https://www.georgiaencyclopedia.org/articles/history-archaeology/andersonville-prison/.

Davis, Stephen. "Atlanta Campaign." New Georgia Encyclopedia. Updated September 17, 2018. https://www.georgiaencyclopedia.org/articles/history-archaeology/atlanta-campaign/.

Davis, Stephen. "The Burning of Atlanta: What Really Happened?" *Civil War News*, September 22, 2022. https://www.historicalpublicationsllc.com/civilwarnews/the-burning-of-atlanta-what-really-happened/article_0f304440-34f0-11ed-acfc-7b4f80e30e60.html.

Demm, Eberhard, and Christopher H. Sterling. "Propaganda: World War I." In *World at War: Understanding Conflict and Society*. ABC-CLIO, 2024.

denton.2@osu.edu. "Who Killed More: Hitler, Stalin, or Mao." Modern Chinese Literature and Culture Resource Center, Ohio State University. February 8, 2018. https://u.osu.edu/mclc/2018/02/08/who-killed-more-hitler-stalin-or-mao/.

Dewan, Angela, and Hilary McGann. "Amnesty International Says US-Led Strikes on Raqqa May Amount to War Crimes." CNN. June 5, 2018. https://www.cnn .com/2018/06/05/middleeast/us-led-coalition-raqqa-war-crimes-intl/index.html.

Diehl, Jörg. "Practicing Blitzkrieg in Basque Country." *Spiegel International*, April 26, 2007. http://www.spiegel.de/international/europe/0,1518,479675,00.html.

Directive to Commander-in-Chief of United States Forces of Occupation Regarding the Military Government of Germany. April 1945. https://usa.usembassy.de/etexts/ ga3-450426.pdf.

Directives for the Treatment of Political Commissars ("Commissar Order") (June 6, 1941). German History in Documents and Images. Accessed September 23, 2024. https://germanhistorydocs.org/en/nazi-germany-1933-1945/directives-for-the-treat ment-of-political-commissars-quot-commissar-order-quot-june-6-1941.

Doenecke, Justus D. "Explaining the Antiwar Movement, 1939–1941: The Next Assignment." *Journalof Libertarian Studies* 8, no. 1 (Winter 1986): 139–62. https:// mises.org/journal-libertarian-studies/explaining-antiwar-movement-1939-1941 -next-assignment?d7_alias_migrate=1.

Douhet, Giulio. "The Command of the Air." In *Roots of Strategy: Book 4*, edited by David Jablonsky, 263–408. Stackpole Books, 1999.

Dower, John W. *War Without Mercy: Race and Power in the Pacific War*. Pantheon Books, 1986.

Downes, Alexander B. "Military Culture and Civilian Victimization: The Allied Bombing of Germay in World War II." In *Civilians and Modern War: Armed Conflict and the Ideology of Violence*, edited by Daniel Rothbart, Karina V. Korostelina, and Mohammed D. Cherkaoui, 72–95. Routledge, 2012.

Duckworth, Tammy, Kirsten Gillibrand, and John Thune. Congressional Record—Senate, no. S5648. November 29, 2023. https://www.congress.gov/118/ crec/2023/11/29/169/196/CREC-2023-11-29-pt1-PgS5648.pdf.

Duelfer, Charles, and United States Central Intelligence Agency. *Comprehensive Report of the Special Advisor to the DCI on Iraq's WMD with Addendums*. Vol. 3. US Government Office, September 30, 2004. https://www.govinfo.gov/content/pkg/ GPO-DUELFERREPORT/pdf/GPO-DUELFERREPORT-3.pdf.

Editorial Team. "The Japanese 'Kill 100 People with Sword' Contest in 1937." *AR-Gunners Magazine*, August 29, 2015. https://www.argunners.com/the-japanese -kill-100-people-with-sword-contest-in-1937/.

Editorial Team. "Understanding Mutually Assured Destruction and Its Implications." Total Military Insight. July 20, 2024. https://totalmilitaryinsight.com/ mutually-assured-destruction/.

Editors of Encyclopaedia Britannica. "Agent Orange." Britannica. Updated March 25, 2025. https://www.britannica.com/science/Agent-Orange.

Editors of Encyclopaedia Britannica. "Causes and Effects of the Spanish-American War." Britannica. Accessed April 26, 2025. https://www.britannica.com/summary/ Causes-and-Effects-of-the-Spanish-American-War.

Editors of Encyclopaedia Britannica. "Geneva Gas Protocol." Britannica. Updated February 28, 2023. https://www.britannica.com/event/Geneva-Gas-Protocol.

Editors of Encyclopaedia Britannica. "Greek Fire." Britannica. Updated January 31, 2025. https://www.britannica.com/technology/Greek-fire.

Editors of Encyclopaedia Britannica. "Iran-Iraq War." Britannica. Updated April 25, 2025. https://www.britannica.com/event/Iran-Iraq-War.

Editors of Encyclopaedia Britannica. "Nationalist Collapse and the Establishment of the People's Republic of China (1949)." Britannica. Updated February 11, 2025. https://www.britannica.com/event/Chinese-Civil-War/Nationalist-col lapse-and-the-establishment-of-the-Peoples-Republic-of-China-1949.

Editors of Encyclopaedia Britannica. "Okinawa." Britannica. Updated April 25, 2025. https://www.britannica.com/place/Okinawa-prefecture-Japan.

Editors of Encyclopaedia Britannica. "Spanish Civil War." Britannica. Updated April 3, 2025. https://www.britannica.com/event/Spanish-Civil-War.

Editors of Encyclopaedia Britannica. "Syrian Civil War." Britannica. Updated April 16, 2025. https://www.britannica.com/event/Syrian-Civil-War.

Editors of Encyclopaedia Britannica. "Tatar." Britannica. Updated April 25, 2025. https://www.britannica.com/topic/Tatar.

Editors of Encyclopaedia Britannica. "Tiananmen Square Incident." Britannica. Updated April 20, 2025. https://www.britannica.com/event/Tiananmen-Square -incident.

Elshakankiri, Hamza. "Prussian Militarism and the German Wars of Unification." *Armstrong Undergraduate Journal of History* 11, no. 2 (October 2021): 40–49. https://doi.org/10.20429/aujh.2021.110203.

Emery, Theo. *Hellfire Boys: The Birth of the U.S. Chemical Warfare Service and the Race for the World's Deadliest Weapons.* Little, Brown, 2017.

Esteban, Rolando. "Cannibalism Among Japanese Soldiers in Bukidnon, Philippines, 1945–47." *Asian Studies: Journal of Critical Perspectives on Asia* 52, no. 1 (2016): 63–102. https://www.asj.upd.edu.ph/mediabox/archive/ASJ_52_1_2016/Cannibal ism_Japanese_Soldiers_Bukidnon_1945_1947_Esteban_Rolando2.pdf.

Ettinger, Shmuel. "The Balfour Declaration of 1917." My Jewish Learning. Accessed August 23, 2024. https://www.myjewishlearning.com/article/the-balfour -declaration/.

Evans, Marissa. "Darfur Genocide (2003–)." BlackPast. March 25, 2009. https://www .blackpast.org/global-african-history/darfur-genocide-2003/.

Evans, Richard J. *The Coming of the Third Reich.* Penguin Books, 2003.

Everts, Sarah. "A Brief History of Chemical War." *Distillations Magazine*, May 12, 2015. https://www.sciencehistory.org/stories/magazine/a-brief-history-of-chemical-war/.

Everts, Sarah. "The Nazi Origins of Deadly Nerve Gases." *C&EN Global Enterprise* 94, no. 41 (October 17, 2016): 26–28. https://doi.org/10.1021/ CEN-09441-SCITECH2.

Executive Order No. 9066. Resulting in Japanese-American Incarceration. 1942. https://www.archives.gov/milestone-documents/executive-order-9066.

Facing History and Ourselves. "America and the Holocaust." Updated February 26, 2021. https://www.facinghistory.org/resource-library/america-holocaust.

Facts and Details. "Chinese Invasion of Tibet in 1950 and Its Aftermath." Updated September 2022. https://factsanddetails.com/china/cat6/sub32/entry-8413.html.

Facts and Details. "Plague Bombs and Gruesome Experiments at Unit 731." Updated September 2016. https://factsanddetails.com/asian/ca67/sub426/entry-5518.html.

Farmer, Sarah. *Martyred Village: Commemorating the 1944 Massacre at Oradour-Sur-Glane.* University of California Press, 1999.

Feifer, George. "The Rape of Okinawa." *World Policy Journal* 17, no. 3 (2000): 33–40. https://doi.org/10.1215/07402775-2000-4009.

Felton, Mark. "Why Were the Japanese So Cruel in World War II?" HistoryNet. November 6, 2017. https://www.historynet.com/a-culture-of-cruelty/.

Feltus, Pamela. "Aerial Reconnaissance in World War I." U.S. Centennial of Flight Commission. Accessed June 6, 2024. https://www.centennialofflight.net/essay/Air_Power/WWI-reconnaissance/AP2.htm.

Fenn, Elizabeth A. "Biological Warfare in Eighteenth-Century North America: Beyond Jeffery Amherst." *Journal of American History* 86, no. 4 (March 2000): 1552–80. https://doi.org/10.2307/2567577.

Ferguson, Niall. *The War of the World: Twentieth-Century Conflict and the Descent of the West.* Penguin Press, 2006.

Ferguson, R. Brian. "War Is *Not* Part of Human Nature." *Scientific American* 319, no. 3 (September 1, 2018): 76. https://www.scientificamerican.com/article/war-is-not-part-of-human-nature/.

Fischer, Fritz. *Germany's Aims in the First World War.* W. W. Norton, 1967.

Fisher, Max. "Americans Have Forgotten What We Did to North Korea." *Vox*, August 3, 2015. https://www.vox.com/2015/8/3/9089913/north-korea-us-war-crime.

Fitz-Gibbon, Jorge. "Arabic Copy of Adolf Hitler's 'Mein Kampf' Found Inside Child's Room in Gaza." *New York Post*, November 12, 2023. https://nypost.com/2023/11/12/news/arabic-copy-of-adolf-hitlers-mein-kampf-found-inside-childs-room-in-gaza/.

Flavion, Gary. "Civil War Prison Camps." American Battlefield Trust. Accessed August 21, 2024. https://www.battlefields.org/learn/articles/civil-war-prison-camps.

Foch, Ferdinand. "Ferdinand Foch 1851–1929." In *Oxford Essential Quotations*, 5th ed., edited by Susan Ratcliffe. Oxford University Press, 2017. https://www.oxfordreference.com/display/10.1093/acref/9780191843730.001.0001/q-oro-ed5-00004492?print.

Fondren, Elisabeth, and John Maxwell Hamilton. "The Universal Laws of Propaganda: World War I and the Origins of Government Manufacture of Opinion." *Journal of Intelligence History* 22, no. 1 (2023): 1–19. https://doi.org/10.1080/16161262.2022.2036498.

Foote, Caleb. *Outcasts! The Story of America's Treatment of Her Japanese-American Minority.* Fellowship of Reconciliation, 1943. https://archive.org/details/outcastsstoryofa00foot/.

Forrest, George. "The First Sacred War." *Bulletin de Correspondance Hellénique* 80, no. 1 (1956): 33–52. https://doi.org/10.3406/bch.1956.2408.

Fraga, Kaleena. "Inside the Chichijima Incident, George H. W. Bush's Harrowing Escape from Cannibal Enemies During World War II." Edited by John Kuroski. All That's Interesting. Updated November 7, 2023. https://allthatsinteresting.com/george-bush-cannibalized-chichijima-incident.

Fraga, Kaleena. "Who Killed the Most People in History? It's Not as Straightforward as You Think." Edited by Adam Farley. All That's Interesitng. July 29, 2022. https://allthatsinteresting.com/who-killed-the-most-people-in-history.

Frischknecht, Friedrich. "The History of Biological Warfare: Human Experimentation, Modern Nightmares and Lone Madmen in the Twentieth Century." *EMBO Reports* 4, no. S1 (June 1, 2003): S47–52. https://doi.org/10.1038/sj.embor.embor849.

Fritz, Stephen G. *Ostkrieg: Hitler's War of Extermination in the East.* University Press of Kentucky, 2011.

Fritzsche, Peter. *Life and Death in the Third Reich.* Harvard University Press, 2008.

Frothingham, Arthur L. *National Security League Handbook of War Facts and Peace Problems*. 4th ed. National Security League, 1919. https://books.google.com/books?id=qVKQAAAAMAAJ&pg=PA1#v=onepage&q&f=true.

Fuller, Thomas. "Crisis in Myanmar over Buddhist-Muslim Clash." *New York Times*, June 10, 2012. https://www.nytimes.com/2012/06/11/world/asia/state-of-emergen cy-declared-in-western-myanmar.html.

Fussell, Paul, and Michael Walzer. "A Defense of the Atomic Bomb and a Dissent." In *Major Problems in American Military History*, edited by John Whiteclay Chambers II and G. Kurt Piehler. Houghton Mifflin, 1999.

Gall, Carlotta, and Andrew E. Kramer. "In the Kyiv Suburb of Bucha, 'They Shot Everyone They Saw.'" *New York Times*, April 3, 2022. https://web.archive.org/web/20220412065516/https://www.nytimes.com/2022/04/03/world/europe/ukraine -russia-war-civilian-deaths.html.

Gambino-Shirley, Kelly. "Napoleon's Missed Opportunities to Maintain Combat Forces Through Medical Innovations and Battling the Hidden Enemy." Air Command and Staff College, December 7, 2011. https://apps.dtic.mil/sti/citations/AD1019074.

Ganguly, Sumit. "Pakistan's Forgotten Genocide—A Review Essay." *International Security* 39, no. 2 (Fall 2014): 169–80. https://doi.org/10.1162/ISEC_a_00175.

Garamone, Jim. "April 1917: America Entered the First World War." U.S. Army. April 7, 2017. https://www.army.mil/article/184897/april_1917_america_entered _the_first_world_war.

Gaughan, Anthony. "Execute Against Japan." Faculty Lounge. December 7, 2018. https://www.thefacultylounge.org/2018/12/execute-against-japan.html.

Gazette. "The Gazette Hall of Fame: Sir Howard Kingsley Wood." Accessed July 1, 2024. https://www.thegazette.co.uk/all-notices/content/224.

Gellately, Robert. *Backing Hitler: Consent and Coercion in Nazi Germany*. Oxford University Press, 2001.

Genos Center. "The Cambodian Genocide: A Look at the Killing Fields." July 21, 2023. https://genoscenter.org/the-cambodian-genocide-a-look-at-the-killing-fields/.

Gergen, Thomas. "The Peace of God and Its Legal Practice in the Eleventh Century." *Cuadernos de Historia del Derecho* 9 (July 2002): 11–27. https://www.researchgate .net/publication/39283239_The_Peace_of_God_its_legal_practice_in_the_Eleventh _Century.

Gershon, Livia. "The US Propaganda Machine of World War I." *JSTOR Daily*, November 17, 2023. https://daily.jstor.org/the-us-propaganda-machine-of-world-war-i/.

Gettleman, Jeffrey. "Disputed Vote Plunges Kenya into Bloodshed." *New York Times*, December 31, 2007. https://www.nytimes.com/2007/12/31/world/africa/31kenya .html.

Geyer, Michael. "German Strategy in the Age of Machine Warfare, 1914–1945." In *Makers of Modern Strategy: From Machiavelli to the Nuclear Age*, edited by Peter Paret, Gordon A. Craig, and Felix Gilbert, 527–97. Princeton University Press, 1986.

Gilbert, Felix. "Machiavelli: The Renaissance of the Art of War." In *Makers of Modern Strategy: From Machiavelli to the Nuclear Age*, edited by Peter Paret, Gordon A. Craig, and Felix Gilbert, 11–31. Princeton University Press, 1986.

Gill, Harold B., Jr. "Colonial Germ Warfare." *Colonial Williamsburg Journal* 26, no. 1 (Spring 2004): 18–23. https://research.colonialwilliamsburg.org/Foundation/jour nal/Spring04/warfare.cfm.

Gilmore, James R. "Our Visit to Richmond." *Atlantic Monthly* 14, no. 83 (September 1864): 372–83. https://www.gutenberg.org/ebooks/20350.

Gladwin, Lee A. "American POWs on Japanese Ships Take a Voyage into Hell." *Prologue Magazine* 35, no. 4 (Winter 2003). https://www.archives.gov/publications/prologue/2003/winter/hell-ships.

Goebbels, Joseph. "Joseph Goebbels Quotes." AZ Quotes. Accessed September 3, 2024. https://www.azquotes.com/author/5626-Joseph_Goebbels.

Goebbels, Joseph. "The Radio as the Eighth Great Power." German Propaganda Archive. Updated January 10, 2024. http://research.calvin.edu/german-propaganda-archive/goeb56.htm.

Goebbels, Joseph. "Two Speeches on the Tasks of the Reich Ministry for Popular Enlightenment and Propaganda (March 15/March 25, 1933)." Edited and translated by Jeremy Noakes and Geoffrey Pridham. German History in Documents and Images. Accessed April 26, 2025. http://germanhistorydocs.ghi-dc.org/docpage.cfm?docpage_id=2431.

Gordon, Gregory S. *Atrocity Speech Law: Foundation, Fragmentation, Fruition.* Oxford University Press, 2017.

Green, David B. "This Day in Jewish History: 1940: Italy Bombs Tel Aviv During WWII." *Haaretz*, September 9, 2013. https://www.haaretz.com/jewish/2013-09-09/ty-article/1940-italy-bombs-tel-aviv-in-wwii/0000017f-f473-d044-adff-f7fb19a10000.

Gregory, Adrian. *A War of Peoples, 1914–1919.* Oxford University Press, 2014.

Gross, Daniel A. "The U.S. Government Turned Away Thousands of Jewish Refugees, Fearing That They Were Nazi Spies." *Smithsonian Magazine*, November 18, 2015. https://www.smithsonianmag.com/history/us-government-turned-away-thousands-jewish-refugees-fearing-they-were-nazi-spies-180957324/.

Grosscup, Beau. *Strategic Terror: The Politics and Ethics of Aerial Bombardment.* Zed Books, 2006.

Gursoy, Berke. "The Eagle's Rise: Napoleon Bonaparte's First Campaign and the Birth of Napoleonic Warfare." *Armstrong Undergraduate Journal of History* 9, no. 1 (April 2019): 17–32. https://doi.org/10.20429/aujh.2019.090102.

Habib, Haroon. "In East Pakistan in 1971: A 'Forgotten' Genocide." *Frontline*, January 30, 2022. https://frontline.thehindu.com/world-affairs/in-east-pakistan-in-1971-a-forgotten-genocide-bangladesh-liberation-war/article38307183.ece.

Hadamovsky, Eugen. "The Living Bridge: On the Nature of Radio Warden Activity." German Propaganda Archive. 2006. http://research.calvin.edu/german-propaganda-archive/hada3.htm.

The Hague Conventions of 1899 (II) and 1907 (IV) Respecting the Laws and Customs of War on Land. Pamphlet no. 5. Carnegie Endowment for International Peace, 1915. https://ia600406.us.archive.org/28/items/hagueconventions00inte_0/hagueconventions00inte_0.pdf.

Hanchett, Ian. "Israeli Spox: Hamas Terrorists Aren't Starving Because They Steal Aid, Pressuring Us Incentivizes Their Evil." *Breitbart News*, April 6, 2024. https://www.breitbart.com/clips/2024/04/06/israeli-spox-hamas-terrorists-arent-starving-because-they-steal-aid-pressuring-us-incentivizes-their-evil/.

Harbi, Mohammed. "Massacre in Algeria." *Le Monde diplomatique*, May 2005. https://mondediplo.com/2005/05/14algeria.

Harris, Robert, and Jeremy Paxman. *A Higher Form of Killing: The Secret History of Chemical and Biological Warfare.* Random House Trade Paperbacks, 2002.

Harris, Sheldon. *Factories of Death: Japanesse Biological Warfare, 1932–1945, and the American Cover-Up.* Rev. ed. Routledge, 2002.

Hasegawa, Guy R. "Proposals for Chemical Weapons During the American Civil War." *Military Medicine* 173, no. 5 (May 2008): 499–506. https://doi.org/10.7205/MILMED.173.5.499.

Haskett, Norm. "Atlantic, Battle of the (1939–1945)." Daily Chronicles of World War II. Accessed April 26, 2025. https://ww2days.com/atlantic-battle-of-1939-1945.html.

Haviland, Jim. "Gothas: The German Bombers of World War I." *Military Heritage* 5, no. 5 (April 2004). https://warfarehistorynetwork.com/article/gothas-the-german-bombers-of-world-war-i/.

Hayward, John. "Algerian President Tebboune Declares Victory with 95% of Vote in Election 'Farce.'" *Breitbart News*, September 9, 2024. https://www.breitbart.com/africa/2024/09/09/algerian-president-tebboune-declares-victory-with-95-of-vote-in-election-farce/.

Hayward, John. "Biden-Harris Admin Finally Notices Sudan Disaster After Doctors Without Borders Bails on Refugee Camp." *Breitbart News*, October 14, 2024. https://www.breitbart.com/africa/2024/10/14/biden-harris-admin-finally-notices-sudan-disaster-after-doctors-without-borders-bails-on-refugee-camp/.

Hayward, John. "Congo Seeks to Execute Americans for Alleged 'Coup' as Joe Biden Soaks Up Sun." *Breitbart News*, August 28, 2024. https://www.breitbart.com/africa/2024/08/28/congo-seeks-to-execute-americans-for-alleged-coup-as-joe-biden-soaks-up-sun/.

Hayward, John. "Taiwan Holds Last Remaining Tiananmen Remembrance in Chinese-Speaking World." *Breitbart News*, June 4, 2024. https://www.breitbart.com/asia/2024/06/04/taiwan-holds-last-remaining-tiananmen-remembrance-chinese-speaking-world/.

Hayward, John. "'Thieves and Parasites': Marxist Uganda Arrests Dozens Protesting Corruption." *Breitbart News*, July 25, 2024. https://www.breitbart.com/africa/2024/07/25/thieves-and-parasites-marxist-uganda-arrests-dozens-protesting-corruption/.

Hayward, John. "Tiananmen Square Massacre Families Tell Xi Jinping: 'We Will Never Forget.'" *Breitbart News*, June 3, 2024. https://www.breitbart.com/asia/2024/06/03/tiananmen-square-massacre-families-tell-xi-jinping-we-will-never-forget/.

Hayward, John. "U.N. Teacher's Passport Found on Hamas Terrorist Mastermind Yahya Sinwar." *Breitbart News*, October 17, 2024. https://www.breitbart.com/national-security/2024/10/17/u-n-teachers-passport-found-hamas-terrorist-mastermind-yahya-sinwar/.

Hayward, John. "W.H.O. Chief Tedros Demands Israel Stop Pursuing Hamas." *Breitbart News*, May 1, 2024. https://www.breitbart.com/middle-east/2024/05/01/w-h-o-chief-tedros-demands-israel-stop-pursuing-hamas/.

Head, Thomas. "The Development of the Peace of God in Aquitaine (970–1005)." *Speculum* 74, no. 3 (July 1999): 656–86. https://doi.org/10.2307/2886764.

Herf, Jeffrey. "Hate Radio: The Long, Toxic Afterlife of Nazi Propaganda in the Arab World." Think-Israel. November 22, 2009. http://www.think-israel.org/herf.nazipropagandainarabworld.html.

Hidden History. "Commander Anthony Miers: War Criminal or War Hero?" Posted August 29, 2024, by Hidden History. YouTube, 11 min., 14 sec. https://www.youtube.com/watch?v=wKTUBQwmnzs.

Hines, Nicola S. "Unit 731 Justice Long Overdue." *State Bar of Texas International Law Section International Newsletter* 1, no. 1 (Fall 2018): 6–10. https://ilstexas .org/wp-content/uploads/2018/09/ILS-Quarterly-3Q18-AE18-F2A_with-links.pdf #page=6.

Hinnershitz, Stephanie. "The Wartime Internment of Native Alaskans." National WWII Museum. June 30, 2022. https://www.nationalww2museum.org/war/articles/ wartime-internment-native-alaskans.

History.com Editors. "120 Emigrants Murdered at the Mountain Meadows Massacre." History. Updated January 31, 2025. https://www.history.com/this-day-in-history/ mormons-and-paiutes-murder-120-emigrants-at-mountain-meadows.

History.com Editors. "Agent Orange." History. Updated February 27, 2025. https:// www.history.com/topics/vietnam-war/agent-orange-1.

History.com Editors. "American-Indian Wars." History. Updated February 27, 2025. https://www.history.com/articles/american-indian-wars.

History.com Editors. "American Indian Wars: Timeline." History. Updated April 15, 2025. https://www.history.com/topics/native-american-history/ american-indian-wars-timeline.

History.com Editors. "Bataan Death March." History. Updated February 27, 2025. https://www.history.com/topics/world-war-ii/bataan-death-march.

History.com Editors. "Ethnic Cleansing." History. Updated April 10, 2025. http://www .history.com/topics/ethnic-cleansing.

History.com Editors. "Khmer Rouge." History. Updated April 15, 2025. https://www .history.com/topics/cold-war/the-khmer-rouge.

History.com Editors. "My Lai Massacre." History. Updated April 1, 2025. https://www .history.com/topics/vietnam-war/my-lai-massacre-1.

History.com Editors. "Russo-Japanese War." History. Updated February 27, 2025. https://www.history.com/topics/asian-history/russo-japanese-war.

History.com Editors. "Tiananmen Square Protests." History. Updated March 6, 2025. https://www.history.com/topics/asian-history/tiananmen-square.

History.com Editors. "Tibetans Revolt Against Chinese Occupation." History. Updated March 2, 2025. https://www.history.com/this-day-in-history/rebellion-in-tibet.

History.com Editors. "Treaty of Versailles." History. Updated April 1, 2025. https:// www.history.com/topics/world-war-i/treaty-of-versailles-1.

History.com Editors. "Union General Sherman's Scorched-Earth March to the Sea Campaign Begins." History. Updated January 30, 2025. https://www.history.com/ this-day-in-history/the-march-to-the-sea-begins.

Historyguy71. "The Spanish Civil War: A Comprehensive Overview." History Bite. February 27, 2024. http://www.historybite.com/european-history/ the-spanish-civil-war-a-comprehensive-overview/.

History Matters. "American Soldiers in the Philippines Write Home About the War." Accessed April 23, 2025. https://historymatters.gmu.edu/d/58/.

HistoryNet Staff. "The Burning of Columbia from the Union and Confederate Perspectives." HistoryNet. September 23, 1998. https://www.historynet.com/the -burning-of-columbia-from-the-union-and-confederate-perspectives-october-1998 -civil-war-times-feature/.

History Skills. "The 4 Most Lethal Chemical Weapons Used in WWI." Accessed August 13, 2024. https://www.historyskills.com/classroom/year-9/wwi-gas-attacks/.

History Skills. "How the Modern State of Israel Was Created in 1948." Accessed August 23, 2024. https://www.historyskills.com/classroom/modern-history/ formation-of-modern-israel-reading/.

Hitler, Adolf. *Mein Kampf*. Translated by Ralph Manheim. Vol. 1. Mariner Books, 1999.
Hoffman, Christopher S. *Major General William T. Sherman's Total War in the Savannah and Carolina Campaigns*. School of Advanced Military Studies, US Army Command and General Staff College, 2018. https://apps.dtic.mil/sti/pdfs/AD1071091 .pdf.
Holland, Elisa Joy. "Massacre at Nogun-Ri." Asia Society. Accessed October 5, 2024. https://asiasociety.org/education/massacre-nogun-ri.
Holland, Eva. "Agony of the Aleutians: The Forgotten Internment." *Anchorage Daily News*, November 9, 2014. https://www.adn.com/we-alaskans/article/ forgotten-internment/2014/11/09/.
Holmes, James R., and Toshi Yoshihara. *Chinese Naval Strategy in the 21st Century: The Turn to Mahan*. Routledge, 2008.
Holocaust Encyclopedia. "The Doctors Trial: The Medical Case of the Subsequent Nuremberg Proceedings." United States Holocaust Memorial Museum. Accessed September 27, 2024. https://encyclopedia.ushmm.org/content/en/article/ the-doctors-trial-the-medical-case-of-the-subsequent-nuremberg-proceedings.
Holocaust Encyclopedia. "The Rwanda Genocide." United States Holocaust Memorial Museum. Updated April 5, 2021. https://encyclopedia.ushmm.org/content/en/ article/the-rwanda-genocide.
Holocaust Encyclopedia. "Treaty of Versailles." United States Holocaust Memorial Museum. Accessed April 27, 2025. https://encyclopedia.ushmm.org/content/en/ article/treaty-of-versailles.
Holzwarth, Larry. "The American Submarine Campaign in the Pacific Changed the Tides of WWII." History Collection. October 26, 2020. https://historycollection .com/the-dramatic-american-submarine-campaign-in-the-pacific-changed-the-tides -of-world-war-ii/.
Homan, Zenobia S. "Unconventional Warfare in the Ancient Near East." *Social Sciences and Humanities Open* 8, no. 1 (2023): 1–10. https://doi.org/10.1016/j .ssaho.2023.100501.
Howard, Martin R. "Walcheren 1809: A Medical Catastrophe." *BMJ* 319, no. 7725 (December 18, 1999): 1642–45. https://www.bmj.com/content/319/7225/1642.
Howe, Amy. "Biden Proposes Supreme Court Reforms." *SCOTUSblog* (blog). July 29, 2024. https://www.scotusblog.com/2024/07/biden-proposes-supreme-court-reforms/.
Hull, Isabel V. *Absolute Destruction: Military Culture and the Practices of War in Imperial Germany*. Cornell University Press, 2005.
Human Rights Watch. "'Break Their Lineage, Break Their Roots.'" April 19, 2021. https:// www.hrw.org/report/2021/04/19/break-their-lineage-break-their-roots/chinas-crimes -against-humanity-targeting.
Human Rights Watch. "China and Tibet." Accessed October 5, 2024. https://www.hrw .org/asia/china-and-tibet.
Human Rights Watch. "Crimea: Persecution of Crimean Tatars Intensifies." November 14, 2017. https://www.hrw.org/news/2017/11/14/crimea-persecution -crimean-tatars-intensifies.
Human Rights Watch. "'I Just Sit and Wait to Die': Reparations for Survivors of Kenya's 2007–2008 Post-Election Sexual Violence." February 15, 2016. https://www .hrw.org/report/2016/02/15/i-just-sit-and-wait-die/reparations-survivors-kenyas -2007-2008-post-election.
Human Rights Watch. "Iran: 1988 Mass Executions Evident Crimes Against Humanity." June 8, 2022. https://www.hrw.org/news/2022/06/08/iran-1988-mass-executions -evident-crimes-against-humanity.

Human Rights Watch. "'They Killed Them Like It Was Nothing.'" October 5, 2011. https://www.hrw.org/report/2011/10/05/they-killed-them-it-was-nothing/need-justice-cote-divoires-post-election-crimes.

Human Rights Watch. "Tibet: Mass Relocations of Tibetans Not Voluntary." May 21, 2024. https://www.hrw.org/news/2024/05/22/tibet-mass-relocations-tibetans-not-voluntary.

Husebo, Wendell. "Report: Israeli Soldiers Find Dead Babies After Palestinian Hamas Attack." *Breitbart News*, October 10, 2023. https://www.breitbart.com/politics/2023/10/10/report-israeli-soldiers-find-dead-babies-after-palestinian-hamas-attack/.

Ienaga, Saburo. "The Glorification of War in Japanese Education." *International Security* 18, no. 3 (Winter 1993–1994): 113–33. https://doi.org/10.2307/2539207.

Imlay, Talbot. "Total War." *Journal of Strategic Studies* 30, no. 3 (June 2007): 547–70. http://www.tandfonline.com/doi/abs/10.1080/01402390701343516.

Imperial War Museum. "Voices of the First World War: Gas Attack at Ypres." Accessed April 26, 2025. https://www.iwm.org.uk/history/voices-of-the-first-world-war-gas-attack-at-ypres.

Imperial War Museum. "Voices of the First World War: Trench Life." Accessed April 26, 2025. https://www.iwm.org.uk/history/voices-of-the-first-world-war-trench-life.

Interfax-Ukraine. "Checkpoints Put at All Entrances to Sevastopol." *Kyiv Post*, February 26, 2014. https://archive,kyivpost.com/content/ukrainepolitics/checkpoints-put-at-all-entrances-to-sevastopol-337655.html.

International Humanitarian Law Databases. "Practice Relating to Rule 76. Herbicides." Accessed September 12, 2024. https://ihl-databases.icrc.org/en/customary-ihl/v2/rule76.

Investor's Business Daily. "Gallup Poll Reveals Obama Has Turned Back Clock on Race Relations." Updated April 12, 2016. https://www.investors.com/politics/editorials/gallup-poll-reveals-obama-has-turned-back-clock-on-race-relations/.

IPT News. "Federal Judge Agrees: CAIR Tied to Hamas." Investigative Project on Terrorism. November 22, 2010. https://www.investigativeproject.org/2340/federal-judge-agrees-cair-tied-to-hamas.

Israel, Steve. "The Roots of Hamas' Terror Attack Can Be Found in Gaza's Schools." Forward. October 25, 2023. https://forward.com/opinion/566841/hamas-schools-indoctrination-antisemitic-textbooks-gaza/.

Jackson, Andrew. "President Andrew Jackson's Message to Congress 'On Indian Removal' (1830)." National Archives. December 6, 1830. https://www.archives.gov/milestone-documents/jacksons-message-to-congress-on-indian-removal.

Janda, Lance. "Shutting the Gates of Mercy: The American Origins of Total War, 1860–1880." *Journal of Military History* 59, no. 1 (January 1995): 7–26. https://doi.org/10.2307/2944362.

Japanese Eye Witness. "Execution of Allied Intelligence Officer by the Japanese." Takeo Tanimizu. Accessed April 26, 2025. https://rjgeib.com/heroes/tanimizu/japanese-execution.html.

Jasiński, Jakub. "Quotes of Ennius." Imperium Romanum. October 28, 2020. https://imperiumromanum.pl/en/roman-art-and-culture/golden-thoughts-of-romans/quotes-of-ennius.

Jewish Telegraphic Agency. "Fourth Day of Palestine Warfare Finds Jewry in Grave Danger Throughout Country." *Jewish Daily Bulletin*, August 27, 1929. http://pdfs.jta.org/1929/1929-08-27_1451.pdf.

Jewish Virtual Library. "The SS (Schutzstaffel): The Waffen-SS." Accessed September 25, 2024. https://www.jewishvirtuallibrary.org/waffen-ss.

Johnson, Amy J. "Algeria: War of Independence, 1954–1962." In *Encyclopedia of African History*, vol. 1: A–G, edited by Kevin Shillington, 101–3. Fitzroy Dearborn, 2005.

Joint Army-Navy Assessment Committee. "Japanese Naval and Merchant Shipping Losses During WWII by All Causes." Naval History and Heritage Command. February 1947. https://www.history.navy.mil/research/library/online-reading-room/ti tle-list-alphabetically/j/japanese-naval-merchant-shipping-losses-wwii.html#pageiv.

Joint Resolution to Provide for Annexing the Hawaiian Islands to the United States. H. Res. 259. 55th Cong. 1898. https://www.archives.gov/milestone-documents/ joint-resolution-for-annexing-the-hawaiian-islands.

Jomini, Antoine Henri de. *The Art of War*. Greenhill Books, 1996.

Jones, Heather. "Prisoners of War." International Encyclopedia of the First World War. Updated October 8, 2014. https://doi.org/10.15463/ie1418.10475.

Keeley, Lawrence H. *War Before Civilization*. Oxford University Press, 1996.

Kennedy, Lesley. "Did Yellow Journalism Fuel the Outbreak of the Spanish American War?" History. Updated February 18, 2025. https://www.history.com/articles/ spanish-american-war-yellow-journalism-hearst-pulitzer.

Kestling, Robert W. "Blacks Under the Swastika: A Research Note." *Journal of Negro History* 83, no. 1 (Winter 1998): 84–99. http://www.jstor.org/stable/2668561.

Khan, Asfandiyar. "The Splitting of East-Pakistan from West-Pakistan in 1971: The Role of India." Modern Diplomacy. February 7, 2020. https://moderndip lomacy.eu/2020/02/07/the-splitting-of-east-pakistan-from-west-pakistan-in-1971 -the-role-of-india/.

Khan, Karim A. A., Mame Mandiaye Niang, Leonie Von Braun, Paolina Massidda, and Sarah Pellet. *The Prosecutor v. Joseph Kony*. International Criminal Court. January 19, 2024. https://www.icc-cpi.int/sites/default/files/CourtRecords/CR2024_00006 .PDF.

Khettab, Djamila Ould. "Q&A: What Really Happened to Algeria's Harkis." *Al Jazeera*, August 22, 2015. https://www.aljazeera.com/news/2015/8/22/ qa-what-really-happened-to-algerias-harkis.

Kifner, John. "Armenian Genocide of 1915: An Overview." *New York Times*. Accessed May 4, 2024. https://archive.nytimes.com/www.nytimes.com/ref/timestopics/top ics_armeniangenocide.html.

King, Byron. "Alaska: 'The Most Important Strategic Place in the World,' Part III." *Daily Reckoning*. February 22, 2008. https://dailyreckoning.com/the-importance -of-alaska-part-ii/.

King, Gilbert. "The Aftermath of Mountain Meadows." *Smithsonian Magazine*, February 29, 2012. https://www.smithsonianmag.com/history/the -aftermath-of-mountain-meadows-110735627/.

Kitazawa, Yuki, and Matthew Allen. "A Story That Won't Fade Away: Compulsory Mass Suicide in the Battle of Okinawa." *Asia-Pacific Journal: Japan Focus* 5, no. 7 (July 12, 2007): 1–4. https://apjjf.org/wp-content/uploads/2023/11/article-2049 .pdf.

Klein, Christopher. "When Japan Launched Killer Balloons in World War II." History. Updated January 27, 2025. https://www.history.com/news/japans-killer -wwii-balloons.

Klein, Joshua. "BBC Mocked for 'Outrageous' Suggestion IDF Should Have Warned Gazans Before Hostage Rescue Mission." *Breitbart News*, June 10, 2024. https:// www.breitbart.com/the-media/2024/06/10/bbc-mocked-for-outrageous-suggestion -idf-should-have-warned-gazans-before-hostage-rescue-mission/.

Klein, Joshua. "IDF Confirms: All 3 Rescued Male Israeli Hostages Held by Gaza 'Journalist.'" *Breitbart News*, June 9, 2024. https://www.breitbart.com/politics/2024/06/09/idf-confirmed-3-male-israeli-hostages-held-gaza-journalist-hamas/.

Klein, Joshua. "'Muslim Terrorists Are in Our Midst': Florida Imam Calls for Annihilation of All Jews." *Breitbart News*, May 13, 2024. https://www.breitbart.com/politics/2024/05/13/muslim-terrorists-are-in-our-midst-florida-imam-calls-for-anni hilation-of-all-jews/.

Kleist, Joseph. "The Battle of Thermopylae: Principles of War on the Ancient Battlefield." *Studia Antiqua* 6, no. 1 (Spring 2008): 75–86. https://scholarsarchive.byu .edu/cgi/viewcontent.cgi?article=1091&context=studiaantiqua.

Klukowski, Ken. "Biden and Harris Propose Abolishing Supreme Court as Independent Branch of Government." *Breitbart News*, July 29, 2024. https://www .breitbart.com/politics/2024/07/29/biden-harris-propose-abolishing-supreme-court -independent-branch/.

Koch, H. W. "The Strategic Air Offensive Against Germany: The Early Phase, May–September 1940." *Historical Journal* 34, no. 1 (March 1991): 117–41. https://doi .org/10.1017/S0018246X00013959.

Kopp, Carlo. "Chemical and Biological Weapons." *Defence Today* 6, no. 3 (2007): 28–30. https://docslib.org/doc/6265356/chemical-and-biological-weapons-pdf.

Koropov, Oleg. "On March 22, 1943, the Nazi Punitive Detachment Destroyed the Belarusian Village of Khatyn." Belarusian Institute for Strategic Research, March 21, 2025. https://bisr.gov.by/en/mneniya/march-22-1943-nazi-punitive -detachment-destroyed-belarusian-village-khatyn.

Korostelina, Karina V. "Devastating Civilians at Home: The Plight of Crimean Tatars and Californians of Asian Descent During World War II." In *Civilians and Modern War: Armed Conflict and the Ideology of Violence*, edited by Daniel Rothbart, Karina V. Korostelina, and Mohammed Cerkaoui, 51–71. Routledge, 2012.

Krebs, Gerhard. "World War Zero? Re-Assessing the Global Impact of the Russo-Japanese War 1904–05." *Asia-Pacific Journal: Japan Focus* 10, no. 21 (May 19, 2012): 1–24. https://apjjf.org/2012/10/21/gerhard-krebs/3755/article.

Krym Media. "Only 3.3% of Crimeans Mention Ukrainian as Their Native Language." March 19, 2015. http://en.krymedia.ru/nationality/3373760-Only-33-of-Crimeans -Mention-Ukrainian-as-Their-Native-Language.

Kushner, Barak. *The Thought War: Japanese Imperial Propaganda*. University of Hawai'i Press, 2006.

Kuzmarov, Jeremy. "The Korean War: Barbarism Unleashed." United States Foreign Policy: History and Resource Guide. 2016. https://peacehistory-usfp.org/ korean-war/.

Lampe, John R. "War Crimes and Trials." Britannica. Updated April 22, 2025. https:// www.britannica.com/event/Bosnian-War/War-crimes-and-trials.

Langenbacher, Eric. "The Allies in World War II: The Anglo-American Bombardment of German Cities." In *Genocide, War Crimes and the West*, edited by Adam Jones, 116–33. Zed Books, 2004.

Legro, Jeffrey W. "Military Culture and Inadvertent Escalation in World War II." *International Security* 18, no. 4 (Spring 1994): 108–42. https://doi.org/10.2307/2539179.

LeMay, Curtis, and McKinlay Kantor. *Mission with LeMay: My Story*. Doubleday, 1965.

LeMay, Curtis E. "The Operational Side of Air Offense." National Security Archive. May 21, 1957. https://nsarchive.gwu.edu/document/20291-national -security-archive-doc-06-operational.

Lepick, Olivier. "France's Political and Military Reaction in the Aftermath of the First German Chemical Offensive in April 1915: The Road to Retaliation in Kind." In *One Hundred Years of Chemical Warfare: Research, Deployment, Consequences*, edited by Breitslav Friedrich, Dieter Hoffmann, Jürgen Renn, Florian Schmaltz, and Martin Worl, 69–76. Springer International, 2017. https://doi .org/10.1007/978-3-319-51664-6_5.

Lerner, K. Lee. "Biological and Chemical Weapons." In *Global Issues in Context*. Gale, 2018. https://www.academia.edu/102704126/Biological_and _Chemical_Weapons_Overview.

Lewis, Susan K. "History of Biowarfare." NOVA Online. Accessed April 26, 2025. https://www.pbs.org/wgbh/nova/bioterror/hist_nf.html.

Lewy, Guenter. *America in Vietnam*. Oxford University Press, 1980. https://archive.org/ details/americainvietnam00lewy/page/450/mode/2up.

Lin, Chan Cheng. "Nanjing Massacre and Sook Ching Massacre: Shaping of Chinese Popular Memories in China and Singapore, 1945–2015." Master's thesis, National University of Singapore, December 18, 2015. https://scholarbank.nus.edu.sg/ entities/publication/c1d85a55-c675-4093-a31b-a24ecc6d260c.

Lindbergh, Charles. "Des Moines Speech—America First Committee." Charleslind bergh.com. September 11, 1941. http://www.charleslindbergh.com/americanfirst/ speech.asp.

Llewellyn, Jennifer, Steve Thompson, and Jim Southey. "Militarism as a Cause of World War I." Alpha History. Updated January 15, 2025. https://alphahistory.com/ worldwar1/militarism/.

Lloyd, Stuart. "Singapore Alexandra Hospital Massacres 1942." Historic UK. February 13, 2022. https://www.historic-uk.com/HistoryUK/HistoryofBritain/ Singapore-Alexandra-Hospital-Massacres-1942/.

Long, Hu, Zhengyu Liao, Yan Wang, Lina Liao, and Wenli Lai. "Efficacy of Botulinum Toxins on Bruxism: An Evidence-Based Review." *International Dental Journal* 62, no. 1 (February 2012): 1–5. https://doi.org/10.1111/J.1875-595X.2011.00085.X.

Lynch, John. "The Lessons of Walcheren Fever, 1809." *Military Medicine* 174, no. 3 (March 2009): 315–19. https://academic.oup.com/milmed/article/ 174/3/315/4333688.

Macdonald, Sally. "He Survived—1,800 Fellow Prisoners Aboard Japanese 'Hell Ship' Died 50 Years Ago Today." *Seattle Times*, October 24, 1994. https://archive.seattle times.com/archive/?date=19941024&slug=1937653.

Machiavelli, Niccolò. *The Prince*. Translated by George Bull. Penguin Books, 2003.

MacKinnon, William P. "Prelude to Civil War: The Utah War's Impact and Legacy." In *Civil War Saints*, edited by Kenneth L. Alford, 1–21. Religious Studies Center, Brigham Young University, 2012.

Maden, Jack. "Mengzi vs. Xunzi on Human Nature: Are We Good or Evil?" Philosophy Break. October 2023. https://philosophybreak.com/articles/mengzi -xunzi-on-human-nature-are-we-good-or-evil/.

Mahan, A. T. *The Influence of Sea Power upon History: 1660–1783*. 12th ed. Little, Brown, 1918. https://www.gutenberg.org/ebooks/13529.

Maizland, Lindsay. "Myanmar's Troubled History: Coups, Military Rule, and Ethnic Conflict." Council on Foreign Relations. Updated January 31, 2022. https://www.cfr .org/backgrounder/myanmar-history-coup-military-rule-ethnic-conflict-rohingya.

Mallmann, Klaus-Michael, and Martin Cüppers. *Nazi Palestine: The Plans for the Extermination of the Jews in Palestine*. Translated by Krista Smith. Enigma Books, 2010.

Margalit, Gilad. "The Uniqueness of the Nazi Persecution of the Gypsies." *Romani Studies* 10, no. 2 (December 2000): 185–210. https://doi.org/10.3828/rs.2000.6.

Margolin, Jean-Louis. "Japanese Crimes in Nanjing, 1937–38: A Reappraisal." *China Perspectives* 63 (January–February 2006): 1–16. https://journals.openedition.org/chinaperspectives/571.

Mark. "General William T. Sherman's Report on the March to the Sea and Capture of Savannah." *Iron Brigader.* Updated August 21, 2018. https://ironbrigader.com/2014/11/22/general-william-t-shermans-report-march-sea-capture-savannah/.

Marr, John S., and John T. Cathey. "The 1802 Saint-Domingue Yellow Fever Epidemic and the Louisiana Purchase." *Journal of Public Health Management and Practice* 19, no. 1 (January–February 2013): 77–82. https://doi.org/10.1097/PHH.0b013e318252eea8.

Marsh, Alan. "POWs in American History: A Synopsis." National Park Service. 1998. https://www.nps.gov/ande/learn/historyculture/pow_synopsis.htm.

Martel, Frances. "Tim Walz Dodges Question on Tiananmen Square Lie: 'I'm a Knucklehead.'" *Breitbart News*, October 1, 2024. https://www.breitbart.com/2024-election/2024/10/01/tim-walz-dodges-question-tiananmen-square-lie-im-knucklehead/.

Martin, Clancy, and Alan Strudler. "Are Humans Good or Evil? A Brief Philosophical Debate." *Harper's Magazine*, October 9, 2014. https://harpers.org/2014/10/are-humans-good-or-evil/.

Mbabuike, Michael C., and Anna Marie Evans. "Other Victims of the Holocaust." *Dialectical Anthropology* 25, no. 1 (March 2000): 1–25. https://doi.org/10.1023/A:1007135521295.

McCurry, Justin. "Told to Commit Suicide, Survivors Now Face Elimination from History." *Guardian*, July 6, 2007. https://www.theguardian.com/world/2007/jul/06/japan.schoolsworldwide.

Melbye, Jerry, and Scott I. Fairgrieve. "A Massacre and Possible Cannibalism in the Canadian Arctic: New Evidence from the Saunaktuk Site (NgTn-1)." *Arctic Anthropology* 31, no. 2 (January 1994): 57–77. https://www.researchgate.net/publication/288244980_A_massacre_and_possible_cannibalism_in_the_Canadian_Arctic_New_evidence_from_the_Saunaktuk_site_NgTn-1.

Mellanby, Kenneth. "Medical Experiments on Human Beings in Concentration Camps in Nazi Germany." *British Medical Journal* (January 25, 1947): 148–50. https://www.ncbi.nlm.nih.gov/pmc/articles/PMC2052883/pdf/brmedj03765-0026.pdf.

Merchant, Zofsha, and Joanna Michalopoulos. "Democratic Republic of the Congo." World Without Genocide. Updated October 2023. https://worldwithoutgenocide.org/genocides-and-conflicts/congo.

Merriam-Webster. "Asymmetric Warfare." Accessed September 3, 2024. https://www.merriam-webster.com/dictionary/asymmetric--warfare.

Merriam-Webster. "Propaganda." Accessed August 18, 2024. https://www.merriam-webster.com/dictionary/propaganda.

Messerschmidt, Manfred. "The Wehrmacht and the Volksgemeinschaft." *Journal of Contemporary History* 18, no. 4 (October 1983): 719–44. http://www.jstor.org/stable/260309.

Messerschmidt, Manfred, and Anne Halley. "The Soldier in the War to Conquer Eastern Europe." *Massachusetts Review* 36, no. 3 (Autumn 1995): 414–20. https://www.jstor.org/stable/25090655.

Meyer, Frederick. "Is Human Nature Inherently Good or Bad? The Buddhist View of Human Nature." Shambhala. June 9, 2023. https://shambhala.org/community/blog/is-human-nature-inherently-good-or-bad-the-buddhist-view-of-human-nature/.

Middle East Media Research Institute. "Hamas' Indoctrination of Children to Jihad, Martyrdom, Hatred of Jews." November 3, 2023. https://www.memri.org/reports/hamas-indoctrination-children-jihad-martyrdom-hatred-jews.

Middleton, Evelyn. "Côte d'Ivoire (Ivory Coast)." World Without Genocide. Updated October 2024. https://worldwithoutgenocide.org/genocides-and-conflicts/ivory-coast.

Mikesh, Robert C. *Japan's World War II Balloon Bomb Attacks on North America.* Smithsonian Annals of Flight, no. 9. Smithsonian Institution Press, 1973. https://reposi tory.si.edu/bitstream/handle/10088/18679/SAoF-0009-Lo_res.pdf?sequence=3.

Military History Matters. "Submarine—The History of Submarine War." March 26, 2011. https://www.military-history.org/feature/submarine-the-history-of-submarine-war.htm.

Miller, Stuart Creighton. *Benevolent Assimilation: The American Conquest of the Philippines, 1899–1903.* Yale University Press, 1982.

Milner, George R. "Warfare in Prehistoric and Early Historic Eastern North America." *Journal of Archaeological Research* 7, no. 2 (June 1999): 105–51. https://doi.org/10.1007/s10814-005-0001-x.

Mitchell, William. "Winged Defense: The Development and Possibilities of Modern Air Power, Economic and Military." In *Roots of Strategy: Book 4*, edited by David Jablonsky, 409–515. Stackpole Books, 1999.

Morris, Benny, and Benjamin Z. Kedar. "*Cast Thy Bread*: Israeli Biological Warfare During the 1948 War." *Middle Eastern Studies* 59, no. 5 (2023): 752–76. https://doi.org/10.1080/00263206.2022.2122448.

Morrison, Samuel Eliot. *History of United States Naval Operations in World War II.* Vol. 4, *Coral Sea, Midway and Submarine Actions May 1942–August 1942.* Little, Brown, 1959.

Mosse, George L. *The Crisis of German Ideology: Intellectual Origins of the Third Reich.* Universal Library, 1971.

Moszynski, Peter. "5.4 Million People Have Died in Democratic Republic of Congo Since 1998 Because of Conflict, Report Says." *British Medical Journal* 336, no. 7638 (February 2, 2008): 235. https://doi.org/10.1136/BMJ.39475.524282.DB.

Mowry, George E. "The First Roosevelt." *American Mercury*, November 1946, 578–84. https://www.unz.com/print/AmMercury-1946nov-00578/.

Mulligan, Bret. "Nepos: Life of Hannibal, the First Punic War." Dickinson College Commentaries. 2013. https://dcc.dickinson.edu/nepos-hannibal/first-punic-war.

Murray, Williamson. "Why Germany's Kriegsmarine Lost the Battle of the Atlantic." HistoryNet. April 28, 2015. https://www.historynet.com/why-germanys-kriegsmarine-lost-the-battle-of-the-atlantic/.

Muxen, Evan. "Okinawa: The Last Battle." U.S. Army. June 22, 2022. https://www.army.mil/article/257747/okinawa_the_last_battle.

My Jewish Learning. "Modern Israeli History: A Timeline." Accessed August 23, 2024. https://www.myjewishlearning.com/article/modern-israeli-history-a-timeline/.

National Archives. "Distribution of Electoral Votes." Reviewed November 4, 2024. https://www.archives.gov/electoral-college/allocation.

National Geographic Society. "The Indian Removal Act and the Trail of Tears." *National Geographic*, updated October 1, 2024. https://education.nationalgeographic.org/resource/indian-removal-act-and-trail-tears/.

National Library Board (NLB). "Alexandra Hospital Massacre." April 3, 2014. https:// www.nlb.gov.sg/main/article-detail?cmsuuid=7d4fd9a0-7bd0-4533-b0ea-3aa5596 73b0e.

National Museum of Australia. "Breaker Morant Executed." Updated September 25, 2024. https://www.nma.gov.au/defining-moments/resources/breaker-morant -executed.

National Museum of the United States Air Force. "The Eight Who Were Captured." Accessed October 11, 2024. https://www.nationalmuseum.af.mil/Visit/ Museum-Exhibits/Fact-Sheets/Display/Article/196770/the-eight-who-were-captured/.

National Park Service (NPS). "The Burning." Accessed March 25, 2024. https://www .nps.gov/articles/000/the-burning-shenandoah-valley-in-flames.htm.

National Park Service (NPS). "History and Legal Status of Prisoners of War." Updated October 26, 2022. https://www.nps.gov/ande/learn/historyculture/history-legal-sta tus-pows.htm.

National Park Service (NPS). "Sand Creek Massacre." Updated March 13, 2017. https://www.nps.gov/sand/learn/historyculture/massacre.htm.

National Park Service (NPS). "Search for Prisoners." Accessed August 21, 2024. https://www.nps.gov/civilwar/search-prisoners.htm.

National WWI Museum and Memorial. "Captured." October 28, 2022–April 30, 2023. https://www.theworldwar.org/captured.

Neill, Celeste. "Otto von Bismarck: Architect of German Unification." History Hit. May 12, 2023. https://www.historyhit.com/1871-unification-germany/.

New Catholic Encyclopedia. "Peace of God." Encyclopedia.com. Accessed April 26, 2025. https://www.encyclopedia.com/religion/encyclopedias-almanacs-transcripts -and-maps/peace-god.

New World Encyclopedia. "Phosgene." Accessed September 8, 2024. https://www.new worldencyclopedia.org/entry/Phosgene.

New York Times. "President Retires Gen. Jacob H. Smith." July 17, 1902. https://times machine.nytimes.com/timesmachine/1902/07/17/101959147.pdf.

Niderost, Eric. "The Fall of Shanghai: Prelude to the Rape of Nanking and WWII." Warfare History Network. November 2003. https://warfarehistorynetwork.com/ article/the-fall-of-shanghai-prelude-to-the-rape-of-nanking-wwii/.

Niemi, Robert. "World War I in Film." *Pop Culture Universe: Icons, Idols, Ideas*. ABC-CLIO, 2024. https://popculture2-abc-clio-com.ezproxy1.apus.ed/Search/ Display/1917732.

Niiya, Brian. "J. Edgar Hoover." Densho Encyclopedia. Updated May 9, 2024. https:// encyclopedia.densho.org/J.%20Edgar%20Hoover.

Nobel Prize. "Fritz Haber." Accessed April 26, 2025. https://www.nobelprize.org/prizes/ chemistry/1918/haber/biographical/.

Norton, Rob. "Unintended Consequences." Econlib. Accessed July 25, 2024. https:// www.econlib.org/library/Enc/UnintendedConsequences.html.

Ochieng, Beverly, Wedaeli Chibelushi, and Natasha Booty. "Sudan War: A Simple Guide to What Is Happening." BBC News. March 21, 2025. https://www.bbc.com/ news/world-africa-59035053.

Office of United States Chief of Counsel for Prosecution of Axis Criminality. *Nazi Conspiracy and Aggression*. Vol. 3. United States Government Printing Office, 1946.

O'Konski, S. R. "The Katyn Forest Massacre: And Five Betrayals of Poland by Its WWII Allies." World War 2 History Short Stories. July 29, 2019. https://www.ww2 history.org/war-in-europe/the-katyn-forest-massacre-and-five-betrayals-of-poland -by-its-wwii-allies/.

Oliver, Mark. "The My Lai Massacre: 33 Disturbing Photos of the War Crime the U.S. Got Away With." Edited by John Kuroski. All That's Interesting. Updated May 20, 2021. https://allthatsinteresting.com/my-lai-massacre-photos#1.

"Operation Downfall: Planned Invasion of the Islands of Japan in World War II." Accessed April 26, 2025. https://i.4pcdn.org/tg/1464699782538.pdf.

Osborne, Samuel. "UN Accuses Russia of Multiple Human Rights Abuses in Crimea." *Independent*, November 16, 2016. https://www.independent.co.uk/news/world/europe/russia-ukraine-crimea-putin-human-rights-abuses-un-accusations-claims-a74 21406.html.

Ostberg, René. "Envy." Britannica. Updated December 3, 2024. https://www.britannica.com/topic/envy.

Ostberg, René. "Gluttony." Britannica. Updated April 22, 2025. https://www.britannica.com/topic/gluttony.

Ostberg, René. "Greed." Britannica. Updated February 21, 2025. https://www.britannica.com/topic/greed.

Ostberg, René. "Lust." Britannica. Updated April 11, 2025. https://www.britannica.com/topic/lust-deadly-sin.

Ostberg, René. "Pride." Britannica. Updated April 14, 2025. https://www.britannica.com/topic/pride-deadly-sin.

Ostberg, René. "Sloth." Britannica. Updated December 17, 2024. https://www.britannica.com/topic/sloth-behaviour.

Ostberg, René. "Wrath." Britannica. Updated August 20, 2024. https://www.britannica.com/topic/wrath.

Ostroff, Maurice. "The Siege of Jerusalem." World Machal. Accessed August 23, 2024. https://www.machal.il/about-machal/the-siege-of-jerusalem/.

O'Toole, Patricia. "How the US Government Used Propaganda to Sell Americans on World War I." History. Updated January 31, 2025. https://www.history.com/news/world-war-1-propaganda-woodrow-wilson-fake-news.

Overcoming the Past: The History and Memory of Nazi Germany. "Ideological Indoctrination." Accessed September 25, 2024. https://nazigermany.lmu.build/exhibits/show/castro/ideological-indoctrination.

Overy, R. J. "Hitler and Air Strategy." *Journal of Contemporary History* 15, no. 3 (July 1980): 405–21. www.jstor.org/stable/260411.

Overy, Richard. "Allied Bombing and the Destruction of German Cities." In *A World at Total War: Global Conflict and the Politics of Destruction, 1937–1945*, edited by Roger Chickering, Stig Förster, and Bernd Greiner, 277–96. Cambridge University Press, 2010.

Pacific Atrocities Education. "The Development of Unit 731." Accessed April 23, 2025. https://www.pacificatrocities.org/the-development-of-unit-731.html.

Pacific Atrocities Education. "Plan Kantokuen and Bacteriological Warfare." Accessed April 23, 2025. https://www.pacificatrocities.org/plan-kantokuen-and-bacteriological-warfare.html.

Padfield, Peter. *War Beneath the Sea: Submarine Conflict During World War II*. John C. Wiley and Sons, 1998.

Palmer, R. R. "Frederick the Great, Guibert, Bülow: From Dynastic to National War." In *Makers of Modern Strategy: From Machiavelli to the Nuclear Age*, edited by Peter Paret, Gordon A. Craig, and Felix Gilbert, 91–120. Princeton University Press, 1986.

Pape, Robert A. "Why Japan Surrendered." *International Security* 18, no. 2 (Fall 1993): 154–201. https://doi.org/10.2307/2539100.

Paret, Peter. "Napoleon and the Revolution in War." In *Makers of Modern Strategy from Machiavelli to the Nuclear Age*, edited by Peter Paret, Gordon A. Craig, and Felix Gilbert, 123–42. Princeton University Press, 1986.

Parkinson, Robert G. "Print, the Press, and the American Revolution." In Oxford Research Encyclopedia of American History. Oxford University Press. September 3, 2015. https://doi.org/10.1093/acrefore/9780199329175.013.9.

Patterson, Ian. *Guernica and Total War*. Harvard University Press, 2007.

Pettinger, Tejvan. "Law of Unintended Consequences." *Economics Help* (blog). September 27, 2019. https://www.economicshelp.org/blog/2381/economics/law-of-unintended-consequences/.

Peyroulou, Jean-Pierre. "Setif and Guelma (May 1945)." SciencesPo. March 26, 2008. https://www.sciencespo.fr/mass-violence-war-massacre-resistance/en/document/setif-and-guelma-may-1945.html.

Pierce, Roy. "Political Power, Technology, and Total War: Two French Views." *Journal of Conflict Resolution* 2, no. 4 (December 1958): 321–28. http://www.jstor.org/stable/172889.

Pimentel, Marc. "Migration of Doom: Bataan Death March." U.S.S. Salt Lake City CA25. Accessed December 20, 2023. https://ussslcca25.com/bataan.htm.

Poll, Richard D. "The Utah War." History to Go. 1994. https://historytogo.utah.gov/utah-war/.

Pollak, Joel B. "Biden Summons Israel for Talks; Says Netanyahu's Position on Hamas Is 'Nonsense.'" *Breitbart News*, March 18, 2024. https://www.breitbart.com/middle-east/2024/03/18/biden-summons-israel-for-talks-says-netanyahus-position-on-hamas-is-nonsense/.

Pollak, Joel B. "Biden Tries to Take Credit for Yahya Sinwar's Killing, but Opposed Israel's Operations; Harris Warned of 'Consequences.'" *Breitbart News*, October 17, 2024. https://www.breitbart.com/middle-east/2024/10/17/biden-tries-to-take-credit-for-yahya-sinwars-killing-but-opposed-israels-operations-harris-warned-consequences/.

Pollak, Joel B. "CAIR Condemns Israeli Hostage Rescue: 'Horrific Massacre.'" *Breitbart News*, June 8, 2024. https://www.breitbart.com/middle-east/2024/06/08/cair-condemns-israeli-hostage-rescue-horrific-massacre/.

Pollak, Joel B. "Hamas Filmed Six Hostages Before Executions; One Clip Released by Family." *Breitbart News*, September 2, 2024. https://www.breitbart.com/middle-east/2024/09/02/hamas-filmed-six-hostages-before-executions-one-clip-released-by-family/.

Pollak, Joel B. "Hamas Leaders to NY Times: No Interest in Helping Palestinians in Gaza, Want 'Permanent' War Against Israel." *Breitbart News*, November 8, 2023. https://www.breitbart.com/middle-east/2023/11/08/hamas-leaders-to-ny-times-no-interest-in-helping-palestinians-gaza-want-permanent-war-against-israel/.

Pollak, Joel B. "IDF: Hamas Is Destroying Shifa Hospital, Firing from Maternity Ward." *Breitbart News*, March 24, 2024. https://www.breitbart.com/politics/2024/03/24/idf-hamas-is-destroying-shifa-hospital-firing-from-maternity-ward/.

Pollak, Joel B. "IDF: Iran Has Fired Missiles at Israel; All Residents Ordered to Bomb Shelters." *Breitbart News*, October 1, 2024. https://www.breitbart.com/middle-east/2024/10/01/idf-iran-has-fired-missiles-at-israel-all-residents-ordered-to-bomb-shelters/.

Pollak, Joel B. "IDF Discovers Hamas, Islamic Jihad Command Center at UNRWA HQ." *Breitbart News*, July 12, 2024. https://www.breitbart.com/middle-east/2024/07/12/idf-discovers-hamas-islamic-jihad-command-center-at-unrwa-hq/.

Pollak, Joel B. "IDF Discovers Subway Tracks Underneath Northern Gaza." *Breitbart News*, September 3, 2024. https://www.breitbart.com/middle-east/2024/09/03/idf-discovers-subway-tracks-underneath-northern-gaza/.

Pollak, Joel B. "IDF Kills 15 Terrorists in 'War Room' at UNRWA School." *Breitbart News*, May 14, 2024. https://www.breitbart.com/middle-east/2024/05/14/idf-kills-15-terrorists-in-war-room-at-unrwa-school/.

Pollak, Joel B. "Israel: UNRWA Is Using Hamas to 'Protect' Aid Trucks." *Breitbart News*, March 6, 2024. https://www.breitbart.com/middle-east/2024/03/06/israel-unrwa-is-using-hamas-to-protect-aid-trucks/.

Pollak, Joel B. "Israel Discovers Weapons Cache in Maternity Ward of Gaza's Shifa Hospital." *Breitbart News*, March 31, 2024. https://www.breitbart.com/middle-east/2024/03/31/israel-maternity-ward-shifa-hospital/.

Pollak, Joel B. "Israel Reveals Hamas Intel Tunnel Underneath UNRWA HQ in Gaza." *Breitbart News*, February 10, 2024. https://www.breitbart.com/middle-east/2024/02/10/israel-reveals-hamas-intel-tunnel-underneath-unrwa-hq-in-gaza/.

Pollak, Joel B. "Official Confirmation: Hamas Leader Yahya Sinwar Is Dead; Killed Trying to Flee Gaza." *Breitbart News*, October 17, 2024. https://www.breitbart.com/middle-east/2024/10/17/official-confirmation-hamas-leader-yahya-sinwar-is-dead-killed-trying-to-flee-gaza/.

Pollak, Joel B. "Palestinian Ruling Party Admits: Hamas Steals Aid, Kills Aid Workers in Gaza." *Breitbart News*, April 21, 2024. https://www.breitbart.com/politics/2024/04/21/palestinian-ruling-party-admits-hamas-steals-aid-kills-aid-workers-in-gaza/.

Pollak, Joel B. "Report: Eyewitness Says Hamas Executed Israeli Woman During Gang Rape." *Breitbart News*, November 9, 2023. https://www.breitbart.com/middle-east/2023/11/09/warning-graphic-content-eyewitness-says-hamas-executed-israeli-woman-during-gang-rape/.

Pollak, Joel B. "Report: Yahya Sinwar Bodyguards Included United Nations Employee; Update: Fake Passport?" *Breitbart News*, October 17, 2024. https://www.breitbart.com/national-security/2024/10/17/report-yahya-sinwar-bodyguards-included-united-nations-employee/.

Pollak, Joel B. "Sinwar's Likely Death Comes Days After Biden, Harris Threatened Israel with Arms Embargo." *Breitbart News*, October 17, 2024. https://www.breitbart.com/middle-east/2024/10/17/sinwars-likely-death-comes-days-after-biden-harris-threatened-israel-with-arms-embargo/.

Pollak, Joel B. "UN Admits: 9 UNRWA Employees May Have Participated in October 7 Terror." *Breitbart News*, August 5, 2024. https://www.breitbart.com/middle-east/2024/08/05/un-admits-9-unrwa-employees-may-have-participated-in-october-7-terror/.

Pollak, Joel B. "UN Report Confirms: Hamas Committed Rape Against Israelis, Hostages." *Breitbart News*, March 4, 2024. https://www.breitbart.com/middle-east/2024/03/04/un-report-confirms-hamas-committed-rape-against-israelis-hostages/.

Pua, Derek, Danielle Dybbro, and Alistair Rogers. *Unit 731: The Forgotten Asian Auschwitz*. 2nd ed. Pacific Atrocities Education, 2018.

Rabson, Steve. "Case Dismissed: Osaka Court Upholds Novelist Oe Kenzaburo for Writing That Japanese Military Ordered 'Group Suicides' in Battle of Okinawa." *Asia-Pacific Journal: Japan Focus* 6, no. 4 (April 1, 2008). https://apjjf.org/wp-content/uploads/2023/11/article-2171.pdf.

Raffaele, Paul. "Uganda: The Horror." *Smithsonian Magazine*, February 2005. https://www.smithsonianmag.com/history/uganda-the-horror-85439313/.

Ralph, William W. "Improvised Destruction: Arnold, LeMay, and the Firebomb-
    ing of Japan." *War in History* 13, no. 4 (November 2006): 495–522. https://doi
    .org/10.1177/0968344506069971.
Ramirez, Jason G., and Douglas R. Bacon. "Modern Chemical Warfare: A History."
    Bulletin of Anesthesia History 22, no. 2 (April 2004): 1, 4–7, 15. https://doi.org/
    10.1016/s1522-8649(04)50015-x.
Ramirez, Jason G., and Douglas R. Bacon. "Modern Chemical Warfare Agents:
    The Anesthesiologist's Perspective." *Seminars in Anesthesia, Perioperative Med-
    icine and Pain* 22, no. 4 (December 2003): 239–46. https://doi.org/10.1053/
    S0277-0326(03)00041-2.
Raphael, Kate. "Mongol Siege Warfare on the Banks of the Euphrates and the Question
    of Gunpowder (1260–1312)." *Journal of the Royal Asiatic Society* 19, no. 3 (July
    2009): 355–70. https://www.jstor.org/stable/27756073.
Ray, Michael. "Executive Order 9066." Britannica. Updated March 6, 2025. https://
    www.britannica.com/topic/Executive-Order-9066.
Reemtsma, Jan Philipp. "The Concept of the War of Annihilation: Clausewitz, Luden-
    dorff, Hitler." In *War of Extermination: The German Military in World War II, 1941–
    1944*, edited by Hannes Heer and Klaus Naumann, 13–38. Berghahn Books, 2000.
Reuters. "Court Charges at Least 42 Ugandan Youths over Anti-Graft Protest."
    July 24, 2024. https://www.reuters.com/world/africa/court-charges-least-42
    -ugandan-youths-over-anti-graft-protest-2024-07-24/.
Reuters. "Ukraine Probing over 122,000 Suspected War Crimes, Says Prose-
    cutor." February 23, 2024. https://www.reuters.com/world/europe/ukraine
    -probing-over-122000-suspected-war-crimes-says-prosecutor-2024-02-23/.
Reuters. "U.N. Judges Overturn Acquittal of Serbian Ultra-Nationalist for Role in
    Wars." April 11, 2018. https://www.reuters.com/article/world/un-judges-over
    turn-acquittal-of-serbian-ultra-nationalist-for-role-in-wars-idUSKBN1HI1WT/.
Reuters. "Witnesses Tell of Congo Massacre." CNN. April 8, 2003. https://www.cnn
    .com/2003/WORLD/africa/04/08/congo.massacre.reut/index.html.
Reynolds, Michael. "Massacre at Malmedy During the Battle of the Bulge." *World
    War II* 17, no. 6 (February 2003): 42–50. https://www.historynet.com/massacre
    -at-malmedy-during-the-battle-of-the-bulge/.
Rhodes, Richard. *Masters of Death: The SS-Einsatzgruppen and the Invention of the Holo-
    caust*. Vintage Books, 2002.
Riechmann, Deb, Mathew Lee, and Jonathan Lemire. "Israel Signs Pacts with 2 Arab
    States: A 'New' Mideast?" Associated Press. September 15, 2020. https://apnews
    .com/article/bahrain-israel-united-arab-emirates-middle-east-elections-7544b322
    a254ebea1693e387d83d9d8b.
Riedel, Stefan. "Biological Warfare and Bioterrorism: A Historical Review." *Baylor Uni-
    versity Medical Center Proceedings* 17, no. 4 (2004): 400–406. https://doi.org/10.108
    0/08998280.2004.11928002.
Riley, Sue Anne. "Loose Lips Sink Ships: American Propaganda in WWII." *Sea Classics*
    45, no. 10 (October 2012): 50–57.
Riñon, Ian. "Another Kristallnacht? Star of David Graffitied on Jewish Homes in Ber-
    lin." Headlines and Global News. Updated October 15, 2023. https://www.hngn
    .com/articles/252954/20231015/another-kristallnacht-star-david-graffitied-jewish
    -homes-berlin.htm.
Roegge, Frederick "Fritz." "December 7th, 1941: A Submarine Force Perspective."
    Sextant. December 7, 2016. https://usnhistory.navylive.dodlive.mil/Recent/
    Article-View/Article/2686191/.

Ronald Reagan Presidential Library and Museum. "Peace Through Strength." Accessed April 27, 2025. https://www.reaganlibrary.gov/permanent-exhibits/peace-through-strength.

Roseman, Mark. *The Wannsee Conference and the Final Solution: A Reconsideration.* Picador, 2002.

Rossino, Alexander B. *Hitler Strikes Poland: Blitzkrieg, Ideology, and Atrocity.* Univeristy Press of Kansas, 2003.

Rothbart, Daniel, Karina V. Korostelina, and Mohammed D. Cherkaoui, eds. *Civilians and Modern War: Armed Conflict and the Ideology of Violence.* Routledge, 2012.

Rotondi, Jessica Pearce. "How Dith Pran's Remarkable Survival Story Exposed Cambodia's Killing Fields." History. Updated March 6, 2025. https://www.history.com/news/dith-pran-killing-fields-cambodia-khmer-rouge.

Rottinghaus, Brandon, and Justin S. Vaughn. *Official Results of the 2024 Presidential Greatness Project Expert Survey.* 2024. https://qzg.wvf.mybluehost.me/wp-content/uploads/2024/05/Presidential-Greatness-White-Paper-2024.pdf.

Royal Museums Greenwich. "What Was Greek Fire?" Accessed April 27, 2025. https://www.rmg.co.uk/stories/topics/greek-fire.

Rubenstein, Richard E. "The Role of Civilians in American War Ideology." In *Civilians and Modern War: Armed Conflict and the Ideology of Violence*, edited by Daniel Rothbart, Karina V. Korostelina, and Mohammed D. Cherkaoui, 21–50. Routledge, 2012.

Rummel, R. J. "20,946,000 Victims: Nazi Germany, 1933 to 1945." In *Democide: Nazi Genocide and Mass Murder.* Transaction, 1992. https://www.hawaii.edu/powerkills/NAZIS.CHAP1.HTM.

Russell, Edmund P., III. "'Speaking of Annihilation': Mobilizing for War Against Human and Insect Enemies, 1914–1945." *Journal of American History* 82, no. 4 (March 1996): 1505–29. https://doi.org/10.2307/2945309.

San Diego Union-Tribune. "Korea Bloodbath Probe Ends; US Escapes Much Blame." Updated August 31, 2016. https://www.sandiegouniontribune.com/2010/07/10/korea-bloodbath-probe-ends-us-escapes-much-blame/.

Sanbar, Sarah. "Twenty Years On, Iraq Bears Scars of US-Led Invasion." Human Rights Watch. March 19, 2023. https://www.hrw.org/news/2023/03/19/twenty-years-iraq-bears-scars-us-led-invasion.

Sand Creek Massacre Foundation. "The Massacre." Accessed August 21, 2024. https://www.sandcreekmassacrefoundation.org/massacre.

Sass, Erik. "Fall of the South: The Burning of Columbia." Mental Floss. February 18, 2015. https://www.mentalfloss.com/article/61743/fall-south-burning-columbia.

Schecter, Anna. "'Top Secret' Hamas Documents Show That Terrorists Intentionally Targeted Elementary Schools and a Youth Center." NBC News. October 13, 2023. https://www.nbcnews.com/news/investigations/top-secret-hamas-documents-show-terrorists-intentionally-targeted-elem-rcna120310.

Schmaltz, Florian. "Chemical Weapons Research on Soldiers and Concentration Camp Inmates in Nazi Germany." In *One Hundred Years of Chemical Warfare: Research, Deployment, Consequences*, edited by Breitslav Friedrich, Dieter Hoffmann, Jürgen Renn, Florian Schmaltz, and Martin Worl, 229–58. Springer International, 2017. https://doi.org/10.1007/978-3-319-51664-6_13.

Scianna, Bastian Matteo. "A Predisposition to Brutality? German Practices Against Civilians and *Francs-Tireurs* During the Franco-Prussian War 1870–1871 and Their Relevance for the German 'Military *Sonderweg*' Debate." *Small Wars and Insurgencies* 30, nos. 4–5 (2019): 968–93. https://doi.org/10.1080/09592318.2019.1638551.

Scott, James M. "Battlefield as Crime Scene: The Japanese Massacre in Manila."
  HistoryNet. January 12, 2019. https://www.historynet.com/worldwar2-japanese
  -massacre-in-manila/.
Searle, Thomas R. "'It Made a Lot of Sense to Kill Skilled Workers': The Firebombing of Tokyo in March 1945." *Journal of Military History* 66, no. 1 (January 2002):
  103–33. https://doi.org/10.2307/2677346.
Searles, Harry. "Utah War Facts." American History Central. Updated June 10, 2024.
  https://www.americanhistorycentral.com/entries/utah-war-facts/.
Searles, Harry. "Utah War Summary." American History Central. Updated February 2,
  2024. https://www.americanhistorycentral.com/entries/utah-war/.
Sempa, Francis P. "The Geopolitical Vision of Alfred Thayer Mahan." *Diplomat*, December 30, 2014. https://thediplomat.com/2014/12/the-geopolitical
  -vision-of-alfred-thayer-mahan/.
Seo, Yoonjung, Andrew Raine, and Gawon Bae. "Torture, Forced Abortions and Insects
  for Food: Life Inside North Korean Jails, Says This NGO." CNN World. Updated
  March 24, 2023. https://www.cnn.com/2023/03/23/asia/north-korea-torture-prison
  -report-intl-hnk-dst/index.html.
Sherman, William T. *Memoirs of Gen. W. T. Sherman*. Vol. 2. 4th ed. Charles L. Webster, 1891.
Sherman, William T. *Sherman's Civil War: Selected Correspondence of William T. Sherman, 1860–1865*. Edited by Brooks D. Simpson and Jean V. Berlin. University of
  North Carolina Press, 1999.
Shirer, William L. *The Rise and Fall of the Third Reich*. Simon and Schuster Paperbacks,
  1988.
Shorrock, Tim. "Can the United States Own Up to Its War Crimes During the Ko-
  rean War?" *Nation*, March 30, 2015. https://www.thenation.com/article/archive/
  can-united-states-own-its-war-crimes-during-korean-war/.
Shulatov, Yaroslav A. "Russia as a 'Trauma': The Rise and Fall of Japan as a Great
  Power." *Russia in Global Affairs* 17, no. 4 (October–December 2019): 78–108.
  https://doi.org/10.31278/1810-6374-2019-17-4-78-108.
Siegphyl. "1937 Battle of Shanghaim, Japan's Brutal Attack on China." War History
  Online. December 15, 2013. https://www.warhistoryonline.com/war-articles/
  1937-battle-shanghai-japans-brutal-attack-china.html.
Simkin, John. "Cuban Missile Crisis." Spartacus Educational. Updated August 2020.
  https://spartacus-educational.com/COLDcubanmissile.htm.
Sledge, E. B. *With the Old Breed: At Peleliu and Okinawa*. Presidio Press, 2007.
Smart, Jeffery K. "History of Chemical and Biological Warfare: An American Perspec-
  tive." In *Medical Aspects of Chemical and Biological Warfare*, edited by Frederick
  R. Sidell, Ernest T. Takafuji, and David R. Franz, 9–86. Borden Institute, Walter
  Reed Army Medical Center, 1997. https://medcoeckapwstorprd01.blob.core.usgov
  cloudapi.net/pfw-images/borden/chembio/Ch2.pdf.
Smith, Matt, and Alla Eshchenko. "Ukraine Cries 'Robbery' as Russia Annexes
  Crimea." CNN World. Updated March 18, 2014. https://www.cnn.com/
  2014/03/18/world/europe/ukraine-crisis/index.html.
Smith, R. Jeffrey. "Srebrenica Genocide." Britannica. Updated March 28, 2025. https://
  www.britannica.com/event/Srebrenica-massacre#ref294001.
Smith, Saphora, and Emily R. Siegel. "'Never Again': Anti-Semitism Surges as Memories of Holocaust Fade." NBC News. Updated January 25, 2020. https://www
  .nbcnews.com/news/world/never-again-anti-semitism-surges-memories-holocaust
  -fade-n1122081.

Smith-Spark, Laura, Phil Black, and Frederik Pleitgen. "Russia Flexes Military Muscle as Tensions Rise in Ukraine's Crimea." CNN World. Updated February 26, 2014. https://edition.cnn.com/2014/02/26/world/europe/ukraine-politics.

Smits, Gregory. "Examining the Myth of Ryukyuan Pacifism." *Asia-Pacific Journal: Japan Focus* 8, no. 37 (September 13, 2010): 1–20. https://apjjf.org/gregory-smits/3409/article.

SOFREP News Team. "The Chichijima Incident: Japanese Soldiers Ate US Pilots That Fell into Their Hands." SOFREP. February 13, 2022. https://sofrep.com/news/the-chichijima-incident-japanese-soldiers-ate-us-pilots-that-fell-into-their-hands/.

Solly, Meilan. "Remembering the Khatyn Massacre." *Smithsonian Magazine*, March 22, 2021. https://www.smithsonianmag.com/history/how-1943-khatyn-massacre-became-symbol-nazi-atrocities-eastern-front-180977280/.

South African History Online. "Women and Children in White Concentration Camps During the Anglo-Boer War, 1900–1902." Updated September 1, 2023. https://www.sahistory.org.za/article/women-and-children-white-concentration-camps-during-anglo-boer-war-1900-1902.

Southgate, Mandy. "Rwandan Genocide: The Hutu Ten Commandments." *A Passion to Understand* (blog). August 13, 2011. https://passiontounderstand.blogspot.com/2011/08/rwandan-genocide-hutu-ten-commandments.html.

Spaight, J. M. *Aircraft in War*. Macmillan, 1914. https://ia902807.us.archive.org/35/items/aircraftinwar00spai/aircraftinwar00spai.pdf.

Spangler, Jonathan. "Dukes of Saxe-Coburg and Saxe-Gotha, Families of Two British Consorts." Dukes and Princes. July 2, 2020. https://dukesandprinces.org/2020/07/02/dukes-of-saxe-coburg-and-saxe-gotha-families-of-two-british-consorts/.

SparkNotes. "The Spanish American War (1898–1901)." Accessed April 27, 2025. https://www.sparknotes.com/history/american/spanishamerican/context/.

Specter, Matthew, and Varsha Venkatasubramanian. "'America First': Nationalism, Nativism, and the Fascism Question, 1880–2020." In *Fascism in America: Past and Present*, edited by Gavriel D. Rosenfeld and Janet Ward, 107–40. Cambridge University Press, 2023.

Spector, Ronald H. *Eagle Against the Sun*. Vintage Books, 1985.

Sperber, Jonathan, and Brian Vick. "From Vormärz to Prussian Dominance (1815–1866)." German History in Documents and Images. Accessed April 18, 2024. https://germanhistorydocs.org/en/from-vormaerz-to-prussian-dominance-1815-1866/introduction.

The Standard. "Alive and Safe, the Brutal Japanese Soldiers Who Butchered 20,000 Allied Seamen in Cold Blood." April 12, 2012. https://www.standard.co.uk/hp/front/alive-and-safe-the-brutal-japanese-soldiers-who-butchered-20-000-allied-seamen-in-cold-blood-6636703.html.

Steer, George. "The Tragedy of Guernica: Town Destroyed in Air Attack: Eye-Witness's Account." *Sunday Times*, April 27, 1937. https://www.thetimes.com/article/bombing-of-guernica-original-times-report-from-1937-5j7x3z2k5bv.

Steinberg, John W. "Was the Russo-Japanese War World War Zero?" *Russian Review* 67, no. 1 (January 2008): 1–7. https://www.jstor.org/stable/20620667.

Stevenson, Struan. "The Forgotten Mass Execution of Prisoners in Iran in 1988." *Diplomat*, July 31, 2013. https://thediplomat.com/2013/07/the-forgotten-mass-execution-of-prisoners-in-iran-in-1988/.

Stewart, Ken. "Zyklon-B." Britannica. Updated October 26, 2024. https://www.britannica.com/science/Zyklon-B.

Stewart, Richard W., ed. "Emergence to World Power 1898–1902." In *American Military History: The United States Army and the Forging of a Nation, 1775–1917*, vol. 1, 341–64. Center of Military History, United States Army, 2005.

Stewart, Richard W., ed. "The Mexican War and After." In *American Military History: The United States Army and the Forging of a Nation, 1775–1917*, vol. 1, 177–98. Center of Military History, United States Army, 2005.

Stewert, Richard W., ed. *The United States Army in a Global Era, 1917–2008*. 2nd ed. Center of Military History, United States Army, 2008.

Stewart, Richard W., ed. "Winning the West: The Army in the Indian Wars, 1865–1890." In *American Military History: The United States Army and the Forging of a Nation, 1775–1917*, vol. 1, 321–40. Center of Military History, United States Army, 2005.

Stilwell, Blake. "Why the US Used Agent Orange in Vietnam and What Makes It So Deadly." Military.com. August 1, 2022. https://www.military.com/history/why-us-used-agent-orange-vietnam-and-what-makes-it-so-deadly.html.

Stone, Richard. "Seeking Answers for Iran's Chemical Weapons Victims—Before Time Runs Out." *Science*, January 4, 2018. https://www.science.org/content/article/seeking-answers-iran-s-chemical-weapons-victims-time-runs-out.

Storry, Richard. *Japan and the Decline of the West in Asia, 1894–1943*. St. Martin's Press, 1979.

Strachan, Hew. "Total War: The Conduct of War, 1939–1945." In *A World at Total War: Global Conflict and the Politics of Destruction*, edited by Roger Chickering, Stig Förster, and Bernd Greiner, 33–52. Cambridge University Press, 2010.

Streeter, Jon, and Joe Parker. "The Khmer Rouge: 'To Destroy You Is No Loss.'" *15-Minute History Podcast*. March 25, 2024. https://www.15minutehistorypodcast.org/episodes/the-khmer-rouge.

Streicher, Julius. "How to Tell a Jew." Translated by Randall Bytwerk. German Propaganda Archive. 1999. http://research.calvin.edu/german-propaganda-archive/story3.htm.

Streicher, Julius. "The Poisonous Mushroom." Translated by Randall L. Bytwerk. German Propaganda Archive. 1999. http://research.calvin.edu/german-propaganda-archive/story2.htm.

Sturma, Michael. "Atrocities, Conscience, and Unrestricted Warfare: US Submarines During the Second World War." *War in History* 16, no. 4 (November 2009): 447–68. https://doi.org/10.1177/0968344509341686.

Sturma, Michael. "Japanese Treatment of Allied Prisoners During the Second World War: Evaluating the Death Toll." *Journal of Contemporary History* 55, no. 3 (July 2020): 514–34. https://www.jstor.org/stable/27067639.

Suny, Ronald Grigor. "Armenian Genocide." Britannica. Updated April 26, 2025. https://www.britannica.com/event/Armenian-Genocide.

Sweetman, Jack. "The Battle of the Atlantic." *Naval History* 9, no. 3 (June 1995). https://www.usni.org/magazines/naval-history-magazine/1995/june/battle-atlantic.

Sydney Morning Herald. "From the Archives, 1950: China Invades Tibet." October 30, 1950. https://www.smh.com.au/world/asia/from-the-archives-1950-china-invades-tibet-20201014-p56560.html.

Syrian Network for Hurman Rights. "Documentation of 72 Torture Methods the Syrian Regime Continues to Practice in Its Detention Centers and Military Hospitals." October 21, 2019. https://snhr.org/blog/2019/10/21/54362/.

Takashi, Tsuchiya. "The Imperial Japanese Medical Atrocities and Its Enduring Legacy in Japanese Research Ethics." *International Congress of History of Science* (August 29, 2005). https://www.researchgate.net/publication/228480362_The_Imperial _Japanese_Medical_Atrocities_and_Its_Enduring_Legacy_in_Japanese_Research _Ethics.

Tanaka, Yuki. *Hidden Horrors: Japanese War Crimes in World War II*. Westview Press, 1996.

Tenorio, Rich. "Documentary on October 7 Supernova Festival Massacre Makes US Debut." *Times of Israel*, February 23, 2024. https://www.timesofisrael.com/ documentary-on-october-7-supernova-festival-massacre-makes-us-debut/.

Times of Israel. "Arabic Annotated Copy of 'Mein Kampf' Found Among Possessions of Terrorist in Gaza Home." November 12, 2023. https://www.timesofisrael.com/ liveblog_entry/arabic-annotated-copy-of-mein-kampf-found-among-possessions-of -terrorist-in-gaza-home/.

Their Majesties' Work as Prince of Wales and Duchess of Cornwall. "Saxe-Coburg -Gotha." Accessed July 25, 2024. https://www.royal.uk/saxe-coburg-gotha.

Thirteen PBS. "The Sedition Act of 1918." Accessed April 27, 2025. https://www.thir teen.org/wnet/supremecourt/capitalism/sources_document1.html.

Thoreau, Henry David. "The Savage in Man Is Never Quite Eradicated." FixQuotes. Accessed June 1, 2025. https://fixquotes.com/quotes/the-savage-in-man-is-never -quite-eradicated-35771.htm.

Thucydides. "The Siege of Plataea." In *History of the Peloponnesian War*, translated by Richard Crawley. 1910. https://www.livius.org/sources/content/thucydides -historian/siege-of-plataea/.

Time. "Trials: My Lai: A Question of Orders." January 25, 1971. https://time.com/ archive/6838471/trials-my-lai-a-question-of-orders/.

Toland, John. *The Rising Sun: The Decline and Fall of the Japanese Empire, 1936–1945*. Modern Library, 2003.

Treaty of Peace with Germany [Treaty of Versailles]. *American Journal of International Law* 13, no. S3 (July 1919): 151–56. https://doi.org/10.2307/2213120.

Treaty Relating to the Use of Submarines and Noxious Gases in Warfare. International Humanitarian Law Databases. February 6, 1922. https://ihl-databases.icrc.org/ assets/treaties/270-IHL-34-EN.pdf.

Trevor-Roper, H. R. *The Last Days of Hitler*. Macmillan, 1947.

Trouillard, Stéphanie. "The First Major Massacre in the 'Holocaust by Bullets': Babi Yar, 80 Years On." France24. September 29, 2021. https://www.france24.com/en/ europe/20210929-the-first-major-massacre-in-the-holocaust-by-bullets-babi-yar-80 -years-on.

Turley, Richard E., Jr. "The Mountain Meadows Massacre." Church of Jesus Christ of Latter-Day Saints. September 2007. https://www.churchofjesuschrist.org/study/ ensign/2007/09/the-mountain-meadows-massacre?lang=eng.

Turner, Matthew D., and Jason Sapp. "Failure to Plan: The Disease That Cost an American Empire." *Military Medicine* 188, nos. 7–8 (July–August 2023): 171–73. https:// doi.org/10.1093/milmed/usad161.

Tzu, Sun. *The Art of War*. Translated by Samuel B. Griffith. Oxford University, 1971.

Uboat.net. "HMS Torbay (N 79)." Accessed September 22, 2024. https://uboat.net/al lies/warships/ship/3498.html.

Ukrinform. "UN Documents Torture and Arrests of Crimean Tatars by Russia." December 12, 2017. https://www.ukrinform.net/rubric-society/2362880-un-documents -torture-and-arrests-of-crimean-tatars-by-russia.html.

United Nations Office for the Coordination of Humanitarian Affairs (OCHA). "Humanitarian Situation Update #221: Gaza Strip." September 23, 2024. https://www.ochaopt.org/content/humanitarian-situation-update-221-gaza-strip.

United States Department of State, Bureau of Democracy, Human Rights and Labor. *Syria 2013 Human Rights Report.* United States Department of State, 2013.

United States Holocaust Memorial Museum. "Cambodia 1975–1979." Updated April 2018. https://www.ushmm.org/genocide-prevention/countries/cambodia/cambodia-1975.

United States Holocaust Memorial Museum. "Côte d'Ivoire." Accessed July 11, 2024. https://www.ushmm.org/genocide-prevention/countries/cote-divoire.

United States Holocaust Memorial Museum. "Definitions: Types of Mass Atrocities." Accessed July 11, 2024. https://www.ushmm.org/genocide-prevention/learn-about-genocide-and-other-mass-atrocities/definitions.

United States Institute of Peace. "Truth Commission: South Korea 2005." April 18, 2012. https://www.usip.org/publications/2012/04/truth-commission-south-korea-2005.

The United States Strategic Bombing Survey: Summary Report (European War). United States Government Printing Office, September 30, 1945. http://www.anesi.com/ussbs02.htm.

United States Strategic Bombing Survey: Summary Report (Pacific War). United States Government Printing Office, July 1, 1946. http://www.anesi.com/ussbs01.htm.

University of Cambridge: Research. "Evidence of a Prehistoric Massacre Extends the History of Warfare." January 20, 2016. https://www.cam.ac.uk/research/news/evidence-of-a-prehistoric-massacre-extends-the-history-of-warfare.

University of Central Arkansas, Government, Public Service, and International Studies. "42. Rwanda (1962–Present)." Accessed July 7, 2024. https://uca.edu/politicalscience/home/research-projects/dadm-project/sub-saharan-africa-region/rwanda-1962-present/.

U.S. Army Air Corps Air War Plans Division. "AWPD-1, August 1941." In *Major Problems in American Military History: Documents and Essays*, edited by John Whiteclay Chambers II and G. Kurt Piehler. Houghton Mifflin, 1999.

U.S. Department of Justice. "Justice Department Announces Terrorism Charges Against Senior Leaders of Hamas." Press release. September 3, 2024. https://www.justice.gov/opa/pr/justice-department-announces-terrorism-charges-against-senior-leaders-hamas.

U.S. Department of State Archive. "U.S. Invasion and Occupation of Haiti, 1915–34." Accessed March 31, 2024. https://2001-2009.state.gov/r/pa/ho/time/wwi/88275.htm.

van Creveld, Martin. "Historical Development." Britannica. Updated September 14, 2023. https://www.britannica.com/topic/tactics/Historical-development.

van der Vat, Dan. *The Pacific Campaign: The U.S.-Japanese Naval War 1941–1945.* Touchstone, 1991.

Vanhoutte, Kristof K. P. "'Oh God! What a Lovely War': Giorgio Agamben's Clausewitzian Theory of Total/Global (Civil) War." *Sotsiologicheskoe Obozrenie/Russian Sociological Review* 14, no. 4 (2015): 28–43. https://doi.org/10.17323/1728-192x-2015-4-28-43.

Vere, Christopher. "The Napoleonic Wars and the Birth of Modern Warfare." *Intelligence and National Security* 24, no. 3 (2009): 464–70. http://dx.doi.org/10.1080/02684520903135065.

Vergun, David. "Submarine Warfare Played Major Role in World War II Victory." U.S. Department of Defense. March 16, 2020. https://www.defense.gov/News/ Feature-Stories/Story/Article/2114035/.

Vidal, César. "La Destrucción de Guernica." Translated by Peter Miller. Buber's Basque Page. Accessed April 27, 2025. http://www.buber.net/Basque/History/guernica-ix .html.

von Clausewitz, Carl. On War. Edited and translated by Michael Howard and Peter Paret. Princeton University Press, 1989.

von Moltke, Helmuth. Memorandum on a Possible War Between Prussia and Austria (1866). German History in Documents and Images. Accessed April 5, 2024. https://germanhistorydocs.org/en/from-vormaerz-to-prussian-dominance -1815-1866/helmuth-von-moltke-memoradum-on-the-possible-war-between-prussia -and-austria-1866.

von Moltke, Helmuth. Memorandum on the Effect of Improvements in Firearms on Battlefield Tactics (1861). German History in Documents and Images. Accessed April 6, 2024. https://germanhistorydocs.org/en/from-vormaerz-to-prussian-domi nance-1815-1866/helmuth-von-moltke-memorandum-on-the-effect-of-improvements -in-firearms-on-battlefield-tactics-1861.

Waller, James E. "It Can Happen Here: Assessing the Risk of Genocide in the US." Center for Develoment of International Law. February 24, 2017. https://worldwith outgenocide.org/wp-content/uploads/2017/03/Waller-Assessing-the-Risk-of-Geno cide.pdf.

Walters, John Bennett. Merchant of Terror: General Sherman and Total War. Bobbs-Merrill, 1973.

War Stories with Mark Felton. "U-Boat Massacre—The Case of U-852." Posted June 7, 2020, by War Stories with Mark Felton. YouTube, 10 min., 21 sec. https://www .youtube.com/watch?v=Q9LeA7V77E0.

Washington, George. "Fifth Annual Message of George Washington." Yale Law School, Lillian Goldman Law Library. December 3, 1793. https://avalon.law.yale.edu/18th _century/washs05.asp.

Washington, George. "Washington's Farewell Address to the People of the United States." United States Senate. September 19, 1796. https://www.senate.gov/artandhistory/ history/resources/pdf/Washingtons_Farewell_Address.pdf.

Watanabe, Kazuko. "Trafficking in Women's Bodies, Then and Now: The Issue of Military 'Comfort Women.'" Peace and Change 20, no. 4 (October 1995): 501–14. https://doi.org/10.1111/j.1468-0130.1995.tb00249.x.

Wehner, Greg. "Arabic Copy of Hitler's 'Mein Kampf' Found in Children's Room Used by Hamas: Israeli Officials." Fox News. November 12, 2023. https://www .foxnews.com/world/arabic-copy-hitlers-mein-kampf-found-childrens-room-used -hamas-israeli-officials.

Weigley, Russell F. The Age of Battles: The Quest for Decisive Warfare from Breitenfeld to Waterloo. Indiana University Press, 1991.

Weigley, Russell F. "American Strategy from Its Beginning Through the First World War." In Makers of Modern Strategy: From Machiavelli to the Nuclear Age, edited by Peter Paret, Gordon A. Craig, and Felix Gilbert, 408–43. Princeton University Press, 1986.

Weigley, Russell F. The American Way of War: A History of United States Military Strategy and Policy. Macmillan, 1973.

Weinberg, Gerhard L. "Some Myths of World War II." Journal of Military History 75, no. 3 (July 2011): 701–18.

Weinberg, Gerhard L. *A World at Arms: A Global History of World War II*. 2nd ed. Cambridge University Press, 2008. https://doi.org/10.1017/CBO9780511818639.

Weindling, Paul, Anna von Villiez, Aleksandra Loewenau, and Nichola Farron. "The Victims of Unethical Human Experiments and Coerced Research Under National Socialism." *Endeavour* 40, no. 1 (March 2016): 1–6. https://doi.org/10.1016/j.endeavour.2015.10.005.

Welch, David. *The Third Reich: Politics and Propaganda*. Routledge, 2008.

Welch, Jeanie M. "Without a Hangman, Without a Rope: Navy War Crimes Trials After World War II." *International Journal of Naval History* 1, no. 1 (April 2002). https://ijnh.seahistory.org/wp-content/uploads/sites/2/2012/01/pdf_welch.pdf.

Welch, Spencer Glasgow. *A Confederate Surgeon's Letters to His Wife*. Neale, 1911.

Werrell, Kenneth P. *Blankets of Fire: U.S. Bombers over Japan During World War II*. Smithsonian Institution Press, 1996.

Westermann, Edward B. "Partners in Genocide: The German Police and the *Wehrmacht* in the Soviet Union." *Journal of Strategic Studies* 31, no. 5 (October 2008): 771–96. https://doi.org/10.1080/01402390802197977.

Wheelis, Mark. "Biological Warfare at the 1346 Siege of Caffa." *Emerging Infectious Diseases* 8, no. 9 (September 2002): 971–75. https://doi.org/10.3201/eid0809.010536.

White House. "Fact Sheet: President Biden Announces Bold Plan to Reform the Supreme Court and Ensure No President Is Above the Law." September 29, 2024. https://bidenwhitehouse.archives.gov/briefing-room/statements-releases/2024/07/29/fact-sheet-president-biden-announces-bold-plan-to-reform-the-supreme-court-and-ensure-no-president-is-above-the-law/.

Wiener Holocaust Library. "Science and Suffering: Victims and Perpetrators of Nazi Human Experimentation." Accessed April 27, 2025. https://weinerholocaust library.org/exhibition/science-and-suffering-victims-and-perpetrators-of-nazi-human -experimentation/.

Willick, Jason. "Biden's Supreme Stunt." *Washington Post*, July 30, 2024. https://www .washingtonpost.com/opinions/2024/07/30/joe-biden-supreme-court-reform/.

Wills, Matthew. "How Commonly Was Smallpox Used as a Biological Weapon." *JSTOR Daily*, April 4, 2021. https://daily.jstor.org/how-commonly-was-smallpox -used-as-a-biological-weapon/.

Winer, Stuart. "Fear in Berlin as Star of David Scrawled at Entrances of Buildings Where Jews Reside." *Times of Israel*, October 15, 2023. https://www.timesofisrael .com/fear-in-berlin-as-star-of-david-scrawled-at-entrances-of-buildings-where-jews -reside/.

Woldt, Collin. "Never Again: Revisiting the 1994 Rwandan Genocide." *Columbia Political Review*. June 4, 2021. https://www.cpreview.org/articles/2021/6/never-again -revisiting-the-1994-rwandan-genocide.

Wolk, Herman S., and Richard P. Hallion. "FDR and Truman: Continuity and Context in the A-Bomb Decision." *Air Power Journal* 9, no. 3 (Fall 1995): 56–62. https:// www.airuniversity.af.edu/Portals/10/ASPJ/journals/Volume-09_Issue-1-Se/1995_Vol 9_No3.pdf.

World Peace Foundation. "Algeria: War of Independence." Mass Atrocity Endings. August 7, 2015. https://sites.tufts.edu/atrocityendings/2015/08/07/algeria-war-of -independence/.

World Without Genocide. "Eight Stages of Genocide." Accessed July 11, 2024. https:// worldwithoutgenocide.org/genocides-and-conflicts/background-and-overview-infor mation/eight-stages-of-genocide.

Wright, Gordon. *The Ordeal of Total War 1939–1945*. Harper Torchbooks, 1968.

Wright, Quincy. "The Nature of Conflict." *Western Political Quarterly* 4, no. 2 (June 1951): 193–209. http://www.jstor.org/stable/443101.

Wyszynski, Diego F. "Men with White Coats and SS Boots: The Children's Euthanasia Programme During the Third Reich." *Paediatric and Perinatal Epidemiology* 14, no. 4 (October 2000): 295–99. https://doi.org/10.1046/j.1365-3016.2000.00278.x.

Yamaguchi, Tomomi. "Japan's Right-Wing Women and the 'Comfort Women' Issue." *Georgetown Journal of Asian Affairs* 6 (2020): 45–54. https://repository.digital .georgetown.edu/handle/10822/1059392.

Yenne, Bill. "Fear Itself: The General Who Panicked the West Coast." HistoryNet. July 11, 2017. https://www.historynet.com/fear-itself-the-general-panicked-west-coast/.

Zapotoczny, Walter. "The Rape of Nanking: Reasons and Recrimination." 2008. https:// www.wzaponline.com/yahoo_site_admin/assets/docs/TheRapeofNanking.2921250 34.pdf.

Zar, Dan. "February 27, 1902: Breaker Morant Executed for War Crimes." History and Headlines. Updated February 14, 2020. https://www.historyandheadlines.com/ february-27-1902-breaker-morant-executed-war-crimes/.

Zar, Daniel. "Sétif and Guelma Massacre, French Slaughter Algerians." History and Headlines. May 8, 2020. https://www.historyandheadlines.com/setif-and-guelma -massacre-french-slaughter-algerians/.

Zinn, Howard. "Just and Unjust War." In *Declarations of Independence: Cross-Examining American Ideology*. HarperCollins, 1990.

Zyla, Greg. "What Years Did Car Manufacturers Not Build Cars During the World War II Era?" Auto Round-Up Publications. September 23, 2019. https://www .autoroundup.com/vehicle/what-years-did-car-manufacturers-not-build-cars-during -the-world-war-ii-era-article-1271.aspx.

www.ingramcontent.com/pod-product-compliance
Lightning Source LLC
Chambersburg PA
CBHW031935090426
42811CB00002B/194